Digital Cities

# Digital Cities

*Between History and Archaeology*

Edited by Maurizio Forte

*and*

Helena Murteira

OXFORD
UNIVERSITY PRESS

OXFORD
UNIVERSITY PRESS

Oxford University Press is a department of the University of Oxford. It furthers the University's objective of excellence in research, scholarship, and education by publishing worldwide. Oxford is a registered trade mark of Oxford University Press in the UK and certain other countries.

Published in the United States of America by Oxford University Press
198 Madison Avenue, New York, NY 10016, United States of America.



CIP data is on file at the Library of Congress
ISBN 978-0-19-750112-2 (pbk.)
ISBN 978-0-19-049890-0 (hbk.)

Hardback printed by Bridgeport National Bindery, Inc., United States of America

# CONTENTS

# ABOUT THE CONTRIBUTORS

**Kristin Agcaoili**, Center for Cyber- Archaeology and Sustainability, Qualcomm Institute, University of California, San Diego, USA

**Zeynep Aktüre**, Associate Professor, Department of Architecture, Izmir Institute of Technology (IZTECH), Izmir, Turkey

**Richard Beacham**, King's College, London, UK

**Jean-Yves Blaise**, Senior CNRS researcher, UMR 3495 MAP CNRS/MC, France

**Geoffrey Booth**, Associate Professor, Department of Landscape Architecture and Urban Planning, and Associate Director, CRS Center for Leadership and Management in the Design and Construction Industry, Texas A&M University, USA

**Maria Leonor Botelho**, Assistant Professor DCTP/ FLUP and Resarcher at CITCEM/ FLUP, Universidade do Porto, Portugal

**Caroline Bruzelius**, A. M. Cogan Professor Emerita, Department of Art, Art History and Visual Studies, Duke University, North Carolina, USA

**Stephen Caffey**, Instructional Associate Professor, Department of Architecture, Texas A&M University, USA

**Gabriela Campagnol**, Lecturer, Department of Architecture, Texas A&M University, USA

**Gianluca Catanzariti**, Geostudi Astier, Italy

**Mark J. Clayton**, Professor, Department of Architecture, Texas A&M University, USA

**Nevio Danelon**, postdoctoral associate, AAHVS, Duke University, North Carolina, USA

**Iwona Dudek**, Senior CNRS researcher, UMR 3495 MAP CNRS/MC, France

**Maurizio Forte**, professor of Classical Studies, director of the Dig@LAb, Duke University, North Carolina, USA

**Maria Alexandra Gago da Câmara**, Assistant Professor, Universidade Aberta (Open University), Lisbon, Portugal, and Researcher at Centre for Art History and Artistic Research (CHAIA), University of Évora, Portugal

**Kevin Glowacki**, Associate Professor, Department of Architecture, and Director, Center for Heritage Conservation, Texas A&M University, USA

**Avner Goren**, Milken Institute, Santa Monica, California, USA

**David Johnston**, Associate Professor of the Practice of Marine Conservation Ecology Coastal Environmental Management, Marine Science & Conservation, Duke University, North Carolina, USA

**Julian Kang**, Associate Professor, Department of Construction Science, Texas A&M University, USA

**Anish Kannan**, Center for Cyber- Archaeology and Sustainability, Qualcomm Institute, University of California, San Diego, USA

**Sarah Kenderdine**, Professor, Laboratory for Experimental Museology (eM+), École polytechnique fédérale de Lausanne (EPFL), Lausanne, Switzerland

**Nancy Klein**, Associate Professor, Department of Architecture, Texas A&M University, USA

**Thomas E. Levy**, Center for Cyber- Archaeology and Sustainability, Qualcomm Institute, University of California, San Diego, USA

**Katherine McCusker**, PhD student, AAHVS, Duke University, North Carolina, USA

**Gianfranco Morelli**, Geostudi Astier, Italy

**Helena Murteira**, Senior Researcher, Centre for Art History and Artistic Research (CHAIA), University of Évora, Portugal

**Everett Newton**, PhD student, Marine Science and Conservation, Duke University, North Carolina, USA

**Eva Pietroni**, CNR Institute of Science for Cultural Heritage, Rome, Italy

**Angelika Psenner**, Institute for Urban Design and Landscape Architecture, TU Wien, Vienna, Austria

**Paulo Simões Rodrigues**, Assistant Professor, Department of History, Director of the Centre for Art History and Artistic Research (CHAIA), University of Évora, Portugal

**Jürgen P. Schulze**, Center for Cyber- Archaeology and Sustainability, Qualcomm Institute, University of California, San Diego, USA

**Bill Seaman**, Co-director of the Emergence Lab and Graduate Program Director, Computational Media, Arts & Cultures; Professor, Department of Art, Art History & Visual Studies, Duke University, North Carolina, USA

**Luís Miguel Sequeira**, PhD student at University Trás- os- Montes and Alto Douro (UTAD), Portugal

**Connor Smith**, Center for Cyber- Archaeology and Sustainability, Qualcomm Institute, University of California, San Diego, USA

**Glenn Yago**, Milken Institute, Santa Monica, California, USA

# Digital Cities

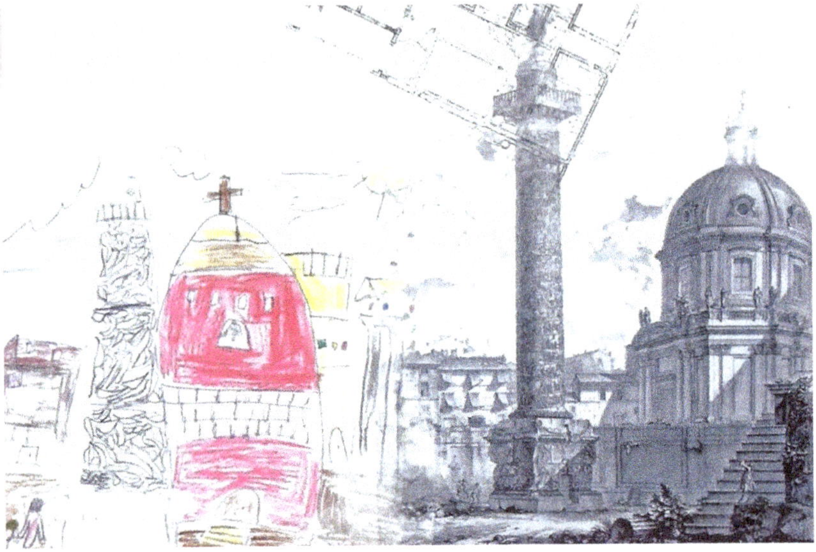

**Figure I.1** Graphics by Claudio Cutrupi (8 years old at the time) for the cover of *From Space to Place: 2nd International Conference on Remote Sensing in Archaeology*, by Stefano Campana and Maurizio Forte (Oxford: BAR Archeopress, 2006).

# Introduction

MAURIZIO FORTE AND HELENA MURTEIRA

Digital cities, intelligent cities, knowledge-based cities, and smart cities are designations that refer to the permeation of the digital into the daily life of cities and the resulting exponential growth and accessibility of urban resources. This digital/technological component can be summarized in the following categories:

- Data capturing (remote sensing, LIDAR)
- Models (3D representations, signs, symbols)
- Diachronic visualizations (4D representations in space and time)
- Operational tools (smart technologies, robotics, networks)

The first three items are related mostly to the digital knowledge and representation of cities through time and space, while the last concerns the functionality of the city in modern and future time.

The complexity of this research topic stems from the transformative nature of the city as a living organism interconnected with the environment (humans, animals, and vegetation), with its geomorphological context and intangible factors such as social, cultural, and political dynamics. The mechanics of this world are based on complex information units that are difficult to describe by internal and external observers. In other terms, the perception and representation of a city are built by a continuous and co-evolving interaction between internal and external communities and by place-making this human knowledge in memories and empirical spaces. This knowledge is created by our spatial embodiment. In fact, the perception of space is a biocultural domain, since the human mind is genetically designed by motor skills for exploring the environment (size, scale, dimensions), while the definition/description of space

Maurizio Forte and Helena Murteira, *Introduction* In: *Digital Cities*. Edited by: Maurizio Forte and Helena Murteira, Oxford University Press (2020). © Oxford University Press.
DOI: 10.1093/oso/9780190498900.003.0001

depends on cultural and chronological contexts. Digital/virtual simulations and the visualization of cities are addressed to highlight the potential attitudes and affordances of the urban context in its multiple relations with the present, past, and future. Therefore, digital cities are able to be experienced as mediated realities at the intersection of empirical observation, multisensory narrative, social memories, and multivocal imagination.

How do history and archaeology come into play when we talk about digital cities? Digital technologies have dramatically changed our relationship with the past. Archaeology, history, art history, and museology have embraced the vast potentialities open to research by digital instruments. Concurrently, our actions and achievements as historians and archaeologists are becoming a fundamental part of the digital network that structures digital cities.

The inexorable impact of the digital world in the humanities gave way to the autonomous discipline of digital humanities more than a decade ago, after having been, for a long period of time, at an intersection of the humanities and computing (Schreibman et al. 2004). The onset of virtual archaeology in the 1990s and its subsequent remarkable development had a crucial effect on the study of past realities—namely on the profound shift that the study, preservation, and display of cultural heritage has suffered in the past two decades (Reilly 1991, Forte and Siliotti 1997, Barceló et al. 2000, Niccolucci 2002, Bentkowska-Kafel and Denard 2012). The fast development of digital technologies and the reinforcement of the relationship between the former and historical research have led to the emergence of cyber-archaeology, a field of study that operates in the digital realm from data capture to research result display (Forte 2010). In cyber-archaeology, information is generated in a dialectical flow from input to interpretation and output, and is translated as simulations open to interaction and, consequently, to diverse readings, continuous updating, and reformulation. Cyber-archaeology is a key example of how the digital world is changing the nature and boundaries of historical research.

Through the intersection of history and computer science, historical evidence is being manipulated and reinvented, thus becoming digital in nature. Digital technologies are not only enhancing the scope of historical, archaeological, and anthropological research but also creating new data and new objects of study.

The study of the city stands out in this context. Cities are microcosms which reflect the structure of societies over time. Archaeologists and historians are faced with the enormous task of understanding the city in all its aspects, in time and space. Lost or profoundly changed urban environments offer a particularly challenging area of research. A vast array of digital tools, such as laser scanning, 3D models, virtual reality, augmented reality, and remote sensing are processing, testing, and displaying archaeological data; architectural remains; built heritage; and written, iconographic, and cartographical sources on a scale and in a fashion never attained before. Multidisciplinary teams, which can bring together historians, architects, geographers, anthropologists, computer engineers, and other experts, are often working in collaborative real-time environments open to interaction with a vast audience of users.

Virtual reconstructions, recreations, or simulations of past urban realities require an extensive amount of information. The data presented rarely reflect similar stages of research and equally consistent evidence. Besides this challenge, models are increasingly more complex in their interactive and immersive features and are thus open to a multitude of interpretations and to different inputs by users. The memory of the past not only is revisited in a new sensory dimension but is also changed by the users' actions and experiences.

The merging of artificial intelligence (AI) and artificial life (AL) simulations in 3D digital environments, increasingly available in the digital game industry, is opening yet another field of possibilities for urban research (Barceló 2007, Bogdanovych et al. 2010, Bordoni et al. 2016). Models are conveyed as rich and multilayered sensory experiences in a social interactive context. As a result, the boundaries between virtual and real environments are becoming increasingly blurred, thus adding a newly significant and contemporaneous view to the past.

We are at the core of a paradigm shift which, although in existence for three decades, is relentlessly changing its contours as a result of the pace and extent of technological development. Therefore, the study of the city is demanding a constant updating not only in terms of data processing and knowledge sharing but also in terms of the methodology in use and the ontological value of the research results. Documents such as the London Charter (2009) and the Principles of Seville (2011) aim to respond to the methodological challenges in place—namely to the requirements of transparency and scientific accuracy. However, the growing complexity of digital tools and the scope of their impact on historical research and display are continuously posing new questions (Parry 2010, 2011; Cameron and Kenderdine 2007 Cooke et al. 2016, among others). It is crucial to understand the hermeneutics of this process, or the epistemological consequences of generating digital data and information in this context, particularly when we are in the presence of a shared social environment.

There are several concepts that pervade our discourse on the relationship between the study of past cities and the digital realm, such as imagination, ubiquity, (cultural) presence, embodiment, and polysensing. They are interconnected and transversal to the real/virtual ambivalence that is increasingly shaping our lives. Imagination is a fundamental part of the scientific research process, whatever its nature or goals. In history and archaeology, virtual simulations are enhancing the role of imagination by exhorting the historical narrative to convey sensory-engaging and scientifically meaningful experiences. Cloud computing, social media, and the proliferation of digital applications and devices are allowing a widespread collection of data, content creation, and sharing. In particular, the increased and massive use of personal devices (e.g., smartphones) develops new and unexplored social attitudes to co-create content for a large-scale consumption, inverting the usual trend of top-down transmission of knowledge. In short, users become content providers.

Scientific research is benefiting from this dynamic. Past urban realities are becoming an important part of contemporaneity through a multitude of channels, from digital repositories, websites, and in situ or ex-situ virtual installations and simulations, to more informal pipelines such as social media. The current demand for open access in the scientific domain is part of this revolution in the information- and knowledge-sharing process that the digital world is allowing and promoting. Information ubiquity is an ongoing phenomenon whose boundaries are yet to be known. The perception of presence ("being there") and cultural presence ("being there and then") is relentlessly pushing its limits in both a sensory and a cognitive dimension, projecting the memory of the past into the present and vice versa. In this context, all-engaging experiences translated as embodied and polysensing simulations of the past or, to be more accurate, of the interpretations of its memory are the ultimate expression of this dialectic relationship among past, present, and future.

The collaborative aspect of this whole process is also being addressed. As Antoine Picon stresses, we are becoming more collective in the way we like to enjoy and share our urban experiences. This assertion includes the way we view the city in its diachronic character. The past is part of our experiences either as residents relying on our city's historical/cultural identity or as tourists wanting to have a glimpse of other cultures and ways of life. This is an experience that, albeit collective, can and should integrate individual singularities:

> From the sensory city, it seems that we are gradually passing to the sentient city. . . . The sentient city might very well represent a first step towards a truly smart city, a city that would not only prove more efficient and livable but be also synonymous with new ways to perceive, deliberate, and ultimately decide about one's collective future. (Picon 2018)

The omnipresence of the digital world in contemporary life has exacerbated the discussion of the advantages and perils of technology. Ethical perspectives and practices are being discussed in the whole spectrum of knowledge and creativity—for example, the Berlin Charter on Citizen Engagement in Cultural Heritage and Humanities Research in the Digital Age (2015). Whatever the perspective, technological advancement is relentless and inescapable.

Where does the study of past cities stand in this technological voracity?

## CONTRIBUTORS AND CONTENT

A building, a complex of buildings, a historic site, or a city is only intelligible as a composite of form, function, and fruition. It is this multifaceted character that gives meaning to architecture and to the inextricable relationship that it has with the urban cosmos. In this book, architecture and the city meet at this crossroads and it is as such that the different case studies in this volume illustrate the subject matter in discussion.

This book considers the city from prehistoric settlements to the present, in different geographical contexts. The authors have different backgrounds—that is, in archaeology, history, urban history/architectural history, architecture, and computer science and technology. The different chapters reflect different levels of engagement with digital tools. However, they all share a common ground: they are at the intersection of the digital world and the study of the city, with the means to advance knowledge and its dissemination.

These are experiences that demonstrate the concurrent relationships between history, archaeology, and digital technology. Some case studies are dated, others are works in progress, but all approaches here share the same interrogation that pervades the study of the past: What is changing? Is it the tools, the methodology, or the subject of research itself? Are we facing a novel paradigm of the narrative on the past, not only by allowing the testing of working hypothesis, collaborative interaction, often in real time, of different experts with various scientific backgrounds but also by making it a contemporaneous reality through sensory experience in a multivocal context? How and to what extent are the epistemological boundaries being expanded? These are questions that do not cease to be asked while technology advances, historical research strives to keep up with the pace, and both interplay and interconnect as digital heritage. These questions do not need to be answered in full; that's an impossible task. It is, however, imperative that we acknowledge them as we go along. The organization of this book's contents into four different parts is more of a practical nature. In one way or another, all the chapters overlap on the questions just mentioned. It is clear that whatever the historic period under study, the interpretation of the past and contemporaneity are today, more than ever, a dialectical reality.

These different chapters thus contribute to a coherent approach to the subject and each provides a suggestive snapshot of the most recent research within its respective field. In part one, "Methodological Challenges," chapter 1, about the Vulci 3000 project, on the Etruscan and Roman site of Vulci, by Maurizio Forte and colleagues, explores the topic of the optimization of scientific research by digital technologies, remote sensing, and 3D visualization. It draws upon the inescapable scientific requirement of merging different technologies in the research process and the resulting methodological and epistemological challenges which archaeologists and historians face. The result is an ongoing process of virtual simulations, predictive models, digital maps, and cityscapes where the core question is the transformation and development of circa 1,500 years of urbanism on the same plateau.

Richard Beacham, in chapter 2, examines the close interaction between digital tools of visualization and simulation and architectural settings of the past. Roman domestic architecture, particularly in its theatrical dimension, is discussed in relation to the potentials and challenges of virtual heritage through the Villa of Oplontis and the Villa of Fannius Synister at Boscoreale projects by the King's Visualisation Lab (KLV).

In chapter 3, Caroline Bruzelius addresses the process of change in architecture and urban design through the project Visualizing Venice. Digital visualization allows this process to be understood within the singularities and fluidity of its pace—as a "flow in time," as the author stresses, adding value to research and heritage display.

Jean-Yves Blaise and Iwona Dudek, in chapter 4, use the case study of the city center of Kraków, Poland, to illustrate how information visualization can be a fundamental tool in dealing with the methodological constraints of handling ill-defined, extensive, and heterogeneous data in urban historical research.

In chapter 5, Eva Pietroni analyzes the impact of digital technologies—namely AI and AL, on the study and display of either tangible or intangible urban heritage from both methodological and epistemological perspectives. The array of digital tools available to researchers, and the diverse topics that may be addressed from an integrative perspective, is allowing a better response to the several constraints of historical research, as in the case of uncertainty. Pietroni argues that in Virtual Heritage, the final result is more complex and ample than the mere sum of its parts, emphasizing the organic character of the process and its multiple significances and readings.

In the final chapter of part 1, Luís Miguel Sequeira underlines the reinforcement of the relationship between digital technology and virtual heritage. Immersive environments allowing a scientific team to test data and a working hypothesis in a collaborative fashion in real time were provided by virtual-worlds technologies such as Second Life and OpenSimulator from 2003 and 2007, respectively. Sequeira stresses the laboratorial character of these models for the study of the physical dimensions of past cities. However, immersion in the memory of a city requires the simulation of also its social and cultural aspects. Scientific research and embodied experiences demand the city be explored as a total organism. Thus, it is crucial to provide the scientific team with an "easy to build" solution for crowd simulation—in other words, an accessible tool for incorporating AI into virtual reality. As a result, the "virtual laboratory of archaeology" can be expanded to the city's multiple components.

Part two, "Conservation, Requalification, and Communication," opens with chapter 7, by Thomas E. Levy and colleagues, on the subject of endangered cultural heritage. The Mar Saba Monastery, in the Judean Desert, is presented in the framework of a cyber-archaeology project that explores the potential of this field of research as a scientific, cultural, and tourism asset. The relationship between the technology used and the goals of the project is emphasized in its multiple aspects—namely its economic value (the creation of a Cultural Heritage Asset District) and its aptitude for research updating and user fruition.

Through the analysis of two projects on Oporto's Historic Center in Portugal, chapter 8 by Maria Leonor Botelho considers the potential of virtual heritage visualization as a means to understanding the past, preserving its heritage and translating its results as a multifaceted sensory experience for a wide and heterogeneous audience.

In chapter 9, Sarah Kenderdine discusses the concepts of "narrative," "place," and "presence" in virtual reality in relation to museum displays of urban memory and heritage. Through a number of projects on ancient cities in a museological context, Kenderdine addresses the need to convey embodied and socially engaging virtual experiences. The shift from "place" to "place and presence" in virtual heritage, as opposed to the focus that the digital industry places on the former, is stressed as the path to be followed. The concept of "deep mapping" in this context is also explored.

Part three, "Hermeneutics and Epistemological Boundaries," begins with chapter 10, where Zeynep Aktüre examines three projects on Çatalhöyük, in Turkey, to address the topic of methodology in virtual archaeology and virtual heritage, particularly in regard to transparency and scientific rigor, as well as the epistemic value of research results and their display. Aktüre uses Umberto Eco's "open work" and Siegfried Kracauer's "cinematic materiality" concepts as reference to reach a better understanding of what an "open multivocality" means—or should mean—in the context of virtual heritage.

Chapter 11's authors Maria Alexandra Gago da Câmara and colleagues explore the concept of virtual simulations of past cities as "memoryscapes" through the example of four different projects on Lisbon, Portugal, before the 1755 earthquake. The city of the past is presented as more than historiography, as also "imaginability" (the merging of different layers of significance that shaped the city throughout time and allows today its perception and understanding). This important dimension of the city is crucial in virtual recreations and simulations allowing users to engage in a simulacrum of the past and to become active participants in the memory that is being conveyed.

In the final part of the book, "Research, Planning, and Learning," chapter 12, by Angelika Psenner, uses the case study of the city of Vienna to examine the potential of urban parterre mapping and urban parterre modeling in the study of the buildings' ground floors and their relationships, as coherent and complete units, with the city's urban layout and fruition. Ultimately, this chapter explores the notion that digital technology is expanding architectural and urban research.

Chapter 13, by Gabriela Campagnol and colleagues, is a suggestive example of how 3D immersive technology is meeting the demands of education in architecture and urban planning. A BIM CAVE, developed by Julian Kang and his students at the Department of Architecture at Texas A&M University, is being used as a core educational tool to facilitate students' perception and understanding of the architectural and urban spatial dynamic, from either a synchronic or a diachronic perspective.

In the final chapter of the book and of this part, Bill Seaman explores the relationship between the contemporary city and computational media. Seaman considers the impact of information ubiquity on the cityscape and emphasizes the need to combine cyber-archaeology and media archaeology to create polysensing and multimodal technological systems that enhance our understanding of the urban context through time.

To summarize, this book explores the idea of digital cities, through a historical, social, and multimodal context, as both past and contemporary entities, thus able to be experienced as memories, as present-day occurrences, and as experiments for a future reality.

## REFERENCES


Barceló, Juan Antonio. 2007. "Automatic Archaeology: Bridging the Gap Between Virtual Reality, Artificial Intelligence, and Archaeology." In *Theorizing Digital Cultural Heritage: A Critical Discourse*, edited by Fiona Cameron and Sarah Kenderdine, 437–455. Cambridge, MA: MIT Press.

Barceló, J., M. Forte, and D. Sanders, eds. 2000. *Virtual Reality in Archaeology*. British Archaeological Reports, International Series 843. Oxford: Archaeopress.

Bentkowska-Kafel, A., and H. Denard, eds. 2012. *Paradata and Transparency in Virtual Heritage*. Farnham, Surrey: Ashgate.

Berlin Charter on Citizen Engagement in Cultural Heritage and Humanities Research in the Digital Age. 2015. http://www.civic-epistemologies.eu/BerlinCharter.pdf.

Bogdanovych, A., J. A. Rodriguez-Aguilar, S. Simoff, and A. Cohen. 2010. "Authentic Interactive Reenactment of Cultural Heritage with 3D Virtual Worlds and Artificial Intelligence." *Applied Artificial Intelligence* 24(6): 617–647. doi: 10.1080/08839514.2010.492172.

Bordoni, Luciana, Francesco Mele, and Antonio Sorgente, eds. 2016. *Artificial Intelligence for Cultural Heritage*. Newcastle Upon Tyne: Cambridge Scholars Publishing.

Cameron, Fiona, and Sarah Kenderdine. 2007. *Theorizing Digital Cultural Heritage: A Critical Discourse*. Cambridge, MA: MIT Press.

Forte, Maurizio, ed. 2010. *Cyber-Archaeology*. British Archaeological Reports, International Series 2177. Oxford: Archaeopress.

Forte, Maurizio, and Alberto Siliotti. 1997. *Virtual Archaeology: Recreating Ancient Worlds*. New York: Abrams.

London Charter for the Computer-Based Visualization of Cultural Heritage. 2009. http://www.londoncharter.org/.

Niccolucci, Franco, ed. 2002. *Virtual Archaeology: Proceedings of the VAST Conference, Arezzo, Italy*. British Archaeological Reports, International Series 834. Oxford: Archaeopress.

Parry, R., ed. 2010. *Museums in a Digital Age*. Abingdon and New York: Routledge.

Parry, R. 2011. "Transfer Protocols: Museum Codes and Ethics in the New Digital Environment." In *The Routledge Companion to Museum Ethics*, edited by J. Marstine, 316–331. London and New York: Routledge.

Picon, A. 2018. "Urban Sensing: Toward a New Form of Collective Consciousness.?" In *Humanizing Digital Reality*, edited by K. De Rycke et al. Singapore: Springer.

Principles of Seville: International Principles of Virtual Archaeology. 2011. http://smartheritage.com/seville-principles/seville-principles.

Reilly, Paul. 1991. "Towards a Virtual Archaeology." In *CAA90. Computer Applications and Quantitative Methods in Archaeology 1990*, edited by S. Rahtz and K. Lockyear, 132–139. British Archaeological Reports, International Series 565. Oxford: Tempus Reparatum.

Schreibman, Susan, Ray Siemens, and John Unsworth, eds. 2004. *A Companion to Digital Humanities*. Oxford: Blackwell.

PART ONE

# *Methodological Challenges*

CHAPTER 1

# Vulci 3000

## *A Digital Challenge for the Interpretation of Etruscan and Roman Cities*

MAURIZIO FORTE, NEVIO DANELON, DAVID JOHNSTON, KATHERINE MCCUSKER, EVERETT NEWTON, GIANFRANCO MORELLI, AND GIANLUCA CATANZARITI

### THE SITE

Vulci (Viterbo, Italy, fl. tenth century BCE–fifth century CE) was one of the largest and most important cities of ancient Etruria and one of the largest pre-Roman cities in the first millennium BCE on the Italian peninsula. The habitation site is an urban context, uniquely stratified and mostly untouched, that includes Iron Age, Etruscan, Roman, and medieval settlements in the same general area. It covered approximately 126 hectares and had an estimated population of thousands of inhabitants in the Classical period (sixth–fifth centuries BCE). It was part of the Etruscan Dodecapolis ("twelve cities"), the Etruscan federation of the most important cities of Etruria. The only historical information available reports that it was conquered by the Romans in 280 BCE, but only archaeological data can reveal urban features, extension, and architectural settings of the pre-Roman and Roman city.

The concentration of archaeological investigations in funeral areas shows the power of elite classes and their impressive capacity to trade with the Attic region (in Greece), but such investigations misinterpret the social, political, and economic aspects of the city. In other words, the entire social body of the city cannot be represented just by aristocratic tombs, funerary art, and ritual

Maurizio Forte, Nevio Danelon, David Johnston, Katherine McCusker, Everett Newton, Gianfranco Morelli, and Gianluca Catanzariti, *Vulci 3000* In: *Digital Cities*. Edited by: Maurizio Forte and Helena Murteira, Oxford University Press (2020). © Oxford University Press.
DOI: 10.1093/oso/9780190498900.003.0002

monuments. Less than 5% of the ancient urban area is known and excavated; no stratigraphic excavation in the city has been published in the last decades.

## METHODOLOGY AND CITYSCAPES

The study of Etruscan cities and urbanism is related to the concept of the city-state and its organization, a very complex and independent socioeconomic unit, self-organized but open to trading and connections in and out of the influence of Etruscan territory (Fulminante 2013). The methodological approach of this project involves: (1) noninvasive technologies for a predictive and extensive interpretation of the site (remote sensing); (2) archaeological excavations (single-context method) in the most important urban areas of the site (micro-scale); (3) accurate and 3D digital documentation of the landscape (macro-scale) and of the excavation trench (micro-scale); (4) lab analyses of archaeological finds and deposits; and (5) digital archiving and dissemination by the collaborative webGIS, remote-sensing data, 3D repositories, and virtual reality (VR) applications.

The combination of these methods creates a new research model for the study and interpretation of ancient cities, applicable to other contexts and sites. The methodological workflow starts with predictive modeling techniques, then based on preliminary results, it proceeds to excavation and finally to dissemination/communication of data through the web and VR applications.

The stratigraphic overlay of about 1,400 years of urban history in the Western Forum (documented by geophysical prospections and by the findings of the 2016–17 excavation seasons), and possibly in most of the southern part of the city, constitutes an extraordinary example of diachronic continuous occupation of a urban space. The chronology can be validated by the Great Temple (first phase, late sixth–fifth century BCE) and by the stratigraphic thickness diagnosed by ground-penetrating radar (GPR), a technology which measures underground densities, and by the recent archaeological excavations in 2016–17. In fact, the foundations of an Augustan building (first century CE) overlay more than 1.5 meters of earlier stratigraphy related to more ancient buildings and pre-Roman wells and cisterns.

## A HISTORY OF EXCAVATIONS AND SURVEYS AT VULCI

The first modern-age excavations at Vulci were begun in 1825 by the Candelori brothers and were quickly followed by wider scale excavations by Lucien Bonaparte, the prince of Canino. These early excavations uncovered dozens of richly decorated tombs, such as the Tomb of the Sun and Moon, in 1830. Thousands of artifacts were removed from various tombs and sold off to European collectors, enriching such families as the Feoli, Candelori, Campanari, and Fossati. Vincenzo Campanari, the "governo pontificio," oversaw excavations at Vulci between 1835 and 1837 (Buranelli 1991, Campanari 1836). In 1857, Alessandro François discovered an extensive late-fourth-century elaborately painted tomb. The frescos removed from the tomb were pinnacle contributions to the construction of Etruscan history and identity. The burial became known as the François Tomb, and it remains one of the few tombs from this century whose location did not become lost and which is still open to the public today. The large burial mound Cuccumella was excavated by Torlonia-Marcelliani between 1875 and 1883. While further excavations of necropolises continued throughout the remainder of the nineteenth century, the lack of controlled and systematic excavation in this century and the early twentieth century has made the study and contextualization of the uncovered tombs and materials impossible.

Fieldwork during the early twentieth century at Vulci continued with a strong focus on the necropolises with excavations by Bendinelli from 1919 to 1924, by Mengarelli from 1925 to 1929, and by Ferraguti from 1928 to 1932 (Falconi Amorelli 1983, 1987; Buranelli 1994; Mengarelli 1929). Excavation continued throughout the 1920s and '30s, ceased during World War II, but resumed during the 1950s after Italy was stabilized. Bartoccini carried out three years of excavation between 1956 and 1958 to completely uncover the Great Temple, the first large excavation inside the city (Bartoccini 1961, 1963). Although Bartoccini's notes are more complete than previous excavations, his unsystematic approach provides only a partial, deficient record. Further unsystematic excavations were carried out in the 1970s and '80s, such as excavation during the early 1980s by German archaeologist Raddatz, focusing on burial grounds and identifying the area of Pozzatella. During the late twentieth century, there were numerous topographical and aerial studies of Vulci, including those of Sgubini Moretti and Giorgio Pocobelli (Sgubini Moretti 1988, 1993, 1994, 2001; see also Pacciarelli 1991 and Gleba 2009). The government officially turned Vulci into a protected natural-archaeological park in 2001. The foundation for a triumphal arch was discovered on the west side of the Roman forum in 2003 and was reconstructed under the auspices of the park. Most recent excavations at Vulci have focused, by necessity, on areas that were either partially looted or that were damaged by natural forces. As rescue and largely unplanned excavations, the contexts were disturbed and damaged, which prevented any systematic excavation. In 2011, Carlo Casi, the director of the Vulci archaeological park, set out to rediscover the lost location of the Tomb of the Sun and Moon, using topographic maps from the nineteenth century. Although he failed to reach that goal, he did uncover over twenty small graves and tombs and two larger funerary complexes, one of which was called the Tomb of Silver Hands, owing to the find of an exquisite set of hands made of silver, located in the tomb (Sgubini Moretti et al. 2014, Carosi and Regoli 2014). A geological survey of Vulci and its landscape was completed by Marchetti in 2013, marking one of the most recent surveys in addition to the Vulci 3000 project of the area (Marchetti et al. 2014).

## THE VULCI LANDSCAPE

Vulci is located atop a plateau, providing a strong natural defense. The city is bordered by the Fiora River running to the west and southwest, with sloping hills in the other directions. People in Vulci used this river to ferry goods down to and up from Regisvilla—a Vulci-controlled port located along the coast of the Tyrrhenian Sea at the mouth of the Fiora River—and, in extension, to the entire region. Use of this port declined after control of Vulci fell into the hands of the Romans, owing to the founding of their own port, Cosa, just up the coast from Regisvilla.

The city itself is surrounded by necropolises, which are vital to an understanding of Vulci, as they are visible and continuous social reminders of the ties that bind and divide societies (Bartoccini 2001). For example, to the west of the city is the Cuccumella Tomb, the largest tomb mound in Italy. This mound is visible from the city center, and it is a monumental status marker. Both old and new aerial images, along with targeted GPR surveys, can offer the location and layouts of these necropolises, which leads to a deeper interpretation of the growth of the city and to its identity and social traditions. These cities of the dead have helped form the early view of the urban portion of the city, although they do not provide the full wealth of information about it as an urban center.

Visibly surrounding the city are large walls built around the late fifth/early fourth century BCE, marking a shift in the city's focus away from the sea and toward the interior (Carandini 1985). The city had five gates, one each in the north, east, south, southwest and west. Today, the north and west gates have been excavated and partially reconstructed for visitors. Inside these walls, only the main east–west decumanus and six structures are visible: the Great Temple, the House of Cryptoporticus, the House of the Fisherman, a Roman-era reconstructed triumphal arch, a medieval building, and part of a Roman basilica (figure 1.1).

It is probable that while the Etruscans used the southern and eastern acropolises as important monumentalized areas, the Romans opted to not use those areas; rather, they centered those activities in the area of the city known as the Western Forum. Since knowledge of the urban plan for the site is limited, owing to a lack of archaeological excavations, the only possible strategy for a large-scale spatial analysis is the use of remote-sensing technologies.

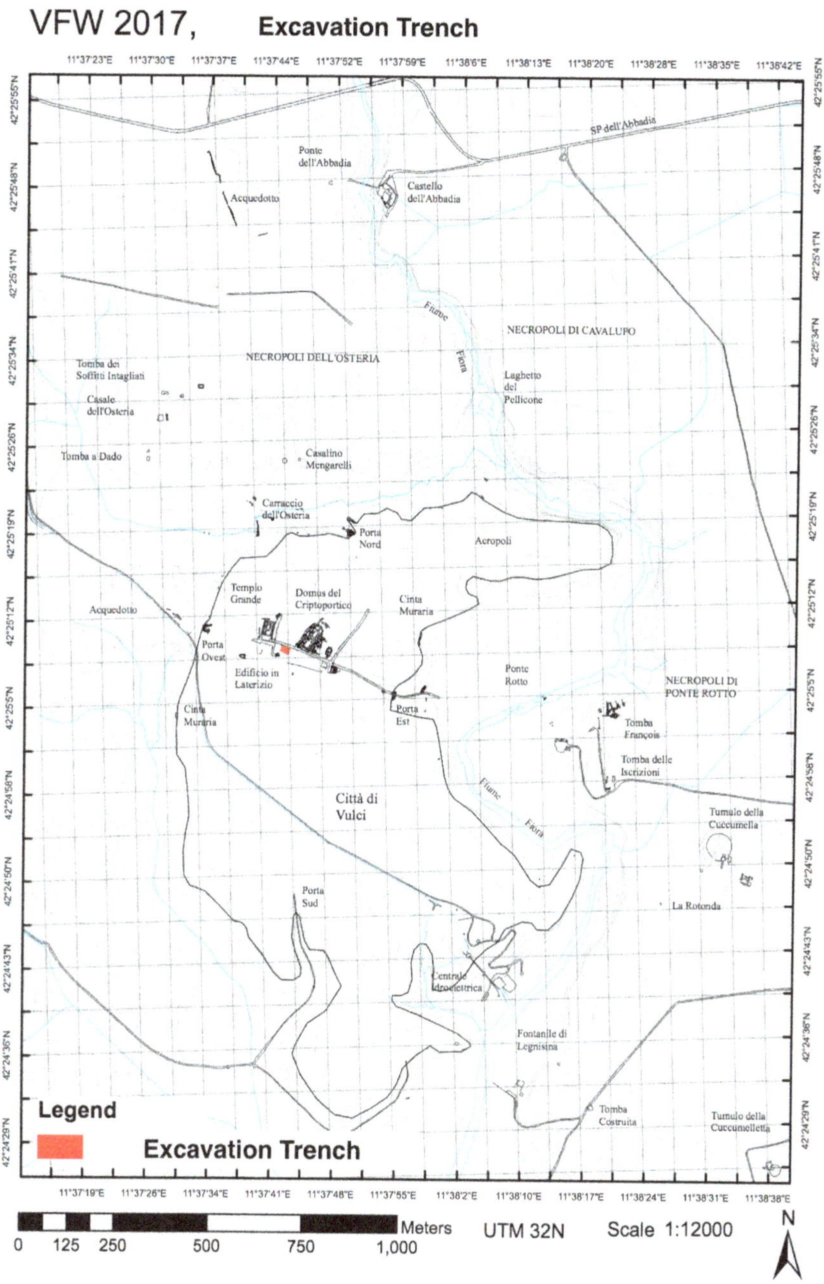


**Figure 1.1** Photogrammetric plan of Vulci's plateau, with the main areas of old and new excavations (in red).

## REMOTE SENSING

Remote sensing is a method of collecting information and data from a distance via use of noninvasive digital techniques, including magnetometry, ground-penetrating radar (GPR), and aerial images via drone (using the primary colors of red, green, blue [RGB] and multispectral), in our case. These technologies have allowed us to see features still hidden beneath the surface and thus to explore the unexcavated areas of the city. While magnetometry did not provide interpretable results, both GPR and drone and aerial images did provide important data for a more comprehensive picture of Vulci's urban center.

The GPR method is a geophysical technique for subsoil investigation involving the transmission of electromagnetic pulses through the subsoil and of receiving, from the surface, the reflections caused by the discontinuities that are present. GPR is sensitive to variations in the lithology, density, and compacting materials, and it is useful in the characterization of both natural and anthropic materials.

The GPR survey has covered a total area of 3.8 hectares, including areas interposed between the Big Temple and the House of Cryptoporticus and a wide area between the west gate and the House of the Fisherman (indicated in yellow, figure 1.2).

Data were acquired with a sixteen-channel IDS Stream-X 200 MHz GPR, towed by a 4x4 vehicle; this system is able to record with a single swath an area of 1.68 meters, composed in a series of fifteen radargrams, equally spaced

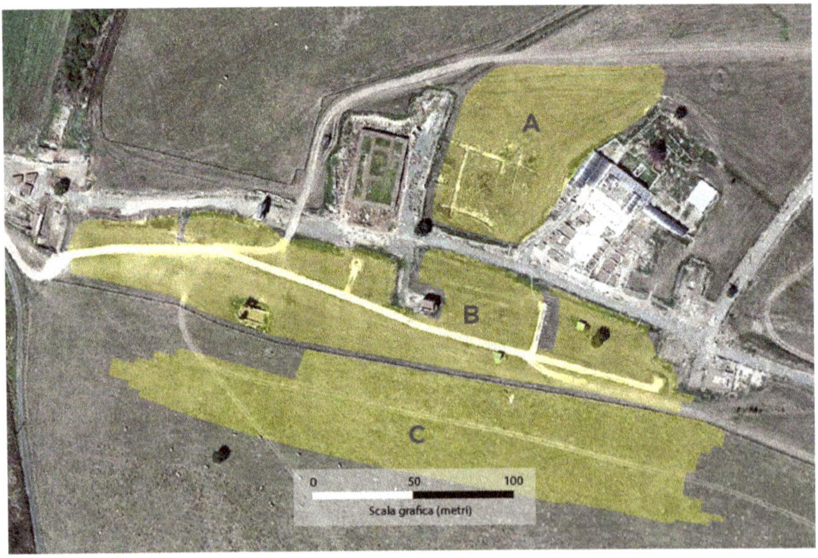


**Figure 1.2** GPR map: in yellow, the area investigated with GPR. Total surface: 3.78 hectares.

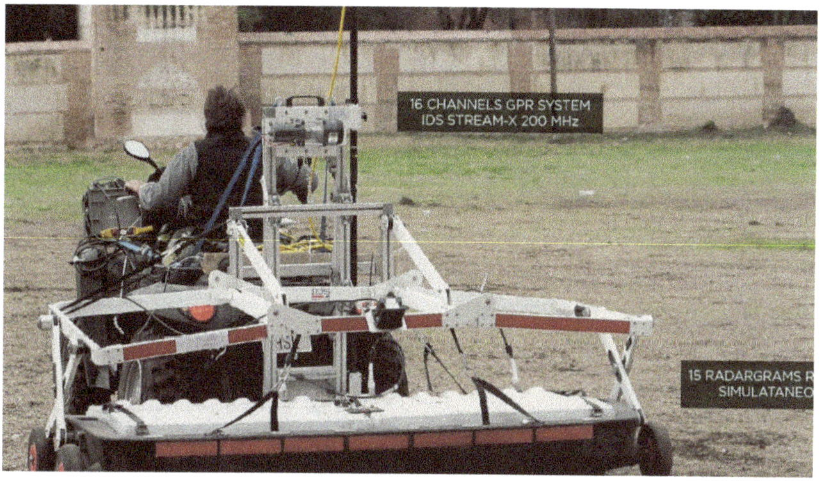


**Figure 1.3** IDS Stream-X 200 MHz, 16 channels.

(12 cm, figure 1.3). The GPR system was interfaced and synchronized with a Trimble R7 GPS device, so all data were geo-referenced in real time with an error of less than 2 centimeters.

Preliminary data processing was performed with the GRED(HD) software of IDS Spa, by applying a standard sequence of filtering to all acquired radargrams; the final post-processing was conducted by the software GPR-Slice of the Geophysical Archaeometry Laboratory of Los Angeles. The radargrams were filtered by applying the following workflow: Dewow -> Move star time -> Gain -> Band-Pass Filter -> Background removal -> Hilbert transform. This data processing has led to the creation of high-definition time-slices (pixel dimensions between 7.5 and 10 cm) of the subsoil in the four investigated sectors.

The final results are shown in figures 1.4–1.9, with the time-slices of each sector superimposed on an orthophoto of the Vulci plateau and a total depth of about 2 meters from the surface. Figures 1.4–1.7 show radar images relative to the northern sector investigated, obtained by projecting on a single horizontal plane all the radar signals detected within four successive horizons of the subsoil, every 0.25 meter in depth (GPR overlay analysis method by GPR-Slice software). Figures 1.8 and 1.9 represent the maps of all the reflectors present from the surface down to an 80 centimeter depth for the central and southern sectors.

The radar data have been represented graphically using a color scale that is able to emphasize the discontinuity of the subsoil. The areas in yellow-brown are associable with the effects produced by the background. The anomalies in black-gray-white are associable to the presence of architectural structures, such as walls, floors, and cavities formed by limestone and volcanic lithology,

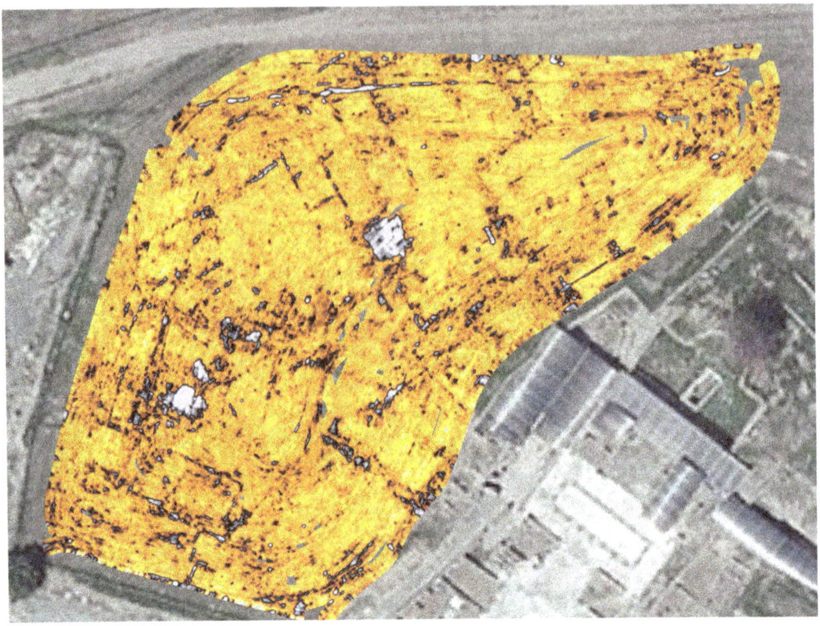

**Figure 1.4** Area D: GPR time-slice representative of 50 cm in depth from the surface.

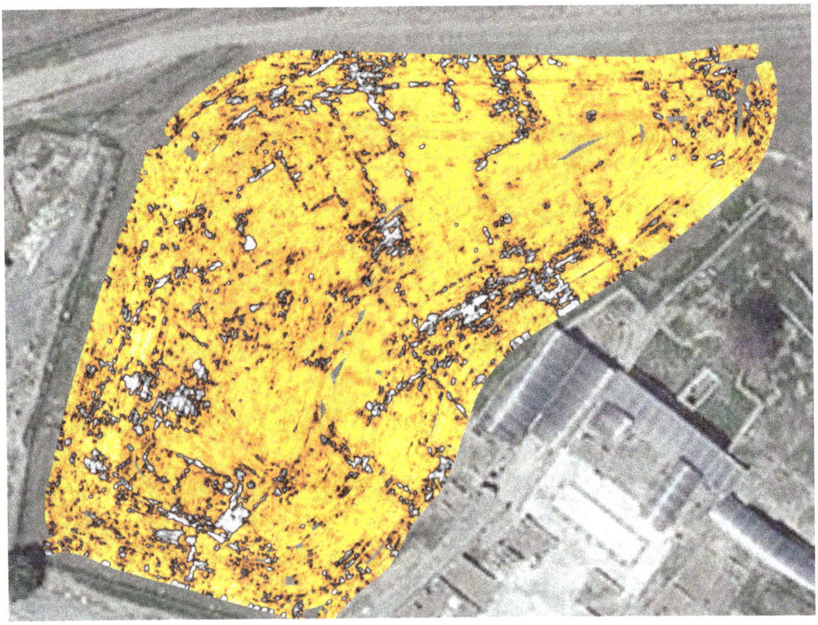

**Figure 1.5** Area D: GPR time-slice representative of 75 cm in depth from the surface.

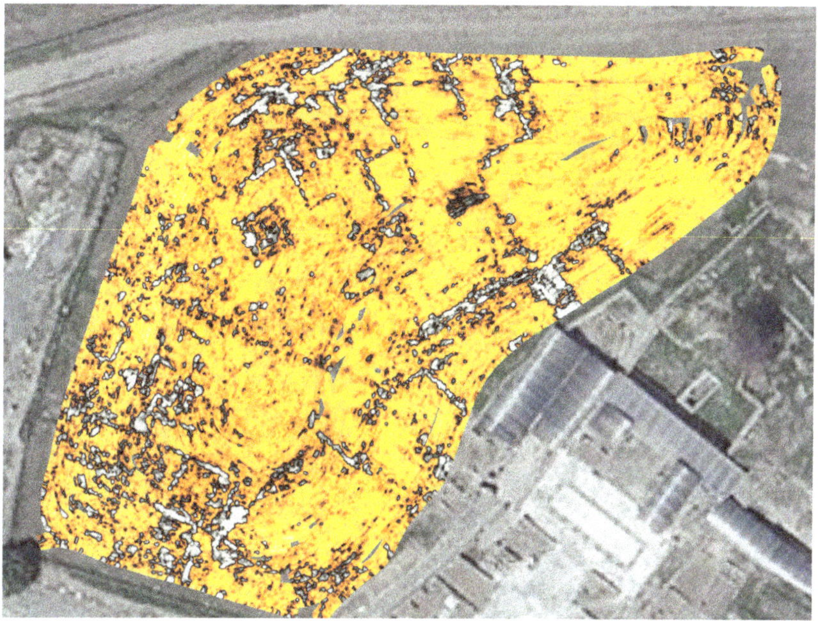

**Figure 1.6** Area D: GPR time-slice representative of 100 cm in depth from the surface.

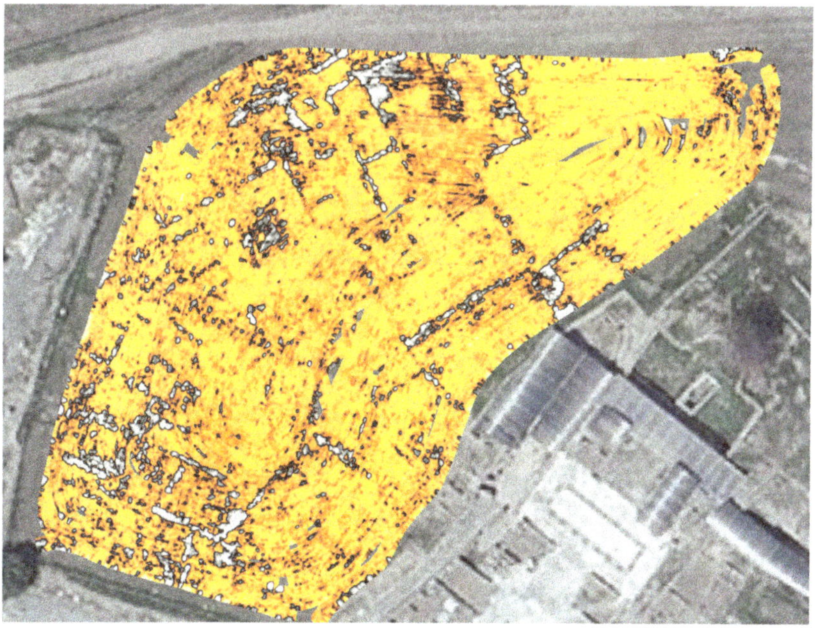

**Figure 1.7** Area D: GPR time-slice representative of 125 cm in depth from the surface.

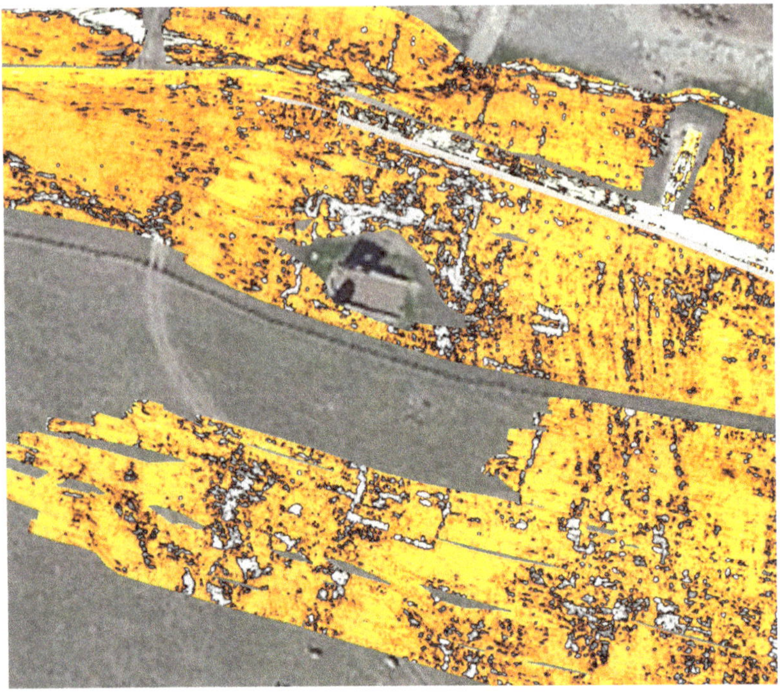

**Figure 1.8** West portion of the areas A, B, and C; GPR time-slice combining all reflectors from surface to 80 cm depth.

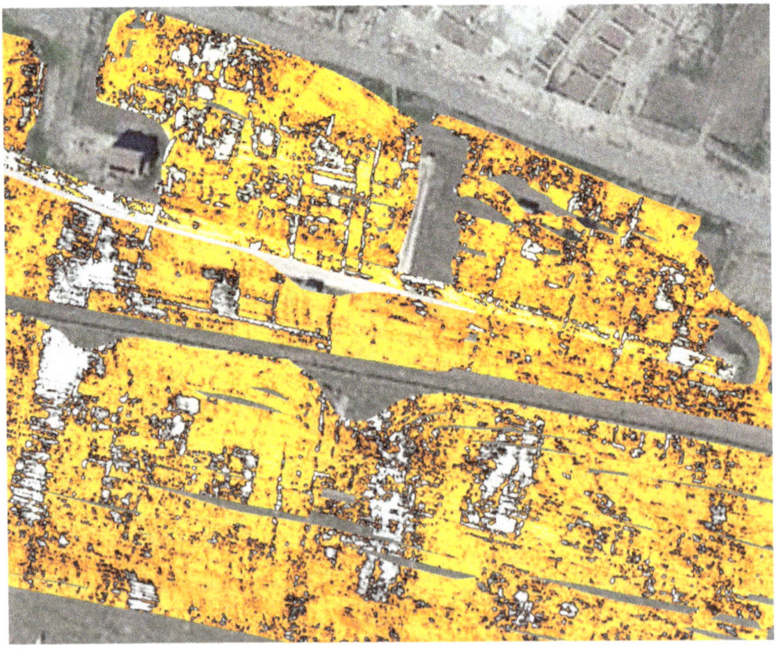

**Figure 1.9** East portion of the areas A, B, and C; GPR time-slice combining all reflectors from surface to 80 cm depth.

which are totally different in respect to the surrounding materials, represented by lithology composed by medium-fine particles.

In the northern part of the investigated area (area D), some masonry structures with orientation axes orthogonal to north–southeast and south–northeast were identified, and they seem related to the Roman grid plan of the city. In the remaining sectors, the orientation of the structures is rotated clockwise from 30 to 40 degrees, which is consistent with the Etruscan temple.

From the combined use of remote-sensing data it is possible to extrapolate the layout of the city, particularly in and surrounding the Western Forum. Figure 1.10 provides an overview of the known features of the city, as well as the preliminary interpreted features through remote-sensing technology. Moving from north to south, the aerial images and GPR data show the visually empty (to the human eye) field between the Great Temple and the House of Cryptoporticus to have numerous iterations of features down through 2.5 meters, signaling a very active and changing area in which features shifted orientations (see area A, figure 1.10). It is reasonable to conclude that this space would have been highly sought after and reused, as it is both physically central and a socially prominent location owing to the adjacent building (figure 1.10).

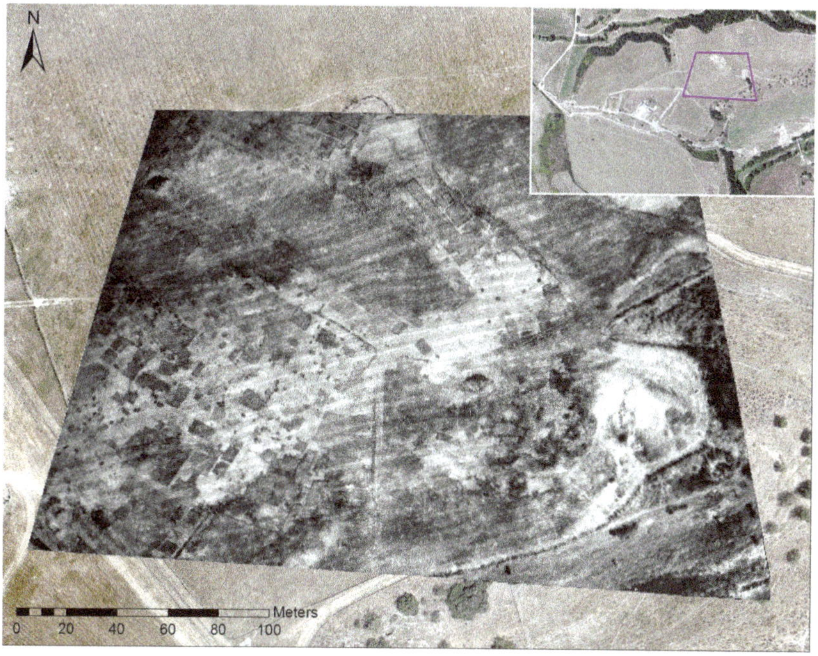


**Figure 1.10** This map outlines the currently known features in black, while showing a selection of new interpreted features based on preliminary conclusions in red. The letters correspond to specific features (further indicated in the section) as follows: A–an impluvium, B–roads, C–oval structures, D–a basilica, E1 and E2–larger buildings in the southern portion of the Western Forum area, and F–a possible theatre.

An examination of the GPR data reveals a long rectangle around 6 meters in width toward the southern portion of the Western Forum, which is aligned with the north–south road to the east of the Great Temple (area B, figure 1.10). While some of the GPR data is piecemeal and allows us only to discern or confirm the presence of multiple features in this area, several structures, including large building footprints, are more evident (area E2, figure 1.10). To the west, several oval buildings, one down 1.1 meters and another at 0.9 meter, are present (area C, figure 1.10). They spatially overlap one another, and we believe them to date to the very late occupation of the site because of their shallow depth, oval shape, and location in an odd position near the west gate, and because they are angled differently from the other features in this area.

Moving eastward, underneath the medieval-era building, the GPR results show evidence beginning at -1.5 meters of a large building with two rounded towers toward the southwestern and northwestern corners (area D, figure 1.10). Further, the remote-sensing data promote the layout of several larger buildings to the south of this feature, although the larger buildings are not visible below 2.3 meters (area E1, figure 1.10). While several buildings appear to be multi-celled or have multiple adjoining rooms, one building reveals a semicircular feature against one wall with radial lines coming out from it (area F, figure 1.10). It is reasonable to place these larger buildings near the Western Forum, as the grander, public buildings of the city. The spatial information gleaned from GPR data does not provide solid conclusions, but it offers several possibilities. It is conceivable that once built, the features and buildings in this area (E1 and F) were longstanding, and if they were reused or reconstructed, then the same foundation was used. On the other hand, it is also possible that the Romans destroyed any features in this area when they remodeled the city and reoriented it around the new Western Forum.

Shifting our focus from the GPR data to the new aerial data, we see a large, multi-room building through the red-edge imagery in the westernmost portion of the city (figure 1.11). Further, these red-edge images show a dark, curved line separating both the eastern and southern acropolises from the main portion of the city.

The 2017 GPR survey was limited in scope to study smaller areas of interest, as well as to document the best methodology using a new model machine. The results thus far have indicated four areas of further interest, three of which are just south of the Western Forum. Two of these areas reveal a long, thick rectangular area of dense material (similar to that seen in the previous GPR data indicating the continuation of a road). In both areas which reveal road-like features, we can also partially see a feature with multiple areas, although the lack of GPR data prevents the emergence of a full picture. The other, third area of interest reveals another multiple-area feature whose

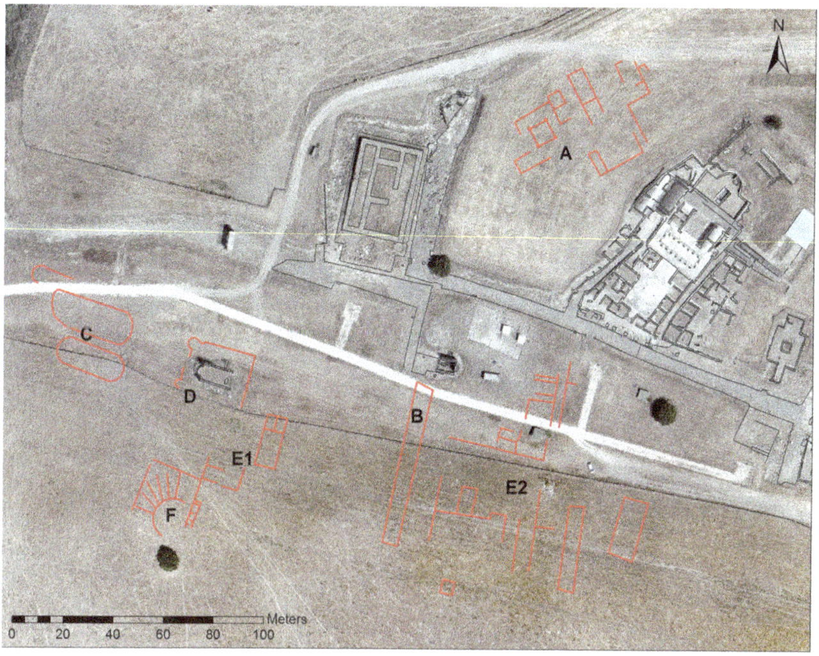


**Figure 1.11** This GPR slice (down 1.05 m) reveals the outline of an impluvium inside a courtyard as part of a larger feature with additional rooms and areas.

footprint is incomplete. The fourth interesting area is a medium-sized t-tomb to the west of the Cuccumella Tomb.

The remote-sensing data sets provide an extended spatial view of Vulci, particularly highlighting the presence of medium and large structures in the southern and southeastern portions of the Western Forum. Black and white oblique aerial images from 1975 provide a unique perspective on this landscape because they offer information from a different point in time, as the visible crop marks in these photos differ from those today (Pocobelli 2004). However, these photos were not geo-tagged images, and thus they had to be geo-referenced before they could be incorporated into the larger pool of data (figure 1.12). Moreover, these images have poor resolution, exacerbated by the distortion created when correctly geo-referencing the oblique angled images. Therefore, new aerial data were collected using drones to supplement and improve the original aerial images.

During the periods of August 4–8, 2016, and July 22–27, 2017, a fixed-wing remotely piloted aircraft was flown over the Vulci plateau. Specifically, it was the eBee, a modular fixed-wing unmanned aircraft system (UAS) produced by the company senseFly. The eBee has a lightweight foam airframe with a single rear-mounted brushless electric motor powered by a lithium polymer battery. The aircraft has a wing span of 96 centimeters and weighs 0.7 kilogram. During the surveys, the eBee followed a pre-programmed three-dimensional

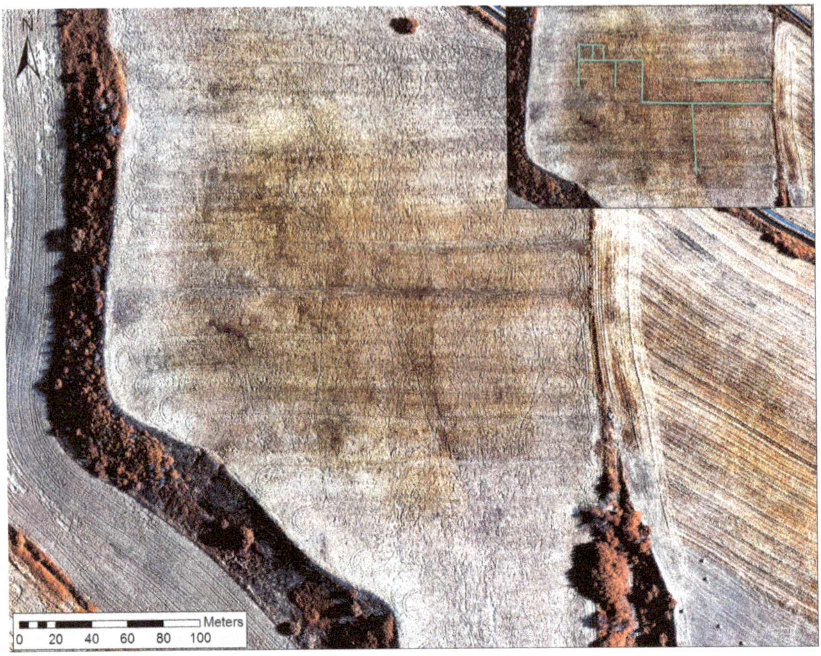


**Figure 1.12** The geo-referenced 1975 black-and-white photograph reveals a pattern of hidden features in the northeast area of the city not seen in the colored aerial image.
Credit: 1975 photography courtesy of Giorgio Pocobelli, 2004.

flight path guided by a precision GPS sensor, a high-resolution barometer, a ground-sensing camera, and wind-speed indicators. Failsafe logic within the autopilot was set to return the eBee to the landing zone if it experienced anomalies in sensor performance or extreme wind conditions, and it telemetered the flight data to the operator over ultra-high frequencies in real time. The instrument was launched by hand and recovered after a linear approach/landing at a predetermined 10 meter radius region. The eBee aircraft supports multiple payloads and post-flight software that can process individual images, orthomosaics, 3D point clouds, digital elevation maps, and multispectral reflectance maps.

For the Vulci surveys, three separate Canon PowerShot S110 cameras were utilized. They were specifically filtered for near-infrared (NIR), red-edge (RE), and red-green-blue (RGB) spectrums, and were electronically adapted for the eBee aircraft to allow the eBee to trigger the camera at precise GPS locations and aircraft attitudes. Radiometrically calibrated NIR and RE remote sensors onboard the drone were used to evaluate plant stress, which may indicate the presence of ancient structures. The NIR camera provides NIR, green, and red bands, while the RE camera provides RE, green, and blue bands. (These spectral bands are frequently used by the agricultural community to analyze crop health.) The team also performed missions with the standard RGB sensor for imagery comparison and to generate a virtual tour of Vulci. The priority ranking for the missions was NIR, then RE, and finally RGB.

Each spectral survey required approximately eight flights of 25 to 30 minutes and 2,900 images to cover the major portions of the Vulci plateau and the eastern necropolis that includes the Cuccumella Tomb (approximately 3.1 sq. km). Takeoff and landing locations were carefully selected to avoid overflight of sensitive areas, and the aircraft remained within the line of sight of the remote pilot so as to ensure safety and avoid conflicts with conventional aircraft. The flights were conducted at approximately 85 meters in elevation, which resulted in an average ground sampling distance of 3.6 centimeters resolution.

The initial processing of flight plans and images was conducted on site so as to ensure good imagery and no gaps in the multispectral maps. An image overlap of 75% laterally and 75% longitudinally was used during the missions to minimize gaps in coverage while maximizing flight efficiency. During post-flight processing in Pix 4D Mapper Pro, the software determined distinct key points for each image and tie points that were common for adjacent images. For the Vulci missions, there was an average of 42,500 key points for each multispectral image and 67,700 key points for each RGB image. These points were utilized to build orthomosaics and 3D point clouds through Structure from Motion (SfM) techniques, as well as reflectance maps for NIR and RE imagery.

## PRELIMINARY INTERPRETATIONS OF REMOTE-SENSING DATA

### Impluvium (Area A)

The spatial analysis of remote-sensing information has to integrate GPR, RGB, and multispectral data in order to produce predictive maps of the urban settlement. It is interesting to consider the difference between crop marks identifiable by the NDVI (normalized difference vegetation index) and active sensors, such as the GPR able to penetrate over 2 meters into the soil. In the first case, the archaeological features are detectable by the different growth of the vegetation, while in the second case, it is by the spectral signature of the archaeological and geological material. In both cases, we have to deal with a mediated information processing, but the overlapping of remote-sensing maps and aerial/drone photography provides a powerful tool for archaeological interpretations.

While many of the features in the area between the temple and domus are nondescript buildings or structures, a clear footprint of an impluvium—a sunken, square pool of water usually in the center of an open-air courtyard—is

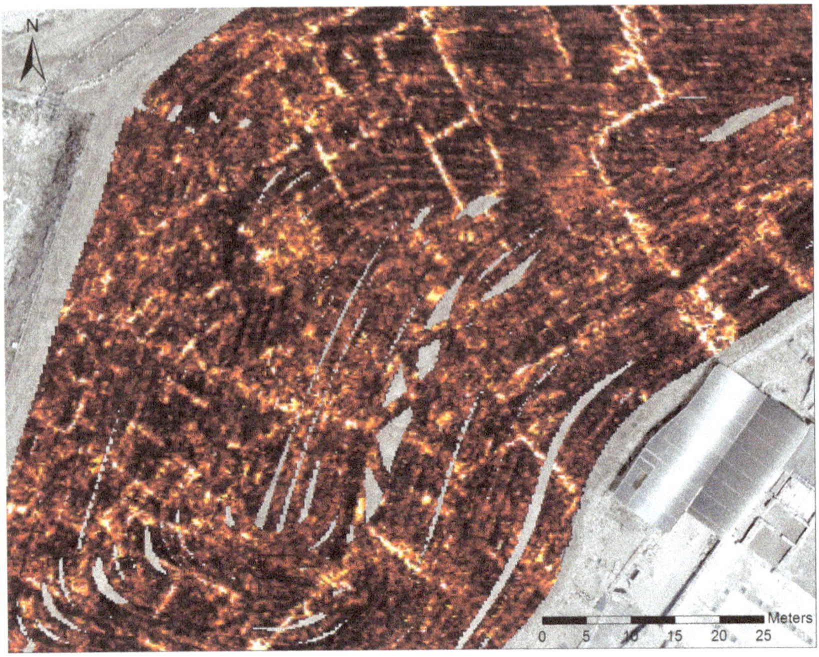


**Figure 1.13** In this view of our GPR data (-0.6 m), one of the main roads can be seen (B) along with indications of larger features to the east of it (E2). The building that is currently being excavated as part of the Vulci 3000 project is pointed out as a reference.

evident (figure 1.13). This appears at 1.3 meters, with the clearest outline of the feature at 0.9 meter and disappearing by 0.7 meter.

In figure 1.13, the bright yellow square signifies the bottom of the pool, likely made from stone, measuring 7.5 meters on each side with 1 meter columns at each corner. Surrounding this square is the outline of a larger square, indicating the limits of the courtyard. A peristyle courtyard is also partially visible to the east of this feature around the depth, furthering the hypothesis of the features in this area being connected and not autonomous.

### Roads (Area B)

The remote-sensing data confirm the location of one of the main north–south roads that can be seen to the east of the excavated Great Temple. While possible isolated portions of the road can be seen 1.1 meters below the surface, it is not until 0.8 meter below that the road in form and shape becomes clear (area B, figure 1.14). This road section is 6 meters in width, matching the average width of the main road beside the temple, and 62 meters in length. There is another possible portion of road visible to the east of the formerly mentioned

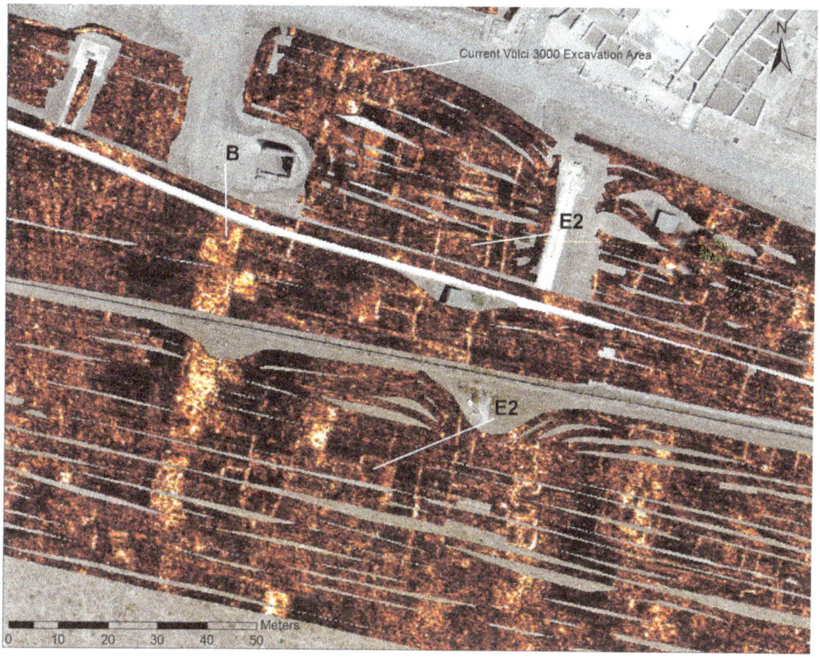


**Figure 1.14** The oval buildings (C), the basilica (D), additional larger structures (E1), and the possible theatre (F) are visible in this GPR slice revealing densities 0.6 meters below the surface.

north–south route, which appears in the GPR data about 0.9 meter down, also measuring 6 meters in width but only 32 meters in length (figure 1.14).

### Oval Structures (Area C)

A clear footprint in the GPR data and old aerial images shows large oval-shaped buildings near the west gate each of different size (area C, figure 1.15). These oval buildings overlap and are located at slightly different depths, suggesting that they were rebuilt over time, one replacing another. Owing to their odd angles and placement, it is probable that they are post-Roman or even later; however, the exact nature and date of these buildings is still being debated.

### Basilica (Area D)

Underneath the still-standing medieval-era building is a large feature whose footprint suggests a basilica. The building measures 26 meters east to west by an estimated 22 meters north to south. Its prominent location in the Western

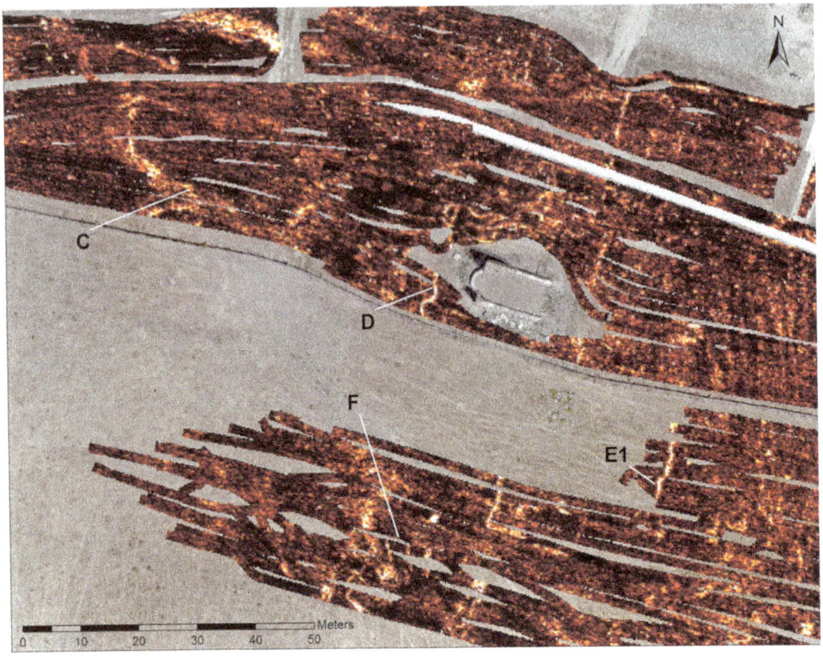


**Figure 1.15** In this far western edge of the plateau, a large footprint of a structure is seen through the red-edge images collected via drone. The smaller image in the upper right-hand corner illustrates our current hypothesis for the possible features based upon this red-edge image.

Forum, along the main decumanus and across from the Great Temple, support the idea of an important public building. Further, looking at the shape created from the reflection of dense materials, we see that there are two rounded sections on the southwest and northwest corners of the building which measure around 5 meters in diameter (area D, figure 1.15). The other walls of the building appear to connect in a large rectangle, although the southern wall is lost, owing to a lack of GPR data in that area.

## Larger Buildings to the South (Areas E1 and E2)

While lacking the breadth of data to make a conclusive argument, we can discern the presence of several larger buildings just to the south of the Western Forum area. One in particular can be seen in both GPR slices and aerial images, and it depicts a larger building with multiple rooms or cells, indicating another possible religious or public building (area E1, figure 1.15). To the east in this area are several partial outlines of what could be larger buildings, particularly possible large platforms just to the east of the main north–south road (see area E2, refer back to figure 1.14) Even farther to the east of these features is a building, also appearing to contain multiple areas, which is the first feature to appear in this area and is at the deepest depth of 2.1 meters below the surface. These larger buildings indicate an area that could have been used for public or community-based purposes.

### Theatre (Area F)

One of the most intriguing buildings to be seen in both aerial images and through GPR data is what appears to be a semicircular architectural feature indicating radial seating inside a medium-sized building (area F, figure 1.15). This semicircular center, only 15 meters in diameter from north to south, is small for a typical Roman theatre stage. The radial lines, measuring between 9 and 10 meters in length, indicate the supports for the stadium seating. The clear double wall with supports on the eastern side points to construction techniques similar to that of Roman theatres, however. The "theatre" building, as well as the larger building to the north of it, appears at the same depth in the GPR data, with the "basilica" to the north of them appearing just slightly higher in stratigraphy. While depth does not equate to precise dates, the relational data suggests an order of events. We hypothesize that the area was cleared of former building and then this building was added to this area in the later period of Roman occupation. This is suggested by the tilted axis on which it is built, which does not line up with the other buildings in the area, as if it were designed at that size and at that angle in order to fit into a space that was previously used for something else.

### Large, Multi-Room Feature in the West

In a far western portion of the city, processed red-edge aerial images reveal the footprint of a large, multi-roomed building, as mentioned above (figure 1.11). The orientation of the building matches that of the rest of the city, supporting its co-existence during Etruscan-Roman times; however, it is far removed from the city center used during either Roman or Etruscan periods, suggesting a temporal displacement. Owing to the restricted area covered by the GPR surveys and the lack of any connecting features from aerial images, it is difficult to interpret this building's relationship to the rest of the city. Through the drone images, we know the feature has a large footprint of around 158 meters north to south by 180 meters east to west. A further GPR survey of this area would provide an indication if this area was used in a similar way multiple times, was used differently, or was never again used at all.

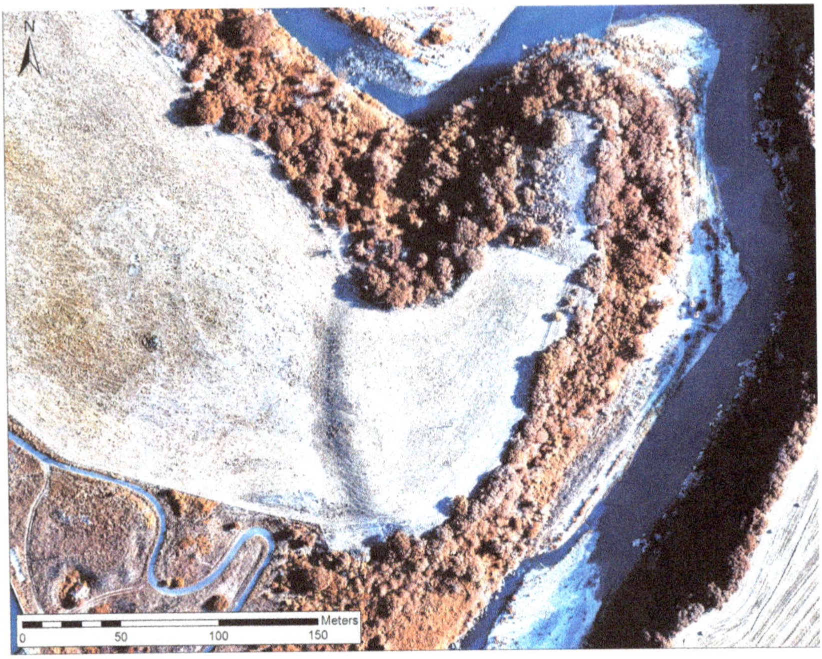


**Figure 1.16** The southern acropolis of Vulci is shown here in red-edge, revealing the presence of a 108-meter long and five-meter wide moat or ditch, outlined in green, which separates this area from the main city center. An identical mark is visible, although slightly lighter in color, in the north separating the eastern acropolis from the city.

### Moats

The two dark, curved lines (outlined in green) seen through the red-edge and infrared images collected via drone in 2016 indicate a trench or defense wall that cuts off the southern and eastern acropolis from the main portion of the city (figure 1.16). This restricted access, as well as the little-published discovery of an Etruscan-era religious building on the eastern acropolis, supports the relatively unsubstantiated claim of a temple on the southern acropolis.

## COLLABORATIVE WEBGIS

The goal of the webGIS (online Geographical Information System; GeoNode in this case) is to share a collaborative platform with all research teams for all excavation and remote-sensing data. GeoNode is an open-source, online platform for sharing geospatial data and maps. Its aim is to provide a collaborative webGIS public platform by which it is possible to visualize, integrate, download, or simply cross geospatial data collected in the field. GeoNode integrates a sophisticated permission system that allows the users to edit, visualize, or download data according to their specific level of access (full, intermediate, or public). This interface allows for public interaction with and visualization of current data and interpretations, but the owner maintains control over its full dissemination. This virtual, collaborative, web-based environment allows for wide-ranging access—a place where scholars from varying fields and locations can incorporate data into their project or approach and, in return, suggest new ideas, theories, or approaches. The platform also allows for the management of different types of geospatial information covering historical aerial photography, geo-referenced archaeological maps, high-resolution orthophotography, digital elevation models (DEMs), remotely sensed data elaborations—such as the NDVI—and vector graphics related to the interpretative layout. GeoNode also provides the project website with server support in order to easily visualize high-resolution geo-referenced imagery.

## THE 3D EXCAVATION

Vulci 3000 was born as an archaeological excavation project that has used fully integrated digital 3D recording techniques since its inception. The need to adopt a digital workflow for the stratigraphic documentation in archaeology is obvious, despite the fact that standard procedures and best practices have not yet been fully established. This is because stratigraphic excavation is, by definition, a destructive and unrepeatable practice aimed at investigating past material culture. The three-dimensional survey is playing an increasingly important role in archaeological documentation. In fact, it provides a precise documentation of the entire surface (interface) of each stratigraphic single-context unit (SCU). This allows the integration of the initial interpretation with a retrospective analysis of the removed archaeological layers and their stratigraphic relationships. Moreover, once geo-referenced, digital 3D models allow for accurate linear and volumetric measurements that would otherwise be difficult to estimate from 2D archaeological photographs or drawings.

Another important aspect concerns access to primary (raw) information rather than providing the scholarly community with peremptory interpretations of the archaeological finds. To a certain extent, 3D recording represents a major qualitative leap forward because it permits both following and revising the interpretative process that leads to the generation of knowledge in archaeology.

The documentation of each SCU is conducted using both conventional methods of representation (photography and manual drawings based on total station measurements) and the newer photogrammetry. Subsequently, the data sets are collated in a GIS for an immediate assessment of the accuracy of the survey. Specifically, SCU plans in shapefile format are superimposed onto the orthophotos generated from the photogrammetric model in order to readjust the outline of each unit. Contour lines and hachures are instead readjusted on the base of a DEM model), which represents altitude values as a 2D raster image (height map). DEMs are also derived from the photogrammetric 3D model (figure 1.17).

These methods are neither alternatives nor are redundant but, rather, are complementary, resulting in better accuracy of the final graphic documentation. On the one hand, entrusting SCU documentation exclusively to the photogrammetric survey would lead to interpretative difficulties that could be unraveled only with direct observation of the site, before the context removal. On the other hand, instrumental survey alone is too sketchy owing to the limited number of measurements that can be made during the documentation

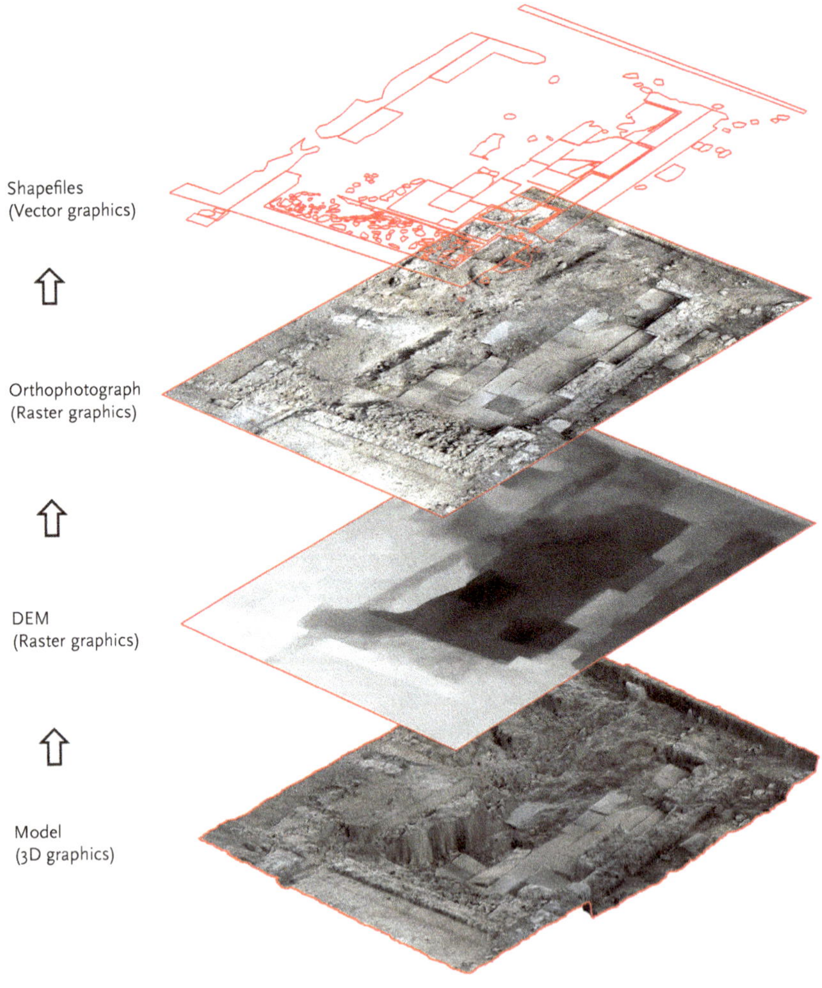


**Figure 1.17** Generation of orthophotos and DEMs from the photogrammetric 3D model and vector outlining of the single context units boundaries.

process. Once in the GIS, however, the raw shapefiles can be edited to match with the orthophotos and DEMs.

The photogrammetric survey is conducted at the multiscale level: micro-scale for the documentation of stratigraphic sections or individual architectural elements; medium-scale for one or more SCU interfaces; and macro-scale for the entire surface of the excavation trench. Depending on the size and type of surface, the survey entails different approaches to photo shooting. Flat features require a mosaic of photographs taken mostly perpendicular to the surface, while complex objects, such as walls and columns, require a greater number of photographs taken mostly obliquely. For the project, the entire excavation trench extent (20 × 15 m) required about 800 photographs and several small checkerboard targets (3 × 3 cm), evenly distributed throughout the surface. These targets marked the position of the ground control points used for geo-referencing the model.

Photographs are taken using a Canon EOS 6D camera with a fixed lens (generally 28 mm; 35 mm for details) of 5472 × 3648 pixels in size. Before starting a sequence of shots, white balancing is performed in order to obtain the most correct representation of the color regardless of the environmental conditions of brightness. Once the photographic session is completed, the documentation phase proceeds through the instrumental survey stage. The coordinates of the targets, the reference points for drawings, and the elevation points are recorded by a robotic total station.

These photographs are then processed using Agisoft PhotoScan SfM software. After a first photo alignment by means of automatic recognition of tie points, a sparse point cloud is generated. As a marker is inserted in the photographs in correspondence with its target, the total station coordinates values are entered. The X (Easting), Y (Northing), and Z (Altitude) values are recorded in the WGS84 global datum and projected in the UTM (32N) geographic coordinate system (figure 1.18). The model is then optimized via bundle adjustment algorithms to minimize the re-projection error. The resulting error is expected to be less than 2 centimeters.

At this stage, a dense point cloud is generated. The DEM is subsequently processed in PhotoScan and exported in raster format (geo-TIFF). The DEM also serves as a support for the generation of very high-resolution orthophotos (up to 2.12 mm/pix) that are exported in a tiled multiresolution geo-TIFF format for an optimized display in the GIS. These very high-resolution images allow for the visualization of the minutest details of the surface, providing archaeologists with visual feedback on the information relating to the consistency and color of the terrain, as textually described in a standard SCU sheet.

Finally, a three-dimensional model of the unit (Tiled Model) is generated and exported for real-time visualization. Unlike the two-dimensional orthophotos and DEMs, the three-dimensional model also allows documentation of vertical surfaces such as wall claddings, stratigraphic sections, and

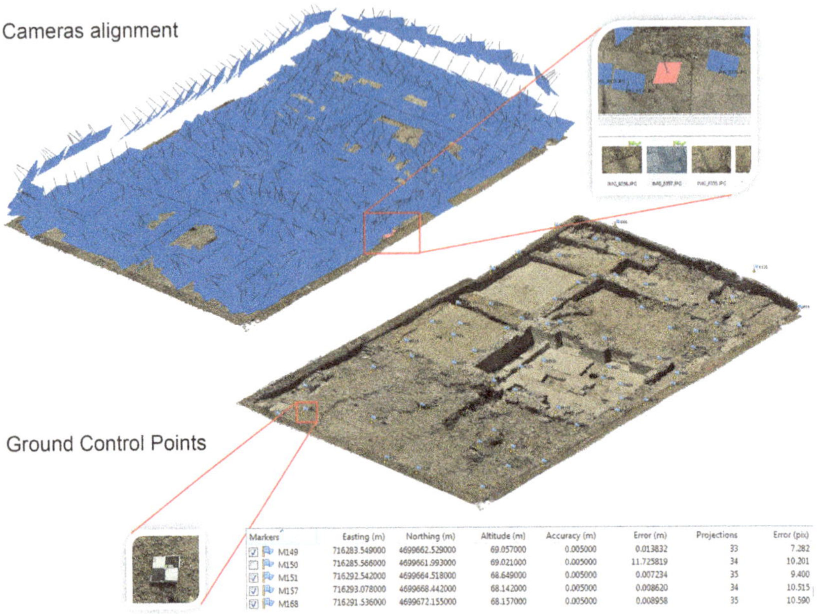


**Figure 1.18** The workflow of photogrammetric processing in PhotoScan software showing the location of the cameras (above) and the ground control points (below).

architectural details in their entirety. The multiresolution Tiled Model generated by PhotoScan allows for a real-time dynamic visualization with greater spatial resolution than a single static mesh, since both the polygonal mesh and the texture of the model are hierarchically subdivided into blocks at different levels of detail.

Besides spatial resolution, temporal resolution must be taken into account. While the first relates to the pixel size—the smaller the pixel in an image, the larger the number of details that can be captured—the latter indicates how frequently the excavation data are recorded and, consequently, how many stratigraphic units can be documented using photogrammetry.

Excavation data sets are archived in different GIS formats ranging from vectorial shapefiles to geo-referenced raster imagery. At any rate, a GIS database alone is not sufficient to store the entire excavation documentation that also includes un-geo-referenced photos and drawings, as well as SCU sheets. Therefore, the GIS must be supported by a full-fledged database management system that incorporates all the data sets as they are produced in the workflow. Based on the experience of the ÇatalHöyük digital excavation project (Lercari et al. 2017), it was decided to adopt a different approach to 3D implementation and visualization that involves the use of height maps (DEMs) and orthophotos, instead of polygonal meshes.

The Vulci 3D webGIS serves as a platform for visualizing archaeological excavation data sets in a geo-referenced 3D environment. It is meant for archaeologists, scholars in the field, and students who need to analyze complex sequences of stratigraphic layers at glance and in a more visually effective manner than the schematic excavation matrix.

The 3D visualization is achieved by automatic extrusion of polygonal surfaces on the base of underlying elevation data stored in DEM format. Vectorial features (points, lines, polygons) and raster imagery can also be projected onto the 3D surface. Compared to 3D meshes, which need to be drastically reduced in the number of polygons in order to reduce the GPU's workload, height maps are more manageable in terms of graphic performance. Yet, tiny details emerging from the surface, such as small artifacts, pebbles, and ossicles, are better preserved than in meshes. This makes multiresolution TIFF images more suitable for real-time visualization.

Furthermore, DEM-based 3D surfaces should be compatible with volumetric analysis tools. One of the main goals of this prototypal 3D webGIS platform is its capability to display stratigraphic sections as volumetric components and not just as unit surfaces. When working with raster data sets, it is easier to compare elevation values in order to calculate the volume between two consecutive DEMs.

To this end, the Vulci 3D webGIS platform will incorporate spatial analysis tool sets, including linear and volumetric measurement tools. Specifically, a section plane will enable the user to see through the archaeological strata in a given location, while a selection tool will allow for executing queries in order to access metadata and other information associated with each SCU.

## CONCLUSIONS AND FUTURE PERSPECTIVES

The study of Etruscan cities and urbanism is related to the concept of the city-state and its organization—a complex and independent socioeconomic unit that is self-organized but open to trading and connections in and out of the influence of Etruscan territory (Riva 2009). Evolution and transformations of cities require a multimodal approach in order to understand the mechanics and dynamics of these changes across time and space.

In our case, the methodological approach involved: (1) noninvasive technologies for a predictive interpretation of the site; (2) archaeological excavations (single-context method) in the most important urban areas of the site; (3) accurate and 3D digital documentation of the landscape (macro-scale) and of the excavation trench (micro-scale); (4) lab analyses of archaeological finds and deposits; and (5) digital archiving and dissemination by collaborative webGIS, VR applications, and 3D repositories.

The combination of these methods allows for creation of a new research template for the study and interpretation of ancient cities, applicable to other contexts and sites. The methodological workflow starts with predictive modeling techniques, then on the basis of these preliminary results, proceeds to excavation and finally to the dissemination/communication of data through the web and VR applications

The Vulci 3000 Project is, and has been since its beginning, an innovative and multidisciplinary project producing *hundreds of times* the amount of data that could be generated just a few years ago. This jump in volume became possible through the use of digitally integrated technologies such as 3D laser scanners, remote-sensing systems, drones, image modeling, virtual reality, geophysical prospections, and immersive systems.

The stratigraphic overlay of approximately 900 years of urban history in the Western Forum, and possibly in most of the southern part of the city, constitutes an extraordinary example of diachronic continuous occupation in a public space. This chronology can be validated through the Great Temple (first phase late sixth–fifth century BCE) and by the stratigraphic thickness diagnosed by GPR and by the last excavations (2016–17). In fact, the foundations of an Augustan building (first century CE) overlay more than 1.5 meters of earlier stratigraphy related to more ancient buildings (presumably Etruscan).

The identification of a pattern—pre-Roman, then Roman—can contribute significantly to the study of urban archaeology on the Italian peninsula. The spatial articulation and the features of the buildings in the Etruscan agora–Roman Forum will reveal also the religious and ritual models of the city and their impact on the general urban organization (orientation, plans, symbolic meanings, ranking). Finally, archaeological and anthropological work on late imperial abandonment levels and deposits will show how a Roman urban community contracted, with specific reference to changing social relations, artifact use, and diet.

A detailed chronology can finally explain the evolution of the Etruscan urban plan in relation to rituals and religious buildings and to the actions/plans of the Romans after their occupation of the city. Take, for example, the construction and planning of the Forum in front of the Great Temple (reused in Roman times). The study of monumental continuity–discontinuity between Etruscan and Roman traditions can explain the persistence or interruption of rituals and the role of religion in the translation of the city. Was the religious perception of the temples the same? Was the visual ranking of public monuments different in the Etruscan and Roman interpretation of the public space? Were the Romans trying to preserve a sort of former consolidated social memory of the core region of the city, or were they willing to erase completely the Etruscan cultural identity?

Archaeological excavations can also reveal the manpower used in the construction of monumental buildings and its social implications in the formation of the city-state first and of the Roman urban administration later. For example, the study of monumentalization and de-monumentalization of temples and public buildings (Warden 2012) can highlight the social relations between power and social identity, visual meaning and rituals. In fact, a public space, such as an agora/square or a forum, is first of all a dynamic entity of performances (political, social, religious) where perceptual factors are essential. The technological approach to the case study is, thus, innovative and able to integrate different methods: remote sensing (aerial surveys and penetrating radar), digital photogrammetry (by digital camera and laser scanners), and digital archives (webGIS).

### Remote Sensing and Geophysics

The most successful technique so far has been the GPR survey. GPR data provide information about features still buried beneath the surface down to 3 meters and thus is an important method concerning a largely unexcavated urban center. Analysis of the GPR data was able both to confirm suggested features from old aerial photographs and to expand our understanding of the fields between the Great Temple and House of Cryptoporticus and the fields to the south of the Western Forum. While there appears to be a shift in orientation and alignment of features in the area between the temple and the house, the area to the south seems to be more uniform in any changes and more continuous in orientation. The GPR data further illustrated the increasing density and complexity of features over the period of the site's development. When the interpreted features from the GPR data were traced, they were combined with traced interpretations from aerial images from 1984, 2015, and 2016 in order to create a more complete picture of Vulci's urban center. The newly collected aerial surveys from summer 2016–17 provided aerial coverage of the entire settlement (including the eastern necropolis) with color, near-infrared, and red-edge bands. What is invisible in a single perspective becomes visible when these three types are combined into an NDVI image. The produced NDVI extenuates differences in vegetation, which are indicators of different features and past activity in an area.

### Intra-site Photogrammetric Survey

The documentation of each single context unit was carried out using traditional methods of representation (archaeological photography and total-station–based drawings) and innovative approaches through photogrammetric survey. Both the direct and indirect surveying methods were conducted independently of each other. They proved to be neither mutually exclusive nor redundant but, rather, complementary, resulting in better accuracy of the archaeological drawings. Once the vector plans (shapefiles) of the units had been superimposed on the photogrammetric orthophotographs and DEMs, increasing the accuracy of the drawings became possible. Since the archaeological excavation is a destructive and irreversible process, the 3D documentation is increasingly playing a critical role in the archaeological record. In fact, it allows for a precise 3D recording of the single-context unit's surface, which also yields a posteriori interpretation, helping to clarify particular circumstances not immediately observed in the course of excavation. The digital model, when properly geo-referenced, allows us to perform accurate linear and volumetric measurements, which are difficult to obtain from simple archaeological photography.

In conclusion, the entire project employs different digital ontologies that are not usually integrated in the same research pipeline in archaeology. In fact, the interpretation and visual reconstruction of ancient cities require a multimodal methodology which has to deal with predictive, empirical, and recorded data sets but also with the subjectivity of the simulation process. This approach deals also with the transformative, diachronic nature of the urban development across time, space, and different societies.

## REFERENCES

Bartoccini, G. 2001. "Le necropolis." In *Veio, Cerveteri, Vulci. Città d'Etruria a Confronto*. Catalogo della Mostra, edited by A. M. Moretti Sgubini, 29–31. Rome: L'Erma di Bretschneider.

Bartoccini, R. 1961. "Tre anni di scavi a Vulci (1956–1985)." In *Atti VII Congresso Internazionale di Archeologia Classica (Roma—Napoli, 1958)*, edited by V. Catalano, 261–278. II: Rome: "L'erma" di Bretschneider.

Bartoccini, R. 1963. "Il tempio grande di Vulci." In *Études étrusco-italiques. Mélanges pour le XXVe anniversaire de la chaire d'étruscologie à l'Université de Louvain*, edited by Franz de Ruyt et al., 9–12. Louvain: Publications Universitaires.

Buranelli, Francesco. 1991. *Gli Scavi a Vulci della Societa Vincenzo Campanari-Governo Pontificio (1835–1837)*. Rome: L'Erma di Bretschneider.

Buranelli, Francesco. 1994. *Ugo Ferraguti. L'Ultimo Archeologo-Mecenate: Cinque anni di scavi a Vulci (1928–1932) attraverso il fondo fotografico Ugo Ferraguti*. Rome: Giorgio Bretschneider Editore.

Campanari, D. 1836. *Sepolcri Vulcenti—Scavi di Vulci.*

Carandini, A. 1985. *La Romanizzazione dell'Etruria: il territoria di Vulci.* Milan: Electa.

Carosi, S., and C. Regoli. 2014. "Vulci, lo scavo nella necropolis dell'Osteria, Area C: campagna 2012." *Etruria in Progress: La Ricerca Archeologica in Etruria Meridionale*, edited by Rosella et al., 112–118. Rome: Gangemi Editore.

Falconi Amorelli, Maria Teresa. 1983. *Vulci: Scavi Bendinelli (1919–1923)*. Rome: Paleani Editrice s.r.l.

Falconi Amorelli, Maria Teresa. 1987. *Vulci: Scavi Mengarelli (1925–1929)*. Rome: Borgia Editore.

Fulminante, Francesca. 2013. *The Urbanization of Rome and Latium Vetus: From the Bronze Age to the Archaic Era*. New York: Cambridge University Press.

Gleba, M. 2009. "Archaeology in Etruria 2003–2008." In *Archaeological Reports,* edited by Doyle and Coulthard, e55:103–121. Hertford, England: The Society for the Promotion of Hellenic Studies and the British School at Athens.

Lercari, N., E. Shiferaw, M. Forte, and R. Kopper. 2017. "Immersive Visualization and Curation of Archaeological Heritage Data: Çatalhöyük and the Dig@IT App." *Journal of Archaeological Method and Theory* 25(2): 368–392. https://doi.org/10.1007/s10816-017-9340-4.

Marchetti, Marco, Vincenzo Sapia, Adriano Garello, Donatella De Rita, and Alessandra Venuti. 2014. "Geology and Geophysics at the Archeological Park of Vulci (Central Italy). *Annals of Geophysics* 57(1): 1–11. http://doi.org/10.4401/ag-6361.

Mengarelli, R. 1929. "Un anno di scavi a Vulci." *StEtr (1929)* 3: 103–110.

Pacciarelli, M. 1991. "Ricerche topografiche a Vulci: dati e problem relative all'origine delle città medio-tirreniche." *Studi Etruschi* 56: 11–48.

Pocobelli, Giorgio. 2004. "Vulci: Il Contributo Della Fotografia Aerea Alla Conoscenza Dell'Area Urbana." In *Archeologia Aerea: Studi di Aerotopografia Archeologica*, edited by G. Ceraudo and F. Piccarreta, 127–143. Rome: Instituto Poligrafico e Zecca Dello Stato.

Riva, Corinna. 2009. *The Urbanization of Etruria. Funerary Practices and Social Change, 700–600 BC.* Cambridge: Cambridge University Press.

Sgubini Moretti, A. M. 1988. "Note di topografia vulcente." In *Un arista etrusco e il suo mondo. Il pittore di Micali,* edited by M. A. Rizzo, 111–116. Rome: De Luca.

Sgubini Moretti, A. M. 1993. *Vulci e il suo territorio*. Rome: Quasar.

Sgubini Moretti, A. M. 1994. "Ricerche archeologiche a Vulci 1985-1990." In *Tyrrhenoi Philotechnoi*, edited by M. Martelli, 9–49. Rome: Gruppo Editoriale Internazionale.

Sgubini Moretti, A. M., ed. 2001. *Veio, Cerveteri, Vulci: Città d'Etruria a confronto*. Rome: L'Erma di Bretschneider.

Sgubini Moretti, A. M., L. Ricciardi, and E. Eutizi. 2014. "Vulci, necropolis dell'Osteria, campagna di scavo 2011-2012: dati preliminary." *Etruria in Progress: La Ricerca Archaeologica in Etruria Meridionale*, edited by Rosella et al., 106–111. Rome: Gangemi Editore.

Warden, P. G. 2012. "Monumnetal Embodiment: Somatic Symbolism and the Tuscan Temple." In *Monumentality in Etruscan and Early Roman Architecture: Ideology and Innovation*, edited by M. L. Thomas and G. E. Meyers, 82–110. Austin: University of Texas Press.

CHAPTER 2

# "I dreamt I dwelt in marble halls"

## *Using Computer-Based Visualization of Roman Domestic Architecture to Evoke the Built and the "Felt" Environments*

RICHARD BEACHAM

We shall not cease from exploration
And the end of all our exploring
Will be to arrive where we started
And know the place for the first time.
—T. S. Eliot. *The Four Quartets*, "Little Gidding," section V

Roman domestic architecture and decor drew upon complex visual strategies to evoke in visitors a wide range of powerful sensual and emotional reactions to what was frequently a carefully and deliberately "staged" environment. The mise en scène of such dwellings was calculated and coordinated to communicate such things as a sense of the power and prestige of the domestic patron; an imaginative access to fantasy realms of mythology or exotic landscapes; and a distortion of space through the painted suggestion of grandiose, often impossible architecture. Computer visualization and representation of such spaces has the challenge—and potential, particularly when configured in a "virtual world" format—to evoke in users not just a sense of the physical structure of such spaces but their sensory qualities as well. This is particularly true when such visualizations—such as that provided by the Oculus Rift and related technologies—are configured to engage the user in

Richard Beacham, *"I dreamt I dwelt in marble halls"* In: *Digital Cities*. Edited by: Maurizio Forte and Helena Murteira, Oxford University Press (2020). 
DOI: 10.1093/oso/9780190498900.003.0003

an immersive environment. This chapter surveys both the nature of the challenges that might be addressed by such applications and their scope and limitations. It also outlines the extent to which they may provide the user with both a perception and an understanding of Roman visual strategies, as well as offers some tentative sense of the sensual impact these might have had on the ancient visitor as she or he encountered and experienced such ancient spaces—now reconstituted virtually, before our eyes.

## ARCHITECTURE AS INSPIRATION

Informing my discussion here is the curious paradox that architecture, the most quintessentially concrete of the arts, whose quiddity resides in its physical presence, is so frequently cited for its capacity to inspire such insubstantial things as poetry, music, psychology, dreams, or history. For example, Friedrich von Schelling (in a suggestion taken up and repeated by Goethe) described architecture as "music in space, as it were, a frozen music" (Schelling 1985, 576).

By contrast, the architectural historian Vincent Scully characterized architecture as "visible history" for its capacity to record and preserve not just historical physical remnants but, concomitant with these and immanent within them, the history of thought, perception, memory, and imagination as well (Scully 1969, 257).

For Denis Diderot, ruins evoked a sense of eternity and a deeply personal awareness and contemplation of his own ephemeral nature: "Time stands still for those who admire."[1]

In thinking about such things, and in particular while pondering the role of ancient architecture in providing an avenue for perceiving and reflecting upon both our collective and our individual existence, I was struck by an eloquent passage in a work by Ben Hecht:

> I once saw a city that seemed to picture the past that lingers in every aging head. It was the once triumphal city of the Caesars, Leptis Magna, on the African coast. Roofless, half dismantled, its statues awry, its great temples toppled and nameless, this former metropolis stood marble-white and lifeless in the Mediterranean moonlight. I sat on one of its high walls and looked on the ghost of grandeur. . . . So do we all look on our youth. We note windows missing, street signs illegible, faces gone, events distorted. As much debris as monument waits our inspection. But a civilization beckons and a poem lies in these ruins. (Hecht, 1963, Biography & Autobiography. Front Matter.)

Thus, when encountering and reacting to architecture and its decor, we experience a complex and multilayered reaction: partly physical and through bodily awareness, as we perceive space and objects, but also cognitively (we think about what we see) and, ultimately, emotionally as well (the perception arouses feelings and imagination). In the face of such various and often quite personal phenomena, how can computer-generated representations of artifacts strive to encompass some of these elements or evoke in the viewers reactions that are equivalent or analogous to those which a visitor might experience in their actual presence? How can an experience be fashioned that is sufficiently immersive to at least allow us to intimate, if not entirely encompass, such things?

## BODIES IN SPACE

I will return to consider the rich and even bewildering complexity of emotional and imaginative response to spaces and their decor, but I focus first on their "concrete" manifestation. It is important to underscore how the Cartesian division in Western thought, positing a separation between mind and matter, observer and observed, and thus imposing notionally a barrier between bodily and cognitive awareness, has limited our appreciation of how fully corporeal we are. Apart from its impact on our own *primary* immediate experience of ourselves in the world around us (including our "holistic" perception of such spatial and visual elements as Pompeian painting and architecture), at a *secondary* level it has impeded (until very recent studies) our critical assessment of this art and architecture, as well as their deployment in the mental and physical spaces of an ancient society with a radically different understanding of the topic.

We acquire understanding through embodied experience; without sense perception, there is nothing for our intellectual or reflective faculties to respond to or upon which we can meditate. Visual perception, in particular, was a major (even arguably the primary) currency of Roman social and political life, and it is itself an embodied element, rather than an abstract or impersonal process or faculty, dependent upon the body.

Cicero called attention to the ambiguous relationship between visualization and thoughtful understanding:

> The most complete pictures are formed in our minds of the things that have been conveyed to them and imprinted on them by the senses, but the keenest of all our senses is the sense of sight, and consequently perceptions received by the ears or by thinking can be most easily retained if they are also conveyed to our minds by the mediation of the eyes. With the result that things not seen and not lying in the field of visual discernment are earmarked by a sort of outline and image and shape so that we keep hold of as it were by an act of sight things that we can scarcely embrace by an act of thought! (Cic. *De orat.* 2.87.357)

In terms of ancient theories of visualization, seeing something was not a merely passive experience on the part of the observer; additionally, object and eye interacted in a dynamic manner through which the mind did indeed directly encounter the physical object either (as Platonists believed) through the eye's projection of rays outward to the object or (as Epicurean thought conceived it) through the eye's reception of the films of visibility emitted by the object into space which traveled to and impacted it.[2] In terms of both ancient visual theory and modern conceptualizations, such as the theory developed by cognitive psychologist James Gibson (1979) in his account of ecological space, what we see and what we cannot see—our visual domain—is understood in relation to bodily experience.

We react, of course, to spatial objects and events with our bodies, as well as our eyes. We respond, for example, to architecture both as coercion—forcing us to move in certain directions and prohibiting movement in others—and as protection. In effect, as we move about a space and its decor which have been designed to engender our reactions to them, our movements may be thought of as directorially "blocked"—choreographed—to determine what objects, views, or images we encounter and in what sequence, while at the same time our emotional, psychological, and cognitive reaction and understanding are similarly programmed. We experience something analogous to participation in a staging in which we both witness the performance as it unfolds before us and actively take part in it: audience participation.

At the same time, and above all, within any spatial environment we react physically to the presence of others; our capacity to relate to other human beings begins with our knowledge, derived from our own bodies, of what will cause them pain and pleasure. We do not merely observe; we also physically and in turn imaginatively react to the performances of dancers and athletes, our muscles twitching in sympathy with their exertions, a factor which has been widely developed in recent "performance art" theory and practice.

Moreover, in the case of the object of my discussion here—Roman domestic architecture and décor—the array of potential reactions and associations is particularly rich because the Romans were clearly so deeply aware of these and their exploitation played such a decisive role in their artisans' deployment of them. Wherever we look we find evidence of a deeply sophisticated visual culture, in which the organization and furbishing of space were deliberately and self-consciously exercised. Even the most cursory contemplation of the manner in which painted, fantasy architecture and vistas ubiquitously merge with actual architectural structure and its organization makes this strikingly clear. As I have discussed elsewhere, in the case of Roman houses, we can observe this visual culture as evident through the creation of what may be understood as veritable stage sets (Beacham 2013).

## MIND, IMAGINATION, MIMESIS, AND MEMORY

The Romans created spaces in which the imaginative and corporeal experience of the visitor was constantly being modulated by often complex visual strategies. What might be thought of as a sort of "kinetic cognition" underlays these. As one walks through the house, one is challenged to distinguish between real and painted architecture, between actual structural elements and views contrived to trick or disorientate the eye. At the same time, as with painted scenery, the surfaces of the walls display scenes from mythology or depict artificial structures, plants, and vistas which evoke both ideas and associations through their subject matter, as well as an iterative and incremental direct sensual response to the visual field itself. At its most intense, such deployments can have an almost hallucinogenic effect upon the visitor as one struggles to orient oneself visually, spatially, and cognitively within an encompassing array—at times almost a bombardment—of impacting phenomena.

How such visual strategies were conceived and functioned can be better appreciated by briefly looking at the ancient practice of mnemonics. In addition to the actual manipulation of views and images—both real and painted—inside real space, within the field of rhetorical training, Roman orators and writers had developed a theory and technique for creating *imagined* spaces as an aid to memory. In considering mnemonics, one is reminded of problems similar to the disorientation sometimes produced by contemporary evocations of cyberspace, virtual identity, and virtual presence; namely that in a world of illusions, we hold ourselves in place by memory. In their deployment of mnemonics, however, Romans held memory in place by illusion.

This technique involved conceiving an internal choreography within the mind's eye through which the speaker visualized himself moving through an architectural landscape, each section of which contained a symbolic, frequently spectacular or emotionally evocative "prop" to trigger memory of the desired verbal sequence and argument.

What was notable about mnemonic technique was its use of the mental pictures of either imaginary or actual existing architectural spaces (for example, the various rooms of one's own house) to create for the speaker "places" in which images evoking the ideas, facts, or even individual words relevant to the speech were "stored" until they were accessed and called to mind by the speaker. The Roman orators Cicero and Quintilian discussed the merits of choosing a more or less complex building; awareness of the distance between each element of the space; and even the quality of the lighting. The images through which the speech (or other body of extensive information) was to be remembered were placed by the imagination within the particular locations of the building, which the speaker had "installed" in his mind's eye. "They place the first idea, as it were, in the vestibule, the second, let us suppose, in the *atrium*, and then move around the open areas allocating ideas systematically not only to bedrooms and bays, but to statues and similar things" (Quint. *Inst. orat.* 11.2.20).

In one sense, when imaginary space is being thus employed mnemonically it is also being used "theatrically," since (as I will consider in greater detail) theatre is the art of space (fictively configured), with time, movement, images, language, and thought combined into meaningful synthesis and sequence to evoke the "memory" of past actions or events (whether they ever *actually* took place or not) in the mind of the audience. The ludic realm of the stage was itself a mnemonic space, par excellence. In it, spectators then and now learn "to think with things" instead of words alone (see Beacham 2016).

Cicero urged the orator to "form mental images" (*effingenda animo*) to represent the things he wished to remember and to store these imaginatively in localities (*loci*). He compared this mental process explicitly to the work of a painter using perspective. "These should be as vivid as possible, with a single image conveying an entire concept on the system and methods of a consummate painter distinguishing the positions of objects by modifying their shapes." He then turned to the theatre for a further metaphor: "the memory of things is the particular quality of the orator; we are able to hold them in our minds by a skilful arrangement of their individual masks (*personae*) so that we can comprehend ideas by images, and their order by locations" (Cic. *De orat.* 2.354, 358, 359). He puts the discussion into the mouth of M. Antonius, who shortens his discourse, pointing out (356) that "the subject is well known and familiar."

Elsewhere, one of the aids explicitly cited to assist in memorizing was to link the words to be recalled to the visualization of active agents—"agentes imagines"—which, using modern computer parlance, we might term and think of as "avatars." Of these agents, the best of all might be actors playing their roles in a theatrical performance "in crowns or purple cloaks, for example . . . or by assigning certain comic effects to our images" (Cic.? *Ad C. Herennium* 3.22.37). Earlier, at 3.21.34, the author of this text uses the rehearsal of *Iphigenia at Aulis*, with the Roman actors Aesopus and Cimber donning their costumes, as the subject for such an exercise.

## THE MISE EN SCÈNE OF THE HOUSE

What, then, is the relevance of mnemonics to an assessment of Roman architectural and decorative visual strategy and, in turn, to our contemporary aspirations to draw upon such strategy in devising computer-based representations of our reconstituted artifacts? Broadly, the theory underlying mnemonics—that objects, spaces, and images can be cognitively linked through the mind's eye to concepts, emotional reactions, and memory—also informs the actual "real-life" organization of space and the effect such deployment was calculated and intended to have upon viewer and visitors. Domestic configurations of space and the creation of visual experience were intended and fashioned to be "thought provoking" for those encountering them, and such emotional and cognitive perception was central to the meaning meant to be communicated to such visitors and observers.

As already noted, a core element of such visual strategy was the evocation of elements conditioned by a highly sophisticated theatricality. This theatrical, scenic quality, which is so prevalent in Roman domestic spaces, further extends their meaning and, in turn, the challenge (should we wish to address it) of representing them through virtual technologies. In devising the particular strategies through which theatricality was deployed in Roman domestic settings, the Romans were able—consciously and deliberately—to exploit the essential intertwinement of architecture and scenery, which is fundamental to each.[3] Theatre of course, as it colludes with the imagination, is "make believe," a fictive realm. But it is also a realm contained within and expressed through real spaces and objects; it is scenery, which has been configured to support and extend the imagination by engaging the senses. What the visitor or participant encounters is something which simultaneously has a physical reality (architecture, painting, objects) and engages this in the service of a fictive experience. This is the essential, paradoxical nature of theatre: it comprises real physical and durable elements (people, objects, architecture) acting as a physical framework within and through which ephemeral acts of the imagination are engendered and supported.[4] Such a perception informs the characterization of architecture as a frozen moment which fuses the physical with the insubstantial. It is concrete reality here and now, and a realm of the imagination: immediate space and a captured moment in time.

## VIRTUAL REALITY-BASED RESEARCH

In its evolution thus far, the use of computer-based 3D modeling has inevitably (and invaluably) concentrated on representing, displaying, and making accessible to scholarly perception and scrutiny the historical, cultural heritage artifacts upon which its technologies have focused. This has been its proper and primary role, as investigators have undertaken to record, preserve, or in many cases notionally "reconstruct" lost or damaged artifacts—most frequently in the form of architecture. Over the past twenty years, scholars have made great progress in investigating the potential and the limits of such virtual reality (VR) depiction of objects and, very importantly, have determined guidelines for the research-based pursuit and creation of credible representations, striving for a methodological rigor that can be subjected to the same sort of analysis, criticism, and evaluation as applied to more traditional forms of scholarship. They have established crucial principles and published these in, for example, the London Charter Computer-based Visualisation and Communication of Cultural Heritage, an internationally recognized standard for heritage visualization (www.londoncharter.org), which the King's Visualisation Lab, (the group I founded and led) was centrally involved in creating and advocating.

But crucial as it is, the representation of architectural artifacts through various modes and media in 3D embraces only a small portion of the rich array of such meanings (as I have sketched them here) which constitute the "felt environment," which are immanent in such architecture, and which the enhanced agency of VR technologies might potentially reveal—or at least might cast upon these meanings some suggestive illumination. Of course, the experience afforded by VR technologies will always be derivative and secondary to the sensual, emotional, imaginative encounters that actual, tangible space may provide. At the same time, it is useful occasionally to remind ourselves that the VR-enabled activity is also itself a "real" experience, which may be realized for its own value, not merely as an analogue referencing the actual space upon which its attentions are focused.

> For I dipped into the future, far as human eye could see,
> Saw the Vision of the world, and all the wonder that would be.
>
> —*Alfred, Lord Tennyson, "Locksley Hall," 1842*

So what might be the challenges—some of them currently outside the reach of investigators and available technology—which VR might potentially take on? Beyond the computer-based representation of space and its decor, and the myriad details of construction, lie a repertoire of topics which, because they constitute the array of qualities afforded by *actual* space, might be conceived as legitimate subjects for VR to explore.

Actual space can be *moved through* by those encountering it; and as they do so, they observe an ever-changing visual field which, as it unfolds before their observing eyes, can engender a great variety of cognitive, psychological, and emotional reactions. A space can be *illuminated* variously; the awareness of its *acoustical* qualities constitutes an important part of its sensual perception; *encounters and exchanges* between "visitors" can take place. And finally, in light of the theatricality which, as I have stressed, was a core element in the fashioning and perception of Roman domestic spaces, and was inseparable from its experience by ancient visitors, we ask how far the affordances of VR modeling might extend to evoke such constituent qualities in its computer-confined simulacra.

Those working in VR and conjuring up—at varying levels of fantasy or verisimilitude—"dreams in marble halls" will do well to take contemplate such challenges, even if only to acknowledge that some of them are irrelevant to a particular project or investigation while others, although highly desirable as a "wish list," are only very partially achievable within the current realms and requisites of VR technologies.

Of course, at the most fundamental level, VR modeling, drawing upon architecture-focused computer-aided design (CAD) software, can create designs which accurately convey the fundamental (*relatively* uncontested and nonsubjective) data about a surviving artifact. These record and are drawn to show the dimensions, layout, plans, and elevations of the structure incorporated into a basic schematic model of the building. It is important perhaps to state the obvious: even such rigorously "factual" representations of relatively uncontested data are still, inevitably, "visualizations." Like an architect's blueprints, they may be an accurate and reliable depiction of the data, but they remain of course not the historicists' "das Ding an sich wie es eigentlicht gewessen"[5] but, instead, a separate entity and object derived from it.

In this context, before venturing further into the terra incognita of possible virtual worlds, it may be helpful to draw attention to a useful (if not entirely definitive) distinction between "reconstruction" and "visualization." My frequent collaborator and former colleague Drew Baker has provided some insight into this in a comprehensive forthcoming publication describing the detailed and carefully implemented process of creating the virtual model for the Oplontis Project (about which I will have more to say). He writes:

> It is perhaps salient at this point to try and distinguish the difference between these two terms which have become synonymous with each other over the interceding years. The distinction is a subtle but a crucial one and is wholly dependent on the nature of communication employed in outputting the computer model and its constituent data. A *reconstruction* uses implication to communicate its message whereas *visualization* relies on inference. To put it in terms that are more familiar and less driven by logic theory, a reconstruction presents one view of the underlying data, that of the author/creator, and as such it is opaque in terms of understanding the process by which the model was derived and is illustrative. Conversely a visualisation attempts to be as transparent as possible in terms of allowing the viewer to assess the underlying data and construction process presenting what can be described as a multifocal data-scape, that is presenting multiple or alternative meaning/conclusions of equal probability or validity, from which different data views can be made. (Baker and Blazeby, forthcoming)

In practice, and in the several projects to which I will turn now as illustrative case histories, eventually a visualization may evolve or (depending upon one's view of such a process) transmogrify into a reconstruction. And that is not always a bad thing in "dream land."

## The Villa of Oplontis Project

This is a visualization project, sponsored by the U.K. Leverhulme Trust, for which I directed the work of the King's Visualisation Lab under the overall project leadership of Professor John Clarke, at the University of Texas. The object of our work was to use computer modeling to depict *as accurately as possible* the current state of the 100 rooms of this sumptuous imperial villa, buried by the eruption of Vesuvius in 79 CE. Our initial task was to show the existing state of the remains. The Oplontis Project research team provided most of the professionally shot photographs for all the surfaces throughout the villa, which were then used to provide the texturing upon the 3D computer model, itself meticulously representing the physical layout of the site.

Not only does the model (using conventional modeling software) constitute an outstanding, meticulously research-based visualization of the site but, importantly, it goes far beyond most other VR archaeological reconstructions by employing Unity Technologies' *Unity* 3D video game-development environment to fashion a "virtual world" environment for optimum navigability. It also allows the user (by clicking on an object or area in the model) to link to an extensive database, providing full transparency of the underlying data informing the model.

Unlike most other current representations of VR models, in which the viewer is essentially a passive observer of images or prerecorded animations depicting the artifact, the deployment of the Unity gaming software enables those accessing the Oplontis model freely to navigate to every space and to examine in detail every surface, interior or exterior, at will. Thus the crucial "felt" environment of the actual villa can be potently evoked for the virtual visitor. A user can experience and observe the sensual impact of movement, of changing vistas, and of the capacity to direct one's view to observe how objects, architectonic elements, and painted images gain context and meaning by juxtaposition, and the relationship in space to one another and to the viewer. The viewer can also note the physical and visual relationship between types of space, and the refinement of their decor, to observe how this reflects the status of a particular space (and its users) within the hierarchical system determining the overall layout of the villa. In short, a great deal of the particular, and often peculiar qualities outlined earlier, as characteristic of what I have termed the mise en scène of Roman domestic architecture can readily be perceived and investigated.

Although simply by using an ordinary computer users are able to engage, in the manner described, with and within the virtual model, the project has extended dramatically the experience available to them by incorporating the model into Oculus Rift. This is an acclaimed program which enables viewers, using headsets, to have extraordinarily compelling "lifelike" access, moving freely throughout the entire villa complex in a manner which is closer to actual "presence" than any other single-user viewing system currently available. Oculus Rift brings to the visitor a powerful evocation and apprehension of the "felt environment" with all its manifold sensual and cognitive qualities.

Although conceived and executed primarily as a "visualization," important aspects of the Oplontis model elide into "restoration," the use of which, however, is always made explicit to the viewer. Thus when using the Unity platform (whether on screen or utilizing the Oculus headset) to "visit" the villa, the viewer can toggle when viewing individual walls, or entire rooms, between the representation of the actual existing state ("visualization") and highly accurate "restorations" of lost portions of paintings. In several cases (for example, the hypothetical restoration of the lost upper portions of the huge Second Style painting in the atrium), we pushed such reconstruction to its limits while keeping it moored to what can plausibly be extrapolated from surviving evidence (see figures 2.1a and 2.1b).

One of the important research benefits arising from the creation of the 3D model is that it readily allows various missing pieces of fresco fragments (dispersed around the site) to be studied, and in many cases they are "reunited" virtually with the painted compositions of which they were once part.

As a further extremely valuable extension of visualization into reconstruction, the user of the Unity model can also easily and optionally view the rooms and their decor under different lighting modes and conditions (at any chosen time of the day or night) to explore how variations in illumination might have affected the perception of those spaces (see figures 2.2a and 2.2b).

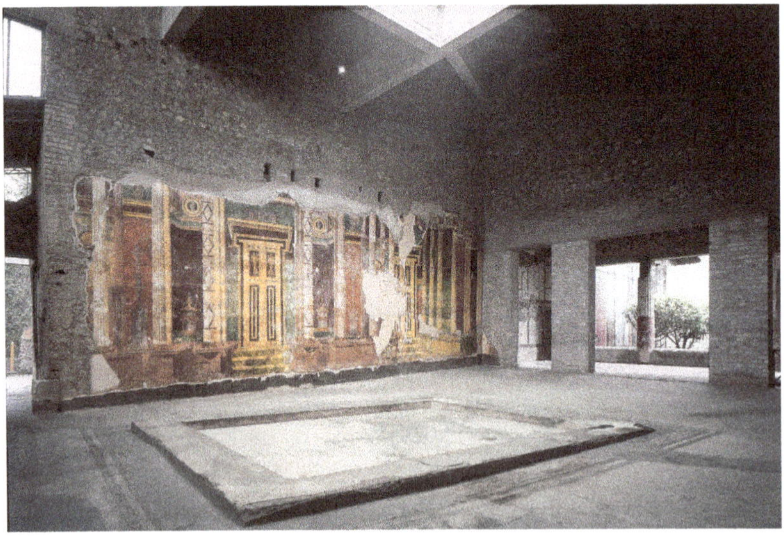

**Figure 2.1a** Room 5, Atrium, of the Villa of Oplontism; west wall existing state.
*Credit:* Photo, Martin Blazeby for King's Visualisation Lab, King's College London.

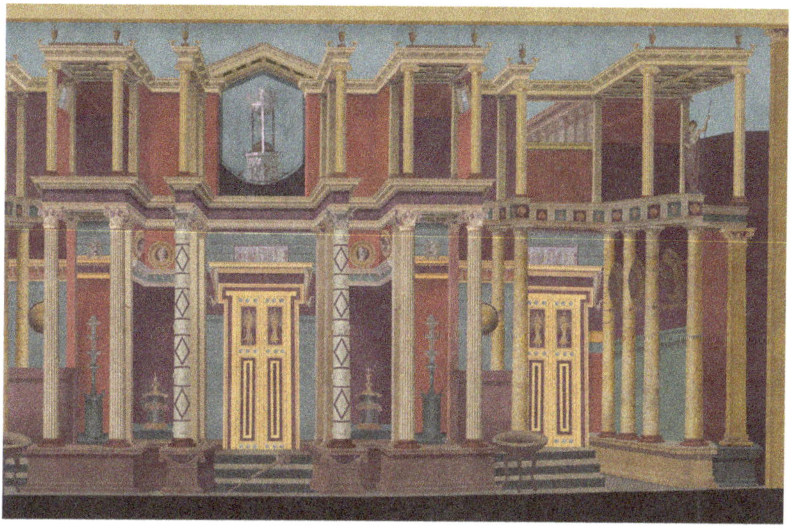

**Figure 2.1b** Digital reconstruction of the lost upper story of the west wall in room 5.
*Credit:* Martin Blazeby for King's Visualisation Lab, King's College London.

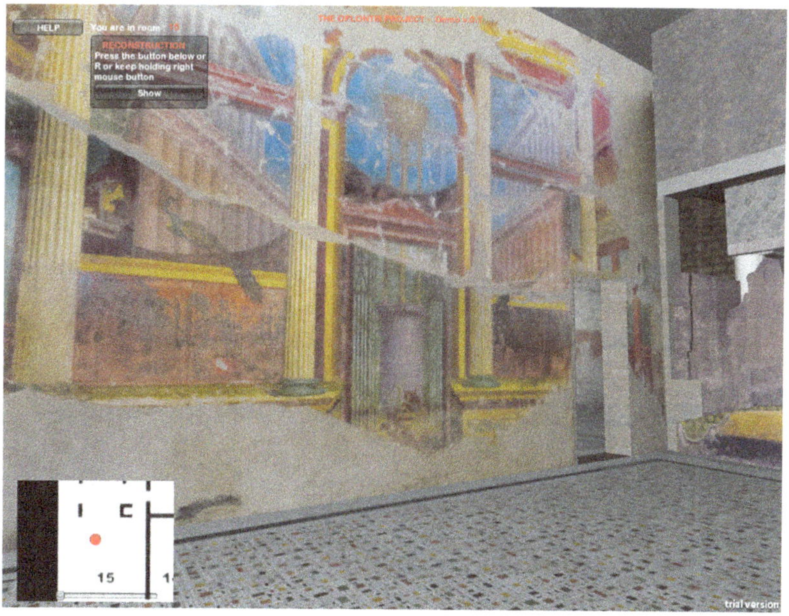


**Figure 2.2a** Screenshot from the interactive Virtual World 3D model of the Villa of Oplontis, room 15; existing state of the east wall.
*Credit:* Model by Marco Bani and Martin Blazeby for King's Visualisation Lab, King's College London.

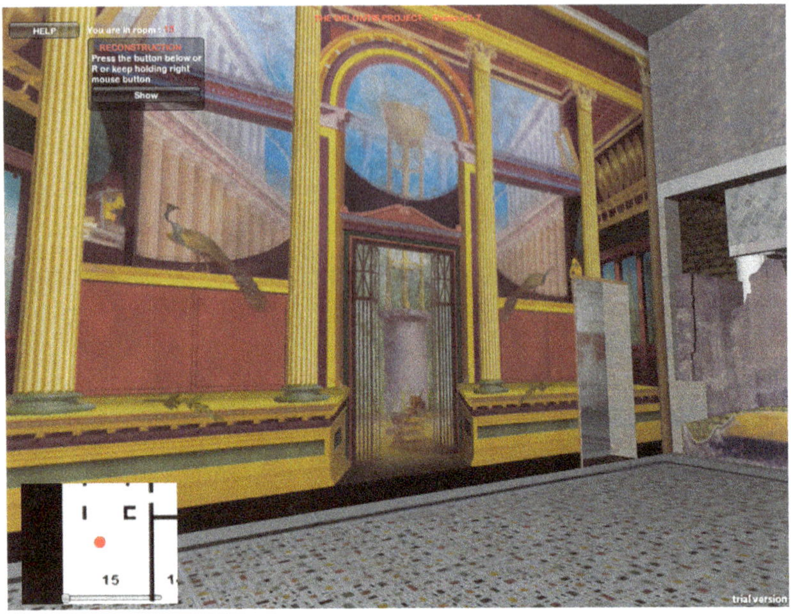


**Figure 2.2b** Screenshot from the interactive Virtual World 3D model of the Villa of Oplontis, room 15; restored state of the east wall.
*Credit:* Martin Blazeby for King's Visualisation Lab, King's College London.

The model, along with an extensive associated database of archival and new research material, forms part of the global publication and dissemination of the work undertaken by the team members and collaborators of the Oplontis Project that includes an e-book in four volumes.

### The Villa of Fannius Synister at Boscoreale Project

By way of marked contrast with the Oplontis Project, the work I led (under the sponsorship of the Metropolitan Museum of Art, New York), creating a virtual model of the Roman villa at Boscoreale, constituted a reconstruction. (See Beacham et al. 2014, which provides an extensive account of our work.) This was because, following its discovery and excavation at the end of the nineteenth century, when many of its paintings were removed, sold at auction, and dispersed to widely scattered museums and collections, the villa itself was reburied and remains inaccessible to the present day. It will only ever again be seen, in its entirety, virtually.

The Met coordinated the cooperation and participation of all the institutions now possessing the villa's paintings and mosaics, and it also prepared new photographs of all those necessary for the modeling process. While using this first-hand evidence to prepare our hypothetical reconstruction of the villa, we were also able to draw extensively on the "official" publication of the villa by Francesco Barnabei (1901), which contains an extensive textual description, complemented by plans, drawings, and several photographs; we supplemented these with a few further archival photographs recording the original state of the finds (see figures 2.3a and 2.3b).

As I have stressed, the fullest meaning and understanding of actual Roman domestic environments only emerges when we are able to consider them as complex, carefully integrated, spatial, decorative, and social ensembles, fashioned by their ancient artists and patrons, often with great subtlety and aesthetic sophistication. To approach an understanding of the lost villa at Boscoreale, it was essential to do our utmost to research and execute a plausible 3D reconstruction of the site.

> When visiting the . . . frescoes from Boscoreale today, we observe them wholly divorced from their original spatial contexts, and objectified, in a late 19th-century museological framework, as discrete, framed, "works of art" rather than as seamless, constituent elements of an all-encompassing visual environment. Even more so than in the case of Oplontis, where the frescoes are displayed in situ, it is easy to forget that the Boscoreale frescoes and mosaics once were part of the living fabric of a working, inhabited building. (Beacham et al. 2014, 16)

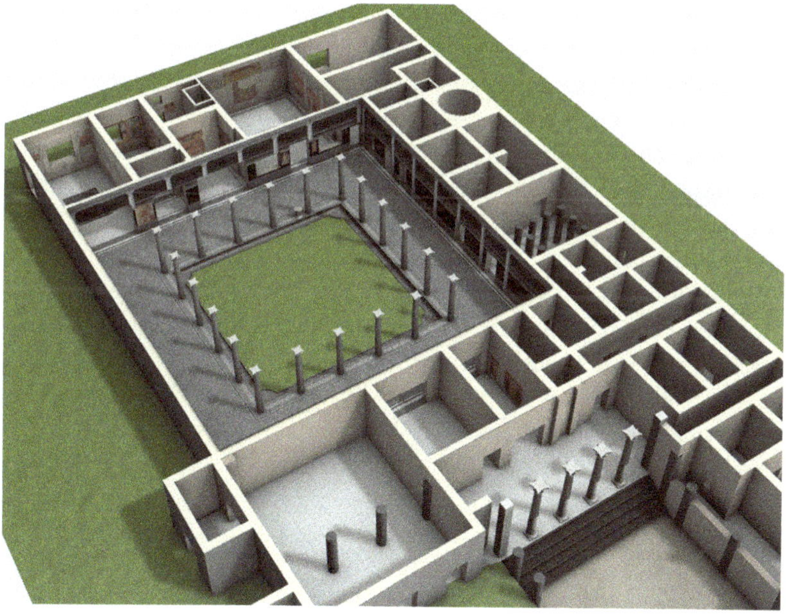

**Figure 2.3a** Virtual model of the Villa Boscoreale, looking northeast; triclinium G is at the upper right corner.
*Credit:* Drew Baker for King's Visualisation Lab, King's College London.

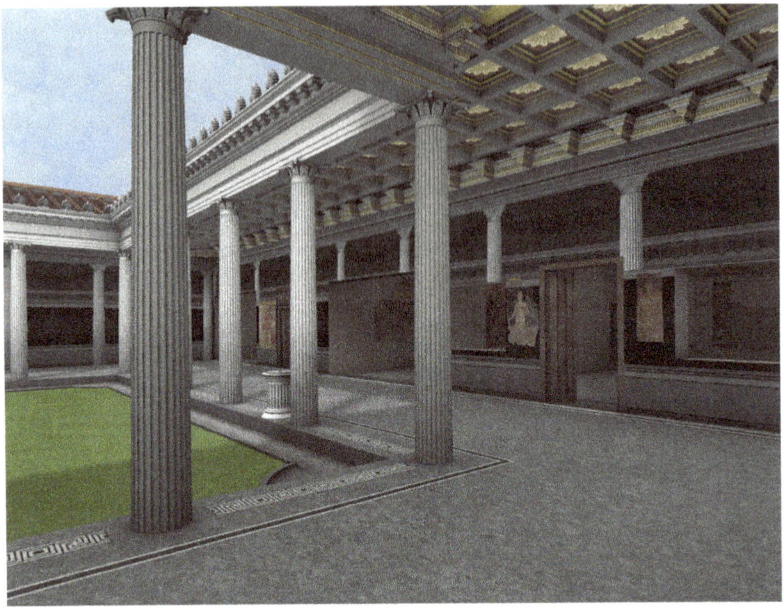

**Figure 2.3b** Peristyle E in the virtual model of the Villa Boscoreale, looking north.
*Credit:* Drew Baker for King's Visualisation Lab, King's College London.

What we prepared was a visual hypothesis based exclusively on the limited direct evidence. We created this initial model of the villa's basic architectural layout showing only those aspects about which our sources were unambiguous. We then hypothetically extended the structures and decor of this partial model through close analysis of the architectural logic of the site and judicious extrapolation both from *comparanda* and from our detailed knowledge of the neighboring villa at Oplontis. Such an approach also informed the hypothetical reconstruction by Drew Baker of the upper stories and roofs of the entire villa for which very little direct evidence exists.

Although inevitably a restoration, we took care to maintain and communicate the scholarly "integrity" of our model by distinguishing between the depiction of actual surviving fresco fragments—which we rendered in full color—and hypothetical restorations of lost elements, which we depicted in more muted shades. The graphical style is thus schematic rather than (as, for example, in our model of the Villa of Oplontis) photorealistic, to avoid the criticism frequently leveled at VR representations for apparently claiming a degree of factual certainty not in fact supported by available evidence.

However, in a few instances (always transparently communicated to viewers) we chose to "fill in the gaps." In some cases, this was quite literally what we did: when the frescoes were removed from their walls and packed for shipping, sale (at auction in Paris in 1903), and dispatch to the various museums that acquired them, they were in several cases enclosed in wooden frames; in this process, some works lost significant portions of the original composition of the frescoes. When we digitally "replaced" such frescoes in our model, positioning them back on the virtual wall of the room where they had originally been executed, we restored the lost portions of the composition. Fortunately, we were enabled to do this not only through careful extrapolation from the designs of the paintings but also by reference to several drawings of the relevant frescoes included in the publication by Barnabei (1901), which he prepared when they were still in situ at the villa. These drawings showed the portions later lost through framing and now restored through our modeling process.

This was a process we carried out (among other rooms) in the large and sumptuous reception space designated "triclinium G." Here it was necessary to restore both the painting now in the Mariemont Museum in Belgium, which had occupied the long south wall of the room, and the painting displayed in the Naples Archaeological Museum, which was on the shorter west wall at the rear of the triclinium (where the diners' couches would have been located) at right angles to the longer wall. Portions of both these paintings had been lost in the framing process, but fortunately each of them had also been recorded in Barnabei's drawings. However, nothing was recorded, nor survives, for the fresco decoration of the other two walls of this room, each of which had a window.

However, there are numerous extant triclinia which, as in room G at Boscoreale, display the proportions suggested by Vitruvius for such spaces, with two-thirds of the floor area reserved for couches and dining (frequently as in room G, demarcated by a mosaic band on the floor), and the rest used to provide access and service. This latter space was also potentially available for the performance of the extensive repertoire of entertainments customarily presented at dinner parties. Invariably within such extant triclinia there are mirrored fresco designs (usually, as at Boscoreale, with Second Style paintings) along the long walls facing each other. Because one of the long walls for room G survives, it was relatively straightforward when undertaking to restore its decor to place the same design upon the opposite long wall.

As part of our analysis of *comparanda*, particularly of examples from Oplontis, we scrutinized the very close affinities between the Mariemont fresco and the paintings on the two long walls in triclinium 14 at Oplontis. Despite some divergent ornamental features, the Boscoreale and the Oplontis walls display almost precisely the same architectural pattern, including the presence in each of them of a circular temple (*tholos*) depicted above the central doorway. The more fully extant frescoes at Oplontis allowed us (as depicted in figure 2.4a) both to create a precise restoration of the damaged fresco from the long south wall now at Mariemont and to extend a hypothetical, but

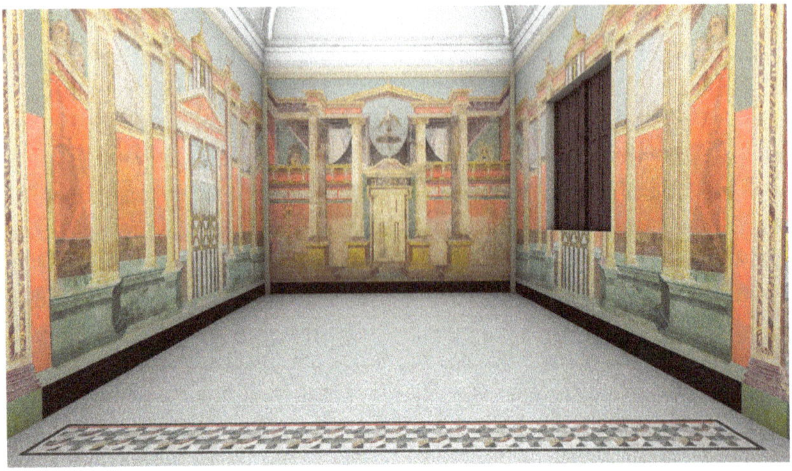

**Figure 2.4a** Triclinium G, as restored in the virtual model of the Villa Boscoreale, looking west.
*Credit:* Drew Baker and Martin Blazeby for King's Visualisation Lab, King's College London.

typologically appropriate, reconstruction of the painted panels now lost from the service area of the room.

So, in pursuit of at least a virtual realization of a dream of marble halls, one can don the Oculus goggles, enter the villa (perhaps, in one's mind's eye, imagining one's attendance there on a warm August evening), and wander along the north side of the columned peristyle E with its "actual" columns mirrored by painted depictions of the same along the inner wall. It is easy to experience a somewhat giddy sense of aesthetic, cognitive, and perceptual blending. What one experiences may even be thought of as analogous to the actual perceptual ambiguity that ancient visitors to the villa were induced to feel.

One might be (in actuality) sitting or standing, perhaps in one's office or computer lab, but experiencing visually, and to some extent corporeally, an extremely vivid and lifelike encounter with a virtual space. This space is depicted in 3D which (in the Roman manner and as evinced ubiquitously in the actual Boscoreale frescoes) prominently features painted trompe l'oeil artificial architecture as part of the decorative scheme depicted on its modeled walls.

After traversing a short distance (peering or venturing into the several rooms along the route), one navigates into triclinium G, passing in the process through several ancillary spaces. This renders this, the most secluded and sumptuous room in the entire villa complex, a special and privileged environment. Once within it, one observes the positioning of the customary three couches (*klinai,* each with three figures lying upon them) and a meal in progress. Within the reconstituted space of this room we have placed reincarnated diners, together with research-based pieces of furniture, braziers, and lamps. The soft illumination plays upon the painted palatial architecture adorning

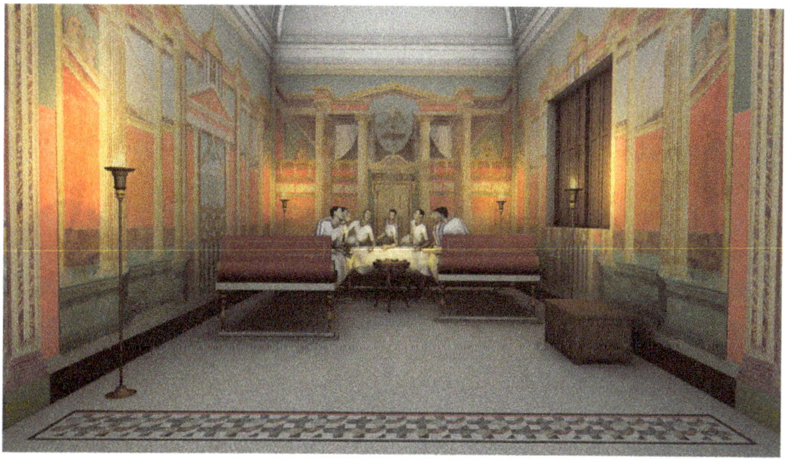

**Figure 2.4b** Triclinium G, in the virtual model of the Villa Boscoreale, looking west at a hypothetical representation of a dining scene.
*Credit:* Drew Baker and Martin Blazeby for King's Visualisation Lab, King's College London.

the room's walls and encompassing those within it, making it easy to elide these artificial, fantasy edifices through one's "oculus mentis" into one's observation and experience of the "real" architecture of the virtual model. Large theatrical masks feature prominently in the decorative scheme, their gaping eyes gazing into the room and, it seems, upon those present with it including now, virtually, oneself (see figures 2.4a and 2.4b).

The virtual guests in our model are, as it were, watching the enactment of a drama. We have extensive references in ancient sources to such performances which, alongside other forms of diversion, were commonplace as dinnertime entertainments. To cite a single example, the Younger Pliny frequently refers in his letters to such dinner entertainments. He mentions that at the home of his greatly admired friend, the general Vestricius Spurinna, then seventy-seven years old, "the dinner is often relieved by actors of comedy, so that the pleasures of the table may have a seasoning of letters" (Plin. *Epist*. 3.1.9).

Accordingly, in our triclinium, two authentically masked and costumed actors are performing both before one's eyes and the "virtual eyes" of the reclining diners a scene from Plautus's comedy *Pseudolus*. As a gesture toward restoring (or at least evoking) some sense of the social and cultural life once led in this long-lost villa, we introduced into triclinium G a video of actors with whom, as part of our previous Body and Mask in Ancient Theatre Space project,[6] we had prepared and recorded performances of such ancient drama (see figure 2.5).

Of course, in this room our model has moved well beyond a vigorously evidence-based visualization into the realm of imaginative reconstruction in order to fashion before the virtual visitor not merely a speculative

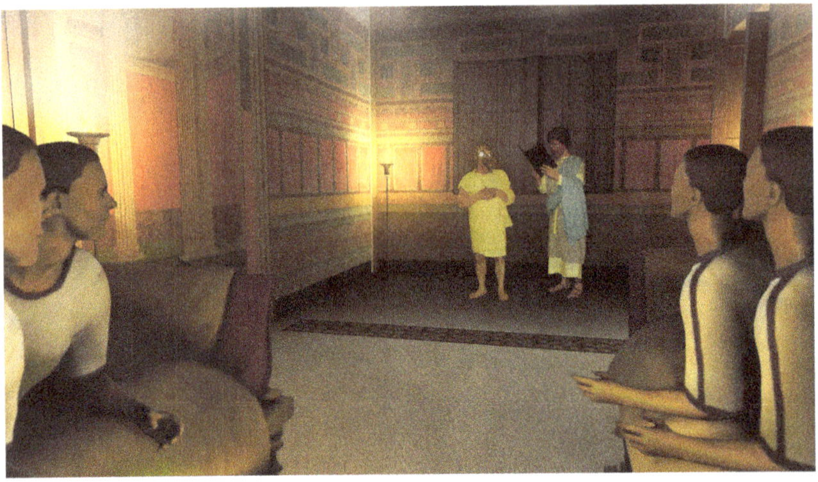

**Figure 2.5** Triclinium G, in the virtual model of the Villa Boscoreale, looking east at a hypothetical representation of a scene from Plautus's *Pseudolus*.
*Credit:* Drew Baker and Martin Blazeby for King's Visualisation Lab, King's College London. Performance by Angelo Crotti and Romans Suarez-Pazos.

representation of physical elements but an evocation of imagined experience as well. One is briefly invited to dwell in a dream of marble halls; one is encouraged, through the agency of one's mind's eye, to contemplate, lingeringly, something of the cognitive resonance which, at the beginning of this chapter, I suggested as immanent in all architecture. Not inappropriately, the viewer's exploration of an ancient architectural space which was itself quintessentially theatrical has transported that viewer into a felt moment of theatre.

## REFERENCES


Baker, Drew, and Martin Blazeby. Forthcoming. "Creating the Virtual Villa." In *Oplontis Villa A at Torre Annunziata*, edited by Michael Thomas and Ivo van der Graaff. *Vol 3: Architecture, Archaeology, and Material Culture, 2006–2010, Part B: "Documentation."* New York: Humanities E-Book Series of the American Council of Learned Societies.

Barnabei, Felice. 1901. *La Villa Pompeiana di P. Fannio Sinistore Scoperto presso Boscoreal*. Rome: Tip. della r. Accademia dei Lincei.

Beacham, Richard. 2013. "Otium, Opulentia and Opsis: Setting, Performance and Perception Within the Mise-en-Scène of the Roman House." In *Performance in Greek and Roman Theatre*, edited by Vayos Liapis and George Harrison, 361–408. Leiden: Brill.

Beacham, Richard. 2016. "Architectural, Pictorial, and Virtual Environments: Making Space for the Oculus Mentis of Ancient Theatre." In *Close Relations: The Spaces of Greek and Roman Theatre*, edited by Paul Monaghan and Jane Griffiths, 63–94. Cambridge: Cambridge Scholars Publishing.

Beacham Richard., D. Baker, M. Blazeby, and H. Denard. 2014. "The Visualisation of the Boscoreale Villa in 3-D." In *La villa romaine de Boscoreale et ses fresques*, edited by

Alix Barbet and Annie Verbanck-Piérard, 165–195. Éditions Errance-Actes Sud et Musée royal de Mariemont.
Diderot, Denis. 1994. "Le paradoxe sur le comédien." Translated by Geoffrey Bremner. In *Denis Diderot, Selected Writings on Art and Literature*. London: Penguin Press.
Eliot, Thomas Stearns. 1943. *The Four Quartets*. New York: Harcourt.
Gibson, James. 1979. *The Ecological Approach to Visual Perception*. Boston: Houghton Mifflin.
Goodman, John. 1995. *Diderot on Art*, vol. 2. New Haven, CT: Yale University Press.
Hecht, Ben. 1963. *Gaily, Gaily*. Garden City, NY: Doubleday.
Johnson, Odai. 2018. *Ruins: Classical Theatre and the Archeology of Memory*. Ann Arbor: University of Michigan Press.
Schelling, Friedrich. 1985. "Philosophie der Kunst, 1802–1805." In *Friedrich Wilhelm Joseph von Schelling, Ausgewählte Schriften*, edited by M. Frank, vol. 6. Frankfurt: Suhrkamp.
Scully, Vincent. 1969. *American Architecture and Urbanism*. New York: Praeger.
Tennyson, Alfred Lord. 1967. *Tennyson: Poems and Plays*. London: Oxford University Press.
Von Ranke, Leopold. 1874. *Geschichte der romanischen und germanischen Völker von 1494 bis 1514*. Leipzig: Dunder & Humblot.

CHAPTER 3

# The Digital Revolution and Modeling Time and Change in Historic Buildings and Cities

## *The Case of Visualizing Venice*

CAROLINE BRUZELIUS

### INTRODUCTION

Cities, like amoebas, change shape. They expand, they contract. They evolve in certain ways at certain times, often for reasons that have little or nothing to do with deliberate or conscious planning and design.[1] Over time, spaces are renegotiated to serve shifting social, environmental, political, or economic purposes,[2] as Jean Sauvaget showed so brilliantly in 1941, illustrating how an ancient colonnaded street evolved into a Muslim souk (figure 3.1).

In Rome, the forum with its great monuments degraded in late antiquity into a cow pasture, but in the early modern period developed into a densely packed residential quarter, only to be voided in the fourth decade of the twentieth century by Mussolini for political and ideological reasons.[3] And at present, the cities that have become destinations for tourists are being modified to suit the demands of leisure travelers: Florence and many other Italian urban centers have closed some streets to enable the flow of tourists from site to site, and local residents often escape to peripheral areas where they can park and use elevators. As a result, the experience of a historic monuments, once immersed in the humming life of a city that was packed with houses and shops, is lost in favor of isolated monumentalization. The pernicious effects of modern tourism exacerbate a separation between historic monument and the

Caroline Bruzelius, *The Digital Revolution and Modeling Time and Change in Historic Buildings and Cities* In: *Digital Cities*. Edited by: Maurizio Forte and Helena Murteira, Oxford University Press (2020). © Oxford University Press.
DOI: 10.1093/oso/9780190498900.003.0004

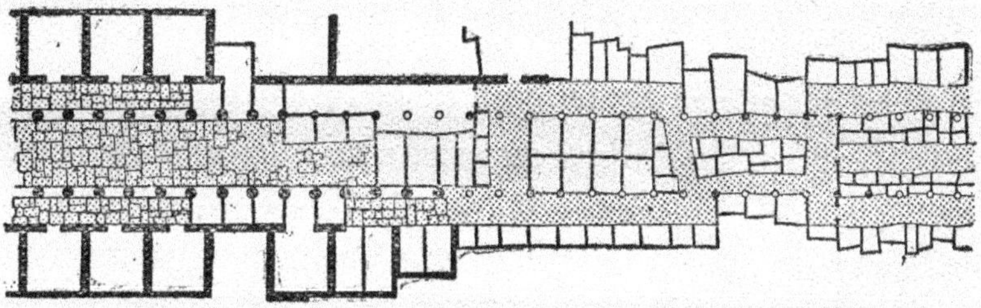

Fig. 25. — Transformation en souk d'une avenue antique : schéma montrant le processus de dégradation de l'ordonnance antique et la localisation topographique des souks médiévaux correspondants.

**Figure 3.1** View of an ancient street in Aleppo as it transforms into a Muslim souk.
*Source:* Sauvaget 1941, 104, Fig. 25.

city that had begun in the nineteenth century, when shops and other buildings that once clustered like beehives around major sites were demolished to create isolated monumental splendor for the building and large open squares, generating vistas: the *parvis* in front of Notre-Dame Cathedral in Paris comes to mind.

## BUILDINGS IN CITIES AND TIME

Change flows from the past into an ineffable and unknowable future. An individual encounters a building or a city in a moment of its ongoing processes of change; her or his experience intersects only briefly with the "long life" of monuments and places, which, after that encounter, will continue to be renegotiated in imponderable ways. There is another aspect of flow and change, as well: in the past, and especially in the medieval period, the construction of large structures was in itself a long-term process that was often characterized by a certain fluidity in the working out of details and design. Medieval buildings were produced in fluid construction environments, frequently subject to change of many kinds *within the period of construction.* These changes could be in building methods, materials, design, and even the plan. For example, in many of the most prominent Florentine churches, façades were left incomplete until the nineteenth century (the Cathedral and Santa Croce), and at the Dominican church of Santa Maria Novella, only the lower part of the façade with its portals and tomb niches existed prior to Alberti's sixteenth-century intervention.[4] The Gothic cathedrals of France—buildings that are viewed as governed by rules of order and proportion—underwent many modifications during construction. Reims Cathedral was lengthened by several bays; at Saint-Denis, had the new choir begun by Abbot Suger (1144) been joined to the abbot's west façade inaugurated in 1140, there would have been an awkward junction between the two very different plans.

There are thus two kinds of time and change that affect our understanding and representation of monuments: the moment in which we ourselves encounter a building or city, and the time (and change) that may have been entailed in its own coming into being in the past.

Representing change as ongoing process in man-made space has been especially challenging for the *longue durée* of ancient and medieval sites. This challenge applies to almost any historic city or monumental architectural enterprise. It is interesting to note that one of the few scholars to actively engage in representing change over time and the process of ongoing transformation, Jean Sauvaget (1901–1950), was an anthropologist, as well as an archaeologist.[5] His interests extended from classical antiquity through the Ottoman period, and ranged from architecture to pottery, to city planning. It is perhaps appropriate that as an anthropologist he focused on the shifting function of space, and his understanding that all history is based on people (and how people change) no doubt informed the creation of his image of the transformation of an ancient colonnaded avenue into a Muslim souk.

## VISUALIZING VENICE: A RESEARCH AND TEACHING INITIATIVE

Prior to the digital revolution, change and process could only be represented as sequential static images that illustrated how buildings and cities were transformed as a series of isolated "frozen moments." For example, images of the medieval walls of Florence show them as a completed project, with no indication of the long, slow process of constructing a defensive system that gradually expanded around the city over multiple decades as property was expropriated and houses were demolished.[6] Although urban defense systems inevitably have the function of defining the edge of a city, marking and fixing borders and emphasizing the transition from countryside to urban space, our traditional methods of representing city walls in static images belie the fact that these (and most other) fortifications came into being over decades, as funds were raised, property was expropriated, and houses were demolished: an animation on the construction of the urban defense system would be a far more accurate method of representing this process than a static image.

The description of sites and monuments in print is static, not fluid. Until now, city plans and representations of buildings have not been able to show the slow *process of becoming in time* of a monumental structure, or urban space. The few new cities that were designed from the outset as a "concept"—either in the past, as in the new towns of the Middle Ages (Aigues-Mortes, for example), or in the modern period (such as Canberra) nonetheless continue to evolve in an ongoing state of flux that constantly renegotiates the original plan to serve new needs. Canberra today echoes only the broadest outlines of Marion Mahony Griffin and Walter Burley Griffin's original designs. Historic cities, such as Siena, crept forward over centuries to cover adjacent ridges and along the main road from Rome to northern Europe (the Via Francigena) as travel and commerce expanded in the twelfth and thirteenth centuries. Often, as in Siena, certain types of institutions, often the convents of the religious orders and adjacent piazzas, "anchored" a new area of town, attracting shops that would serve expanding communities; in medieval Italy and elsewhere, communal governments often supported these anchoring institutions to promote expansion and growth.

The disjuncture between the traditional methods of visual representation and the actual *process of becoming* is analogous in monumental architecture, especially the buildings of the Middle Ages, which were constructed over decades or even centuries, and underwent modification and change of design even while under construction.[7] Our understanding of the evolving process of creating a historic building within a fluid construction environment has been conditioned and restricted by our tools for representation (plans, sections, elevations, photographs, reconstruction drawings), which tend to show the whole rather than a *successive aggregation of parts*. In other words, the nature of our capacities of representation, partnered with a desire for the concepts of harmony, unity, and visual coherence, collaborated with and perhaps stimulated a formalist approach to art and architectural history, especially in the second half of the twentieth century. A desire for aesthetic unity superseded the visible reality of change and process, and this, in part, is because we did not have the tools to represent these phenomena as occurring in the fluidity and unpredictability of ongoing time. Thus, when we consider a historic monument, such as the Cathedral or Santa Croce in Florence, or Santa Maria Gloriosa dei' Frari in Venice, we experience (either in person on site or in publication) these structures in their completed state rather than comprehending them as the aggregation of incremental gestures that slowly brought a building or space into being, often over decades or centuries. One of the most notable and beautifully documented examples of this process of change and design fluidity, as well as its implications for an architect, is Michelangelo's work on the transept arms of Saint Peter's Basilica in Rome, a phenomenon documented by Howard Burns on the basis of Michelangelo's letters.[8]

Slow, ongoing process of construction and modification are also characteristics of urban space in relation to the insertion of large-scale buildings or institutions such as hospitals or convents, which entailed the destruction of preexisting structures to make space, a phenomenon particularly acute (and painful) in the crowded cities of the thirteenth and fourteenth centuries. The making of large-scale historic buildings, such as a cathedral, would have profoundly disrupted surrounding areas and thus, as urban interventions had a ripple effect, entailing decades of upheaval. At Santa Maria Gloriosa dei' Frari in Venice, construction moved forward at an achingly slow pace, as individual donors contributed one column at a time; and as this occurred, we can probably imagine the surrounding piazza and neighboring structures being reconfigured or demolished in response.

The study of medieval cities and buildings can often reveal an ongoing internal dialogue between initial design and subsequent phases of construction; this was evident not only in my work on Notre-Dame Cathedral in Paris[9] but also at the Cathedral of Noyon. Elevation drawings of the chevet (figure 3.2) show heavy upper stories with multiple and layered moldings precariously perched over slender *en délit* columns of the hemicycle below.

Was there a change in architect to account for the profound transformation of design between the arcade at hemicycle level and the gallery above?[10] Was there yet another master builder who took over as construction at Noyon moved from east to west, introducing an entirely new concept with superimposed wall passages in the transept? These disjunctions and their implications have not yet been seriously addressed in the literature on this building.

Representations of buildings in the traditional form of elevations and sections provide vital information, but they cannot address the ongoing nature of disjunction and process. Drawings, plans, sections, elevations, and maps condition historical narratives toward a presentation of coherence and continuity. The traditional tools of representation incline scholars to study buildings and cities as completed projects rather than as an ongoing *process of becoming:* the predilection to see and understand buildings and spaces as total concept and coherent space has superseded that of understanding them as progressively negotiated interventions.

All this is obvious, at least to a medievalist who has spent a lifetime studying historic buildings and spaces. And while of course historians study the process of constructing a building or shaping a city, the tools of representation have privileged interpretations of coherent and complete vision as "ideal" project (in Albertian terms of coherent design), rather than the evolving and emerging and fluid *process* of design and construction. In many ways, the fifteenth century seems to have been a critical moment for this shift, for there were not only the treatises of Alberti and the concretized notion of the "ideal project" but also the invention of printing (and therefore also of "fixing" an image through the multiplication of publication).[11]

Until the recent invention of digital technologies, particularly 3D modeling and mapping tools and animations, modes of representation available to the architectural or urban historian (maps, plans, photographs) were static. They could not easily engage with the creation of urban or architectural space as *a flow in time*.

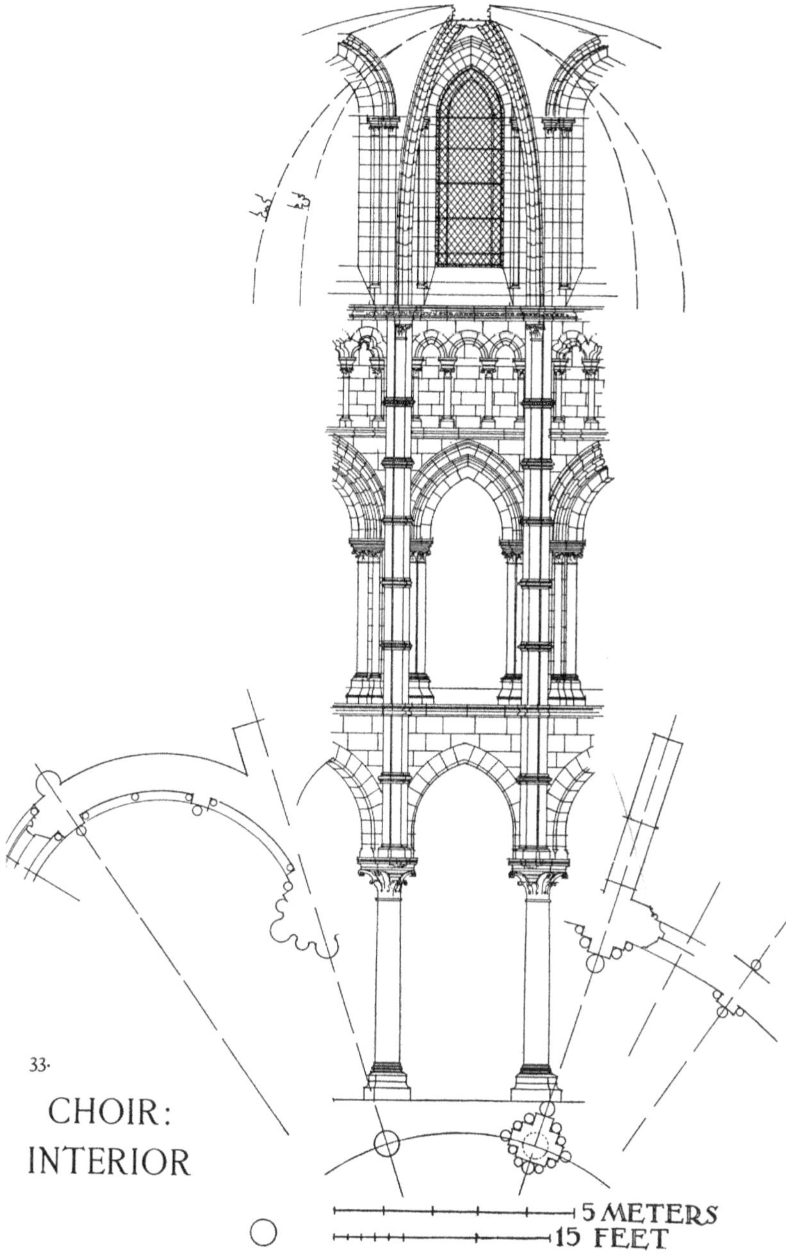


**Figure 3.2** Elevation of the choir of Noyon Cathedral.
*Source:* Seymour 1939, 111.

## VISUALIZING VENICE: A RESEARCH AND TEACHING INITIATIVE

Visualizing Venice (figure 3.3) is an international collaboration created by myself and Donatella Calabi in 2010 to explore the capacities of digital technologies to reflect *change as an ongoing process* and address the concept of cities and buildings as fluid construction environments.

Our primary goal has been to train and engage young scholars in understanding that digital technologies now form a critical and transformative part of the intellectual process, and that new tools stimulate new types of questions about evidence and radically new capacity for *narrating* the lives of buildings and cities. The goal of our initiative is to work against the notion of an idealized, "authorial," and "fixed" understanding of built space, and instead to explore the intermittent adaptive interventions that respond to the vicissitudes of finances, available space, and changing use.[12] Visualizing Venice as a research initiative pushes against the constraints of traditional representation, engaging with the flow of a city and its buildings in time, examining the affordances of mapping and modeling technologies to produce 3D models and

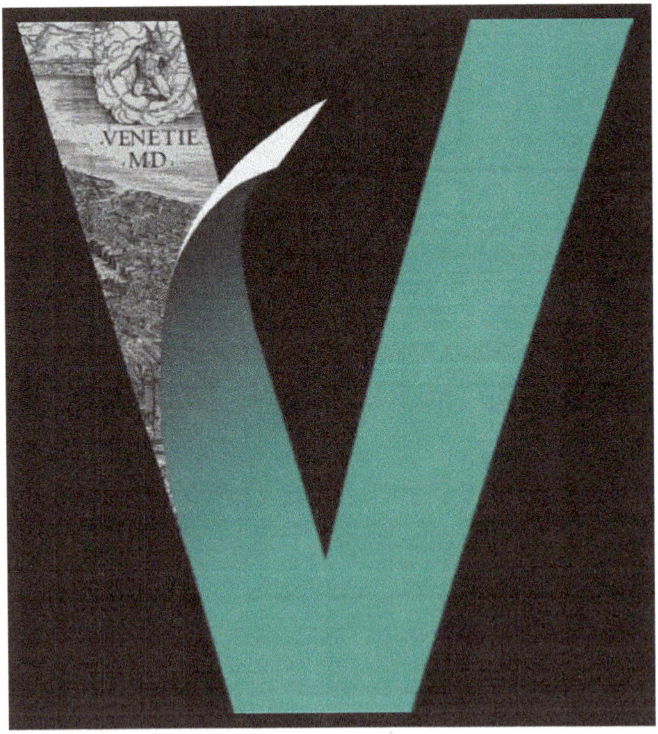


**Figure 3.3** The Visualizing Venice website logo.

animations that reflect the transformation of cities and buildings as part of the organic, "amoeba-like" flow of change and adaptation in time.[13]

Visualizing Venice focuses on different types of interconnected experiments in mapping and modeling, and combines research with teaching and public outreach. Begun as a trial experiment in 2010 to map and model change in Venice over time, and to engage graduate scholars and recent PhDs in new representational tools, Visualizing Venice projects are documented by historic maps and cadasters. The initiative has grown to include annual training workshops, exhibitions, and a series of modeling projects that focus on different regions of the city.[14]

From the start, the research group has been committed to teaching and training, a concept that emerged from the experience and experiments of the Wired! lab at Duke University.[15] In 2009, there were few opportunities for graduate students to learn digital tools in the context of their home universities, as few departments had the combination of equipment, intellectual interest, and in-house expertise to support such integration of new modes of representation and analysis that digital technologies make possible. To achieve this goal, the Wired! group at Duke[16] created a digital laboratory at Venice International University to train students at all levels and teach best-practice technologies for gathering data, mapping, and modeling. We offer training for graduate students and young professionals in the fields of architectural preservation, architecture, art, architectural and urban history. The operational principle for the workshops is to teach a variety of digital technologies through the study of a particular topic—that is to say, through data collected on a certain theme. The first workshop (in 2012), taught in collaboration with Giorgio Gianighian (Università luav di Venezia, or IUAV), focused on the cisterns of Venice: how the system of provisioning fresh water conditioned the shaping of the city. Subsequent workshops concentrated on the relationships between the city and the lagoon (the topic of a major exhibition at Palazzo Ducale in Venice in 2015, curated by Donatella Calabi), the Venetian Ghetto, and the Biennale.[17] In 2016, the workshop returned to the topic of the Venetian Ghetto in commemoration of its founding five hundred years ago.

Visualizing Venice is also committed to research and the 3D modeling of change in buildings and cities. We first focused on the area of Santi Giovanni e Paolo, with a series of projects and collaborations that examined the transformation of this part of the city from circa 1500 to the present (figure 3.4).

Extended research by Gianmario Guidarelli and Ines Tolíc investigated the various unbuilt projects for the development and expansion of the former convent into Venice's primary hospital after the Napoleonic suppression of churches and convents.[18] Other collaborators modeled changes to the exterior of the Dominican basilica, as well as ideal routes of vision for the perspectival reliefs on the façade of the Scuola Grande di San Marco. Additional projects modeled the expansion of the Arsennale and the area around the Academia Museum (formerly the convent of Santa Maria della Carità). At

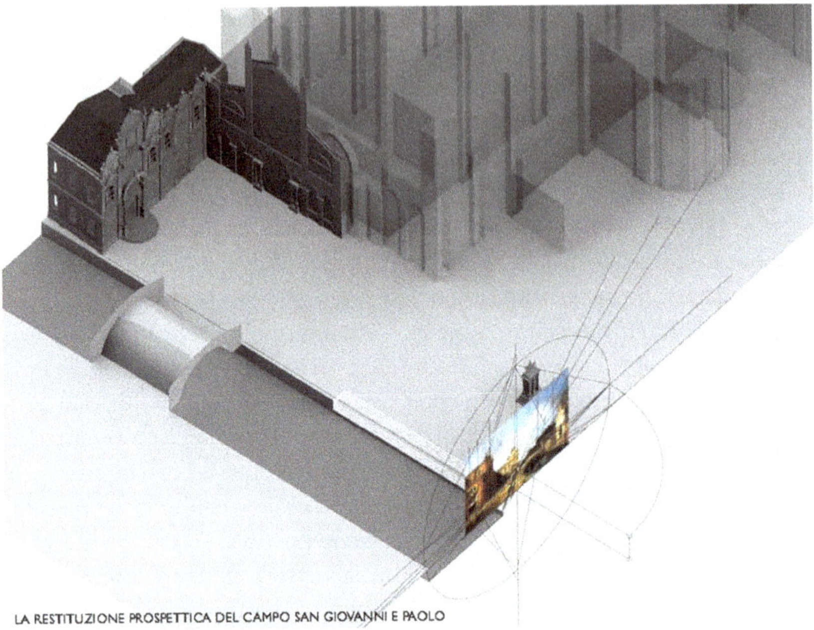


**Figure 3.4** Perspectival study of Santi Giovanni e Paolo based on Canaletto painting.
*Source:* With kind permission of Andrea Giordano, the University of Padua.

Duke University, students working under the guidance of Kristin Lanzoni created interactive displays based on the De' Barbari map of circa 1500. Other important projects produced by members of the group include modeling growth and change at the city of Carpi, by Elena Svalduz and her colleagues at the University of Padua, as well as a wide range of digital interventions on the city and buildings of Padua.

There is considerable cross-fertilization of work, between the collaborators at the different partner universities and between the teaching and research projects. For example, material for the splendid exhibition on the provisioning of Venice (fall of 2014) on view at the Palazzo Ducale in Venice; "Aqua e Cibo" was related to the topic of the 2014 training workshop. Projects developed for the training session held in 2013 on the Venetian Ghetto are informing, and providing part of the data for, an upcoming exhibition and associated apps held in the summer of 2016 that commemorates the five hundredth anniversary of the founding of the first ghetto in Venice.

## CONCLUSION

Learning, research, modeling, and public outreach are woven into the Visualizing Venice enterprises. The team is deeply committed to the notion that digital tools are transformative and essential for modeling space and structures in time. The capacities of these technologies are transforming our ability to engage in questions on man-made space, tracking its modification and transformation as social, political, economic, and religious institutions evolved. The digital option enable scholars to reach out to the public through websites and apps, engaging tourists, visitors, and students in a new way of understanding where they are and why places there look the way they do.

## REFERENCES


Bruzelius, C. 1987. "The Construction of Notre-Dame in Paris." *Art Bulletin* 69: 540–569.

Bruzelius, C. 2014. *Preaching, Building and Burying: Friars in the Medieval City*. New Haven, CT, and London: Yale University Press.

Burns, H. 1995. "Building Against Time: Renaissance Strategies to Secure Large Churches Against Changes to their Design." In *L'Église dans l'architecture de la Renaissance*, edited by J. Guillaume, 107–131. Paris: Picard.

Day, F. 1954. "Jean Sauvaget, 1901–1950." *Ars Orientalis* 1: 259–262.

Ferrighi, A. 2014. "Urban History Through New Technologies. Visualizing Venice, a Pilot Project." In *Digital Urban History. Telling the Story of the City in the Age of the ICT Revolution*, edited by R. Tamborrino, 59–66. Rome: Università degli Studi Roma Tre—CROMA.

Guidarelli, G., and I. Tolíc, I. 2016. "Buildings that Never Were: The Unbuilt Projects for the Civic Hospital of Venice." In *Visualizing Venice: Mapping and Modeling Time and Change in a City*, edited by K. Huffman, A. Giordano and C. Bruzelius, 29–35. Abingdon, Oxon: Routledge.

Huffman, K., A. Giordano, and C. Bruzelius, eds. 2016. *Visualizing Venice: Mapping and Modeling Time and Change in a City*. Abingdon, Oxon: Routledge.

Kallis, A. 2014. *The Third Rome, 1922–1943: The Making of the Fascist Capital*. Houndmills, Basingstoke, Hampshire: Palgrave Macmillan.

Kennedy, C. 2011. *The Evolution of Great World Cities: Urban Wealth and Economic Growth*. Toronto and London: University of Toronto Press.

Sauvaget, J. 1941. *Alep. Essai sur le développement d'une grande ville syrienne, des origines au milieu du XIXe siècle*. Paris: Geuthner.

Seymour, C. 1939. *Notre-Dame of Noyon in the Twelfth Century: A Study in the Early Development of Gothic Architecture*. New Haven, CT: Yale University Press.

Trachtenberg, M. 2013. *Building in Time: From Giotto to Alberti and Modern Oblivion*. New Haven, CT, and London: Yale University Press.

CHAPTER 4

# Exploring Visually the Known and the Ill-Known About Kraków's Center Urban Evolution

## *An Information Visualization Perspective*

JEAN-YVES BLAISE AND IWONA DUDEK

### INTRODUCTION

Studying and depicting the designs and transformations of urban structures have been concerns ever since the nineteenth century, from the point of view of both historians and urban planners. What exactly has been the impact of the "digital humanities" or, more precisely, the digital technologies, on the way we employ our means of investigation today? How can we best support the analyst's reasoning tasks, by visual means, in the context of historical research and on the scale of a city? What methodological pitfalls are likely to be encountered? What can be done when there simply is not enough data to carry out a reconstruction, or when there is a need to formalize hypotheses that can be used to back up a reconstruction process? How can we support analysis tasks when our concern is not a spatial pattern but, rather, a temporal pattern?

An obvious difficulty when attempting to analyze the evolution of a "city" in the broad sense is the fact that the available information is sometimes scarce or questionable, when it is not completely absent. Hence, this chapter raises the question: How can analysts formalize the ill-known and employ it pragmatically in contemporary information systems?

Another difficulty is the heterogeneity of the objects under scrutiny. Consequently, this chapter also presents a key aspect of reasoning about a

Jean-Yves Blaise and Iwona Dudek, *Exploring Visually the Known and the Ill-Known About Kraków's Center Urban Evolution* In: *Digital Cities*. Edited by: Maurizio Forte and Helena Murteira, Oxford University Press (2020). © Oxford University Press.
DOI: 10.1093/oso/9780190498900.003.0005

city: that the step from analyzing a single edifice to that of an urban structure is not *quantitative*—it is primarily *methodological*. That is, an urban structure is a system in which interrelations and interdependencies play major roles; these relationships may be between edifices, between edifices and open spaces, or between the urban fabric and a context, such as wars, floods, laws, or the aesthetics of a time. In that sense, switching to an analysis of an entire urban structure may reveal such interrelations, patterns of change, and possible causal chains, thereby allowing for greater information and subsequent knowledge discovery.

This chapter is not a theoretical discussion of how to "reconstruct the past," however. Instead, it introduces real-life experiments that were carried out on the city center of Kraków, the former capital of Poland, and it uses those case studies to illustrate some key scientific and technological difficulties we are likely to encounter when wanting to better understand, and have others better understand, the way a city changes over time. Through these experiments we illustrate the fundamental problem of imperfect knowledge in the historical sciences; the need to handle ill-defined temporal data and the constraints in doing so; and the necessity to at least complement "classic" virtual reconstruction efforts with visual solutions that better match the nature of the underlying data.

## BACKGROUND INFORMATION

The city of Kraków developed significantly during the first six centuries of the second millennium, during which time it was a center for a renaissance in art and architecture. But in the waning years of the sixth century, the city entered a period of relative decline, culminating in the Swedish invasion, after which Polish King Sigismund III moved the capital of the kingdom to Warsaw. This period of decline and somewhat thereafter offers an opportunity for us today, however. The city experienced little significant change during the subsequent nineteenth and twentieth centuries, compared to other cities of similar size in Europe. Moreover, years of preservation and investigative actions resulted in the production of a large quantity of documents (descriptions, analyses, drawings, photographs, plans, etc.) that need to be collected, organized, and made available for study.

Our research, therefore, has focused on understanding and recording the evolution of Kraków's architectural and urban heritage, as reflected in these documents (341 objects, 885 phases studied). The work has involved primarily historical sources but also elements of context that can help in identifying potential causal chains (legal issues, natural phenomena, fire propagation patterns, demography, wars and rulers, etc.). Specifically, we have been studying the evolution of Kraków's historic center (figure 4.1), on various scales, for over fifteen years now. Consequently, the work has evolved from analyzing solely an edifice to analyzing an entire urban area. This broadening of perspective clearly represents a methodological break,

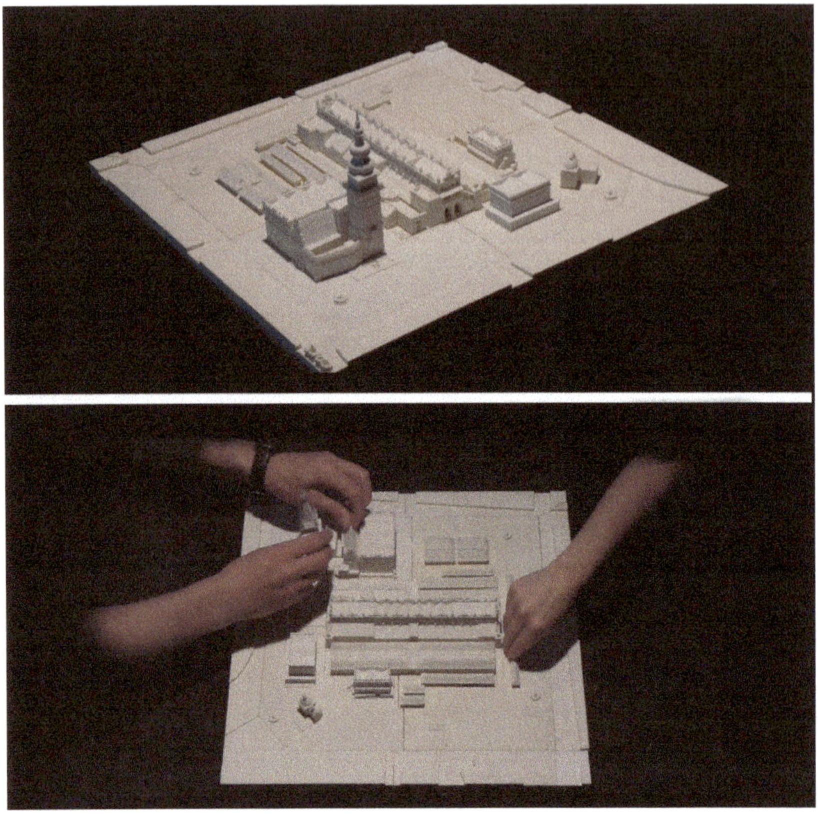

**Figure 4.1** The Tangible Chronology experiment: physical models combined on a master board, and depicting the evolution of Krakow's urban center over time (Blaise and Dudek 2013). Top: layout of the city's main square in 1770, according to what historical hints allow us to understand. Bottom: illustration of a usage scenario of the Tangible Chronology experiment—a guided, didactic combination of physical models matching this or that phase of evolution.
*Credit:* © I. Dudek & J. Y Blaise, UMR CNRS/MCC 3495 MAP.

not only in numbers (more elements) studied but also in heterogeneity (different objects of study, different quantities and/or quality of data, different levels of knowledge).

A commonsense assertion would be to consider that, as time passed, we would get to know more and more about the changes that had occurred in the city. And indeed, we did uncover facts about individual edifices, the urban layout, uses and practices of urban spaces, the defensive system, and the like. But, as it is often the case in historical research, as time went by we also raised more and more questions and identified more and more ill-knowns or unknowns. The reality of such historical research on a large, heterogeneous corpus—that is, an entire city—is this: the more one digs, the more unknowns are uncovered. Thus, our experience established a counterintuitive pattern: the proportion of unknowns or ill-knows grows in relation to the time spent comprehensively analyzing a city's evolution.

But is that pattern *really* so surprising? As we have analyzed more and more individual components of the city, we also have pulled in more and more sources, more and more elements of context. Accordingly, we added up not only the unknowns corresponding to each component or each piece of information but also the doubts about potential relations between the components' context. A classic example of this is Kraków's fire propagation patterns. If a source tells us that a fire *probably* damaged artifact A, chances are that it also had an impact on any B neighbor of A. And so not only is the information concerning A in doubt but it also initiates a whole chain of potential (and questionable) consequences.

Furthermore, understanding a city's changes over time cannot depend on only one or even several "pictures" of its spatial layout at given times $t_1, t_2, \ldots, t_n$ (synchronic analysis). It also requires understanding the temporal patterns that characterize those changes: rhythms, order, cycles, and so on (see figure 4.2). Hence, there is the necessity to complement "assertive" visual results

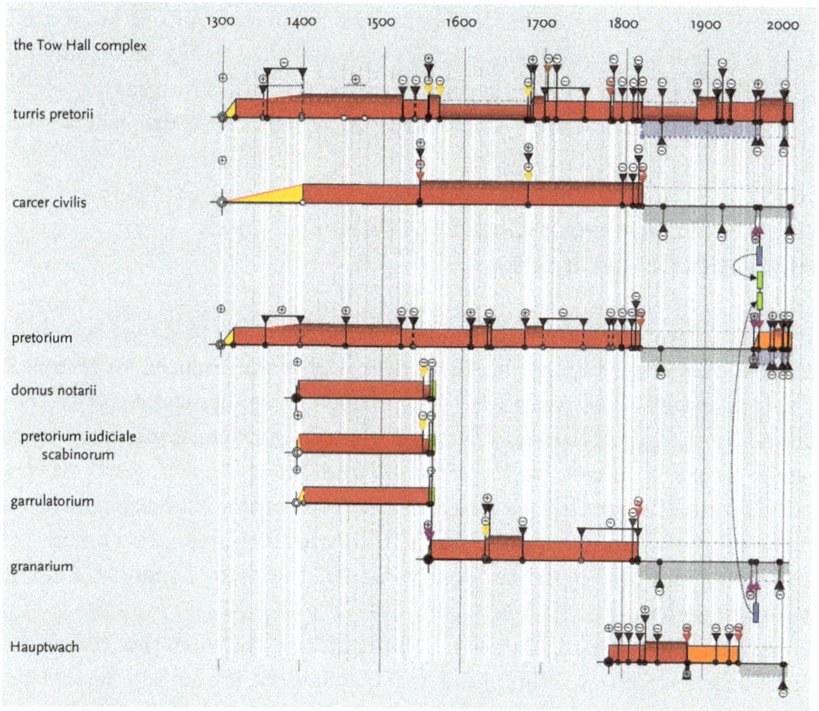


**Figure 4.2** Trying to spot and analyze temporal patterns: the diachrogram visual formalism (Dudek and Blaise 2012). Here, changes on subparts of the Town hall ensemble are distributed in time (time runs from left to right, each diachrogram corresponds to one of the subparts). Colors and glyphs are used to position and differentiate among occurrences of relevance with regard to each subpart. The visual formalism allows the analyst to spot temporal density patterns (highlighted by white lines, vertically) and more generally to cross-examine, in time, the historical evidence on events and processes that impact each of the subparts.
*Credit:* © I. Dudek & J. Y Blaise, UMR CNRS/MCC 3495 MAP.

such as 3D models with other analysis platforms that are better suited to handle massive data sets, doubtful information, and temporal data.

But on what methodological and/or technological basis can such platforms be designed and built? As will be shown, doing analysis visually on large information spaces is at the heart of a subfield of information sciences—namely InfoVis (Information Visualization) and its sister field, visual analytics. Our intent in this chapter is to show, through case studies, that historical research about the pasts of cities can benefit from complementing photorealistic representations with abstract graphics inspired by the InfoVis paradigms and practices.

First, we give an overview of the mainstream practices in terms of virtual reconstructions of cities. We then discuss four major scientific challenges that such initiatives raise: how to deal with the ill-known, how to represent temporal data, how to ensure the reproducibility of experiments, and how to enable the analysis of collections as such.

Finally, we briefly position InfoVis with regard to its legacy and the general context of information technologies, and we detail some of our key experiments. As will be shown, these graphics were intended to help in reading the available data, assist in filtering and reorganizing it, and ultimately aid in exploring visually the known and the ill-known about a city's urban evolution.

## VIRTUAL RECONSTRUCTIONS: PRACTICES AND OPEN METHODOLOGICAL ISSUES

With the increase in digital technologies developed in the past twenty years, a number of new tools (in the broad sense) have been introduced for historical research and its communication. These tools include digitization platforms, online databases, information systems, virtual reconstructions, augmented reality disposals, immersive 3D platforms, and others. Providing an overview of these "new" opportunities is beyond the scope of this chapter; instead, we discuss three significant trends, born in the early 1990s, so as to pinpoint the underlying methodological issues that we consider to be common, and still open for resolution.

Back in 1993, Patricia Alkhoven (1993) published research that remains in many ways a landmark accomplishment. In that work, the author blended historical research on the evolution of the city of Heusden (the Netherlands) with an architectural analysis of morphotypes, and with a significant 3D modeling effort that resulted in a series of "virtual reconstructions." The technology of that time is, of course, obsolete today, but her research early on exemplified the pitfalls, work processes, and potential benefits of using 3D models in the course of such research. These features included heterogeneous and untrustworthy inputs, pieces of general knowledge utilized to counterbalance missing data, case-by-case inferences made, the seductive power of virtual environments, and so on. Today, building such "virtual reconstructions" on the scale of a city has become a popular research goal; however, there are nuances to be noted in the way that activity is conducted.

The initial virtual-reconstruction paradigm can be described as an effort to translate the expert's understanding of a particular city's layout and its components at a given time *t* into a set of geometric shapes distributed in space. The underlying geometric canvas can be obtained by various means, from old maps to contemporary digitization campaigns. Alkhoven's 3D models were partially based on a 1649 map by Joan Blaeu. The Virtual Leodium project (Pfeiffer et al. 2013) is based on a 3D survey of a physical model representing the city of Liège (Belgium) toward 1730. The Virtual Kyoto city model (Takase et al. 2008) integrates automatic generation of a 3D city model using LIDAR data, a 2D digital map, and aerial photography. Significant research efforts are put today toward trying to produce photorealistic city models and publish them on the internet, with the focus quite more often on the technological suite rather than on the scientific added value.

But, as exemplified by Alkhoven's project, the objective is not necessarily to reach to a photorealistic anamorphosis in the virtual world. Rather, it can be on testing different geometric combinations or fostering an educated debate among scientists. Here, non-photorealism (NPR) appears to be a potential alternative approach (e.g., Döllner and Buchholz 2005).

Beyond figuring a city's outlook, 3D models can be used as interfaces—sort of a 3D implementation of the Geographical Information System (GIS) paradigm. An early attempt posted on the internet was carried out by Heinonen et al. (2000), who developed what they call an "information database for VRML cities" (VRML being Virtual Reality Modeling Language), an interactive 3D model that can be queried interactively so as to retrieve information about services, shops, and so on. We also applied this principle in early developments of the Kraków project (Blaise and Dudek 2000), whereby the data we had gathered on individual edifices or urban spaces were available using a 3D, NPR model of the city (figure 4.3). An interactive slider was used to browse through time, monitoring the content of the 3D model in real time. A number of graphic codifications were introduced to convey information about the nature of the documentation behind each component in the model or about the nature of the hypothesis made in the reconstruction.

The Virtual Leodium project previously mentioned reinterprets this principle in a more contemporary technological suite: the objective for the project was not only to reconstruct the city but also to provide access to the historical evidence for each edifice in the 3D model. To this day, however, a solid evaluation of the performance (in terms of information retrieval) of such platforms still needs to be made.

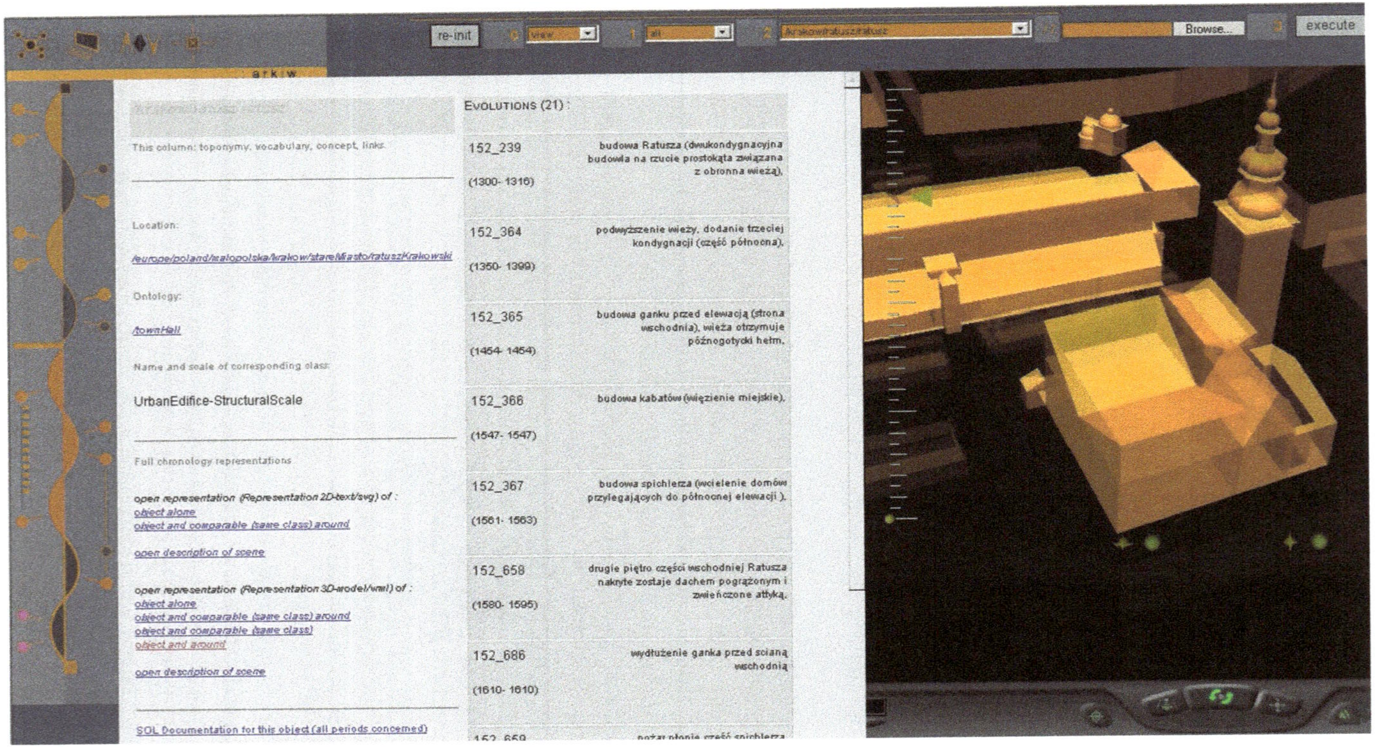


**Figure 4.3** The VIA platform investigates the use of non-photorealistic 3D models as interface: a time slider is included in the model; each edifice is associated with a list of evolutions, and with the corresponding bibliography (shown here is the example of the town hall ensemble).
*Credit:* © I. Dudek & J. Y Blaise, UMR CNRS/MCC 3495 MAP.

With the passage of time and greater levels of accomplishment in the technologies, we should see virtual reconstructions targeting even greater general communication goals. The past decade has shown a clear move from 3D modeling for producing images of lost cities and testing hypothetical reconstructions to what could be called "visual storytelling." Such initiatives, supported by technological breakthroughs such as haptic interfaces and virtual and augmented reality (see, for instance, Dima et al. 2014), will borrow the visual language (when not the actual graphic engine or software suite) of computer games. They will end up on with "visual products" that range from interactive 3D scenes to videos or "serious" games. The issue at stake for scientists, then, is to use these media to explain the scientific basis behind such products.

This approach is exemplified in our test field, the city of Kraków, with research initiated on the occasion of the 750th anniversary of the city's foundation. A significant multidisciplinary research effort, this program has been conducted to back up scientifically the production of a 3D model showing the city at around 1650. Outputs include, on the one hand, a series of convincing communication products (videos, images, CD-Rom) and, on the other, scientific publications summarizing the historical evidence upon which the hypotheses are built (Marek 2011, Firlet 2010, Rydzewski 2006, and others).

Unfortunately, in many cases similar comprehensive efforts to explain the making of a 3D model have been neglected. In short, there are potential problems regarding knowledge sharing in initiatives that focus not on the data and its interpretation process but, rather, on fostering a type of artistic freedom—legitimate in the context of a communication-oriented activity but dangerous for scientific legitimacy. Therefore, a major concern for those involved in creating virtual 3D cities today is the scientific and technological "sustainability" of those end products and, in parallel, a growing dependence, including in terms of work processes, on tools and software suites. To meet this challenge, a number of initiatives have flourished that attempt to apply "open science" approaches—typically, developing standards such as CityGML (Lorenzini 2009) or BIM (building information modeling; see Baik et al. 2015). Yet in this chapter we focus on a more fundamental challenge: the underlying methodological issues. We discuss four main issues (with no attempt at exhaustiveness): how to deal with uncertainties, the role of time parameters, the recording of processing methods, and methods for analysis of complex data sets.

## The Uncertainty Issue

Dealing with uncertainty in the process of reconstructing a "lost city" is regularly cited as a basic issue in the scientific community. Sylaiou and Patias (2004) put it this way:

> the main problem is that advanced graphic systems that are used for computer reconstructions may sometimes be too realistic. Things can be difficult, especially when they are based on partial evidence, but nevertheless suggest an impression of good knowledge of the past. . . . A very good "image" can give the impression to the viewer that archaeologists know more than they actually know. Some products of computer reconstructions can be considered as scientifically accurate, because they seem to be accurate. (p. 6)

Back in 1999, Strothotte and colleagues had proposed to "encode systematically more information about the model than just its geometry" and then to "produce non-photorealistic images which encode this additional information" p. 37). In our first experiment on the Kraków project (in 1999), we differentiated systematically the existing form of the hypothetical by using graphic encoding (textures and transparencies; see figure 4.4). But what we learned over time is that there exists nothing that could be termed a "general," or "overall" uncertainty; there are, in fact, different factors of uncertainty, which include the credibility of sources, degree of precision, quality of temporal density, presence of contradictory data, and so on. Our claim here, therefore, is not that one should avoid representing objects that are ill-known but, rather, that it is vital if a 3D reconstruction is to be reused scientifically that one keep track of all that is *known*—including what is not known, what is partially known, and what is inferred from analogies and generic knowledge (Blaise and Dudek 2005; see figure 4.4).

This effort means making a significant attempt to formalize and convey existing doubts. The following list has definitions of some of the notions that can be used to better differentiate among uncertainty factors (a partial list, including the origin of the uncertainty).

- *Reliability/credibility* (origin: sources). A judgment made by the human consumer of the information about the information source. An information source that supplies data that conflict with other data, has supplied unreliable data in the past, or is otherwise suspect for reasons that imply uncertainty (Skeels et al. 2010).
- *Imprecision* (origin: sources and interpretation): Inexactness in measurement (Thomson et al. 2005). By extension, this can apply to factual eyewitness reports.
- *Incompleteness* (origin: sources): Information available to analysts that is never complete. Even with excellent collection capabilities, observed evidence is likely to be only a small portion of the whole (Thomson et al. 2005).
- *Lineage* (origin: sources): A questionable conduit through which information is passed (number of steps) (Thomson et al. 2005).

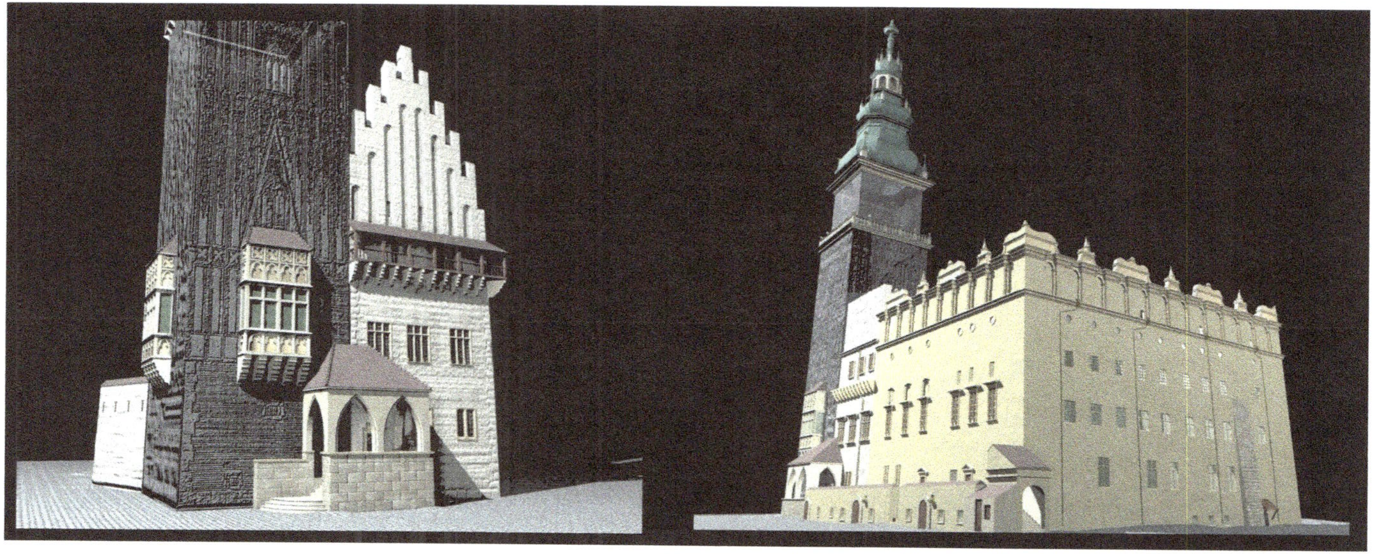

**Figure 4.4** Virtual reconstructions of the town hall ensemble. Existing parts (i.e., the belfry walls as they remain today) are differentiated visually from the hypothetical parts (represented in color).
*Credit:* © I. Dudek & J. Y Blaise, UMR CNRS/MCC 3495 MAP.

- *Approximation* (origin: sources and interpretation): An attempt to come close to measuring or describing a phenomenon, when in fact we cannot measure or describe it with perfect precision (Skeels et al. 2010).
- *Timing* (origin: sources): Temporal gaps that may exist between occurrence, information collection, and use (Thomson et al. 2005).
- *Subjectivity* (origin: interpretation): The amount of private knowledge or heuristics that has been utilized to reach conclusions (Zuk and Carpendale 2007).
- *Inconsistency* (origin: sources): The extent to which information components agree or not ("5 say 'M'; 2 say 'T'") (Thomson et al. 2005).
- *Accuracy* (origin: sources): Any existing differences between observation and reality (Thomson et al. 2005).

Historical research is clearly a relevant test field when we want to investigate the uncertainty issue, yet the topic is rarely at the core of historic research. In contrast, it is a hot topic in the information sciences, as exemplified by Zuk and Carpendale (2006) or Gschwandtner et al. (2014).

### The Time Parameter Issue

Reconstructing "lost cities" is not only (and not even mainly) a matter of spatial distribution, of geometric shapes—it is a matter of *time*. Handling the time parameter as such is, even in the information sciences, an issue that is scarcely addressed, although recent contributions (like Aigner et al. 2011) are pushing it toward the top of the research agenda. In the context of the historical sciences, the issue is, by its very nature, crucial. We mention two potential pitfalls we think those involved in the analysis of "lost cities" should be conscious of avoiding: temporal consistency, and the specificity of temporal data.

*Temporal consistency.* Ensuring the geometrical consistency of a 2D or 3D reconstruction is challenging because of the nature of the underlying data; repositioning old maps is a typical example of this. But the same difficulty exists when attempting to ensure the temporal consistency of that reconstruction. Quite often, the historical "hints" we handle do not permit us to be sure of temporal co-occurrences of the objects we represent in a single, synchronic view of a city (e.g., there is no direct evidence that two artifacts coexisted at time $t$). Even when we do have corroborated evidence that artifacts A and B coexisted at time $t$, we are often forced, owing to lack of information, to represent A based on information we have on A at time $t - \gamma$ and B based on information we have on B at time $t + \delta$. It's like an anachronism we might find in the theatre or the cinema—the classic example is a gladiator wearing a watch.

*Specificity of temporal data.* A second pitfall—and from our view the more prominent one of the two time factors—is not understanding the nature of temporal data, and therefore failing to encourage analysis specific to the time involved. These analysis tasks include spotting cycles, reviewing densities of change, comparing overlapping time slots, seeing potential causal chains, managing uneven time granularities and branching time, uncovering significant trends such as speed of changes, or acknowledging exceptions such as periods of intensive destruction. Avoiding this pitfall implies making primarily a conceptual modeling effort, built on notions derived from the information sciences.

Some precautions can be taken to minimize the impact of uncertain or ill-defined temporal indications at various stages of the research process (conceptual modeling, data management, reconstruction):

- Distinguish the uncertainty concerning geometrical aspects from the uncertainty concerning dating.
- Record the real data, independent of the choices and hypotheses needed to be made before a reconstruction (e.g., if evidence is available for 1658, and the reconstruction shows 1700, record "1658" and indicate that the inference was made that it was "probably" unchanged in 1700).
- Switch from a *synchronic* approach (i.e., corresponding to a time point) to a *diachronic* approach (where the focus is put on transformations, or time intervals, rather than on states (Dudek and Blaise 2010).
- Inform users visually (see figure 4.5) about temporal inconsistencies in the 3D model itself (Blaise and Dudek 2003).

### The Issue of Recording the Process of Analysis

Historical research starts with the cumbersome task of pulling together fragmental and heterogeneous evidence that will act as input for an analysis.

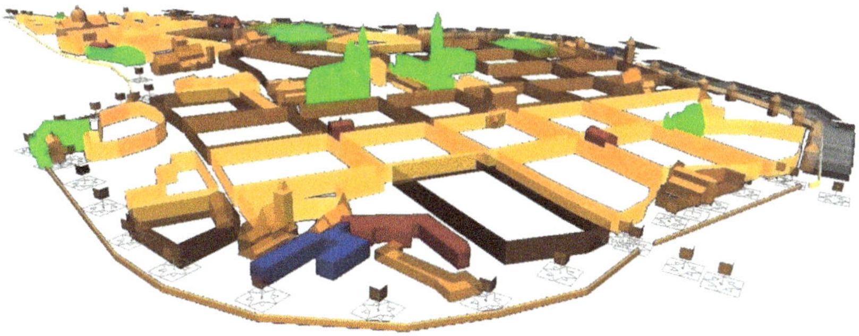

**Figure 4.5** Color and emissivity are used to spot in a VIA scene, corresponding to a period *t*, shapes that correspond to an earlier or later phase of evolution (Blaise and Dudek 2003, Blaise and Dudek 2005).
*Credit:* © I. Dudek & J. Y Blaise, UMR CNRS/MCC 3495 MAP.

But this pulling together of evidence requires numerous interpretative steps. Hence, there is potential information loss if only the results of this process are passed on, especially in the form of a 3D model, without indications as to why the 3D model is what it is and what inferences have been made by the analysts in its preparation (e.g., critical analysis and selection of sources, analogies used, generic knowledge applied, archaeological findings considered).

"Images" are easy to understand, and therefore are simple to share with a wide public. The 3D reconstruction of how a city could have looked like at a particular period becomes an interchange point whereby an interpretation of evidence is presented by "the ones who know" (scholars, researchers) and made available to others (those who want to know) in a form that is expected to be easily readable.

But what exactly does a 3D model "contain" in terms of knowledge? Recreating the spatial layout of a city in its past involves numerous interpretation steps and numerous inferences. Considering a 3D reconstruction (i.e., 3D model) as the *only* result of this process would be like reducing a whole research process to its conclusion. If no precautions are taken in this presentation, the 3D reconstruction may conceal the selections of inputs, assessments of data quality, inferences made, and choices selected. The 3D model may then impose on the analysts involved the responsibility of memorizing and passing on their research process (Dudek et al. 2015)—yet another methodological pitfall. What is a 3D reconstruction beyond a one-shot visual production if we do not give ourselves the means to formalize, record, and visualize more than the result of a process? What is its value if we neglect to mention the way it was achieved?

Indeed, what is at stake here is not only the transfer of knowledge or the clarification of choices and inferences; it is also the cornerstone of scientific practices: *reproducibility* and *intersubjective verification.*

## The Issue of Methods for Wider Analysis

Analyzing an entire city introduces a quantitative leap when compared to analysis of a single monument; more edifices or other urban components now have to be taken into consideration. But the leap is not only quantitative; it is also (and in our view primarily) methodological. That is, not only are we dealing with more data but our methods must be adjusted to encompass a wider structure. Simply put, the information gathered for studying a whole city has heterogeneous urban components (buildings as well as open spaces) and nonlocalized data (e.g., laws enforcing changes on this or that component), contextual information (e.g., a plague and its consequences on the population), and reports of significant events (e.g., sieges, floods). Assembling the evidence and analyzing it requires an approach (and visual tools) that go beyond distributing individual components in a given space.

A number of such approaches, applied to the analysis of a city's territory and its changes over time, have been introduced by the InfoVis community. Slingsby et al. (2009), for instance, reuse a visual formalism called "treemaps," introduced by Johnson and Shneiderman (1991), for "showing spatial and temporal variation in house prices." And indeed their approach does enable a clear understanding of the patterns of change in each district of a city over time.

Focusing on the temporal aspect of the underlying data, the Commuting Scales experiment by Beaude and Guillemot (2012) summarizes in an abstract way the periodically recurring travel times occurring within an entire urban area. Yet another approach is that of Yabuuchi's (2011) Home to School diagram, a visualization inspired by musical staves that turns an urban route into a series of individual scenes. For each scene, components in the surroundings (vegetation, buildings, vehicles, and so on) are visualized through specific icons, thereby helping portray something rather intangible—an atmosphere.

These examples illustrate a fundamental principle: visualization and analysis in regard to a collection of data are allowing not only for the analysis of "lots" of individual items but also for the depiction the data as such.

As mentioned, analyzing an entire city differs from analyzing a "list" of individual components, in the sense that it involves the relations these components may have with one another, as well as within a wider context. InfoVis practices that focus on making sense of such large data is a promising research path to follow. However, most examples of such research use "well-defined" data sets. It is, as mentioned earlier, the analysis of a city's past that requires handling large, heterogeneous data sets, but whose data are riddled with uncertainties. Bridging the InfoVis-originated approaches and the methods of historical research remains an open challenge.

## MAKING SENSE IN HISTORICAL RESEARCH: AN INFORMATION-VISUALIZATION PERSPECTIVE

Information visualization is founded on a specific interpretation of the term *visualization* that encompasses a wide range of practices. This interpretation was expressed by Spence (2014, 1) as follows: "visualisation is a cognitive activity." In other words, visualization targets greater *understanding*—it is not a result but, rather, the means to perform the tasks of analysis.

Information visualization is defined by Kienreich (2006, 9) as "the use of computer-supported, interactive visual representations of abstract data to amplify cognition." Friendly (2006) considers the discipline as evolving from both statistical graphics/data visualization and thematic cartography.

Information visualization is an active research area, with still rather scarce applications in historical research, although hot topics like time-oriented data or the handling of "dirty data" can be seen as potential methodological bridges to historical research.

We do not intend here to discuss information visualization as such but, rather, to illustrate through a chronology of experiments how the visual tools inspired by information visualization can be useful in avoiding the four major pitfalls discussed earlier. The examples noted here share a number of common choices made in terms of conceptual modeling, as well as computer infrastructure:

- Urban components (edifices, urban blocks, streets, squares, etc.) are modeled as one unique "being," with potentially successive morphologies.
- The model includes typologies classifying *states* (a morphology and a time slot during which the city is believed to have remained stable) and *transformations* (changes that occurred and how long they took). See Dudek and Blaise 2008.
- Components are not considered as having disappeared until confirming evidence is found that they have been fully dismantled (underground structures included).
- Qualitative lexical scales are used to formalize various types of doubts concerning the lifeline of components, uncertain morphology, materials, functions, durations, and the like.
- All pieces of data are stored in RDBMS and XML files.
- The infrastructure's user interface is web-compliant (HTML, produced on the fly).
- Graphics (2D or 3D) are produced on the fly, as answers to user queries, in standard formats (mainly scalable vector graphics [SVG]).

### Localizing and Interfacing the Historical Evidence

Uncertainty, as mentioned earlier, affects both the spatial and the temporal determinacy of a component, as well as affecting them both differently. Regarding the former, one of our earliest experiments was an information system called VIA. This system was used to establish a direct link between historical sources and urban components, and to produce 2D and 3D visual displays that could be used as interfaces (Blaise and Dudek 2003). Each component can be queried in order to retrieve the information available in it; vice versa, any historical source can be queried in order to localize its "architectural content" in 2D or 3D. VIA gave us a first opportunity to dig into uncertainty factors (figure 4.6, right). Each urban component was described as a set of "justifiers"—that is, values representing our estimates of the reliability of

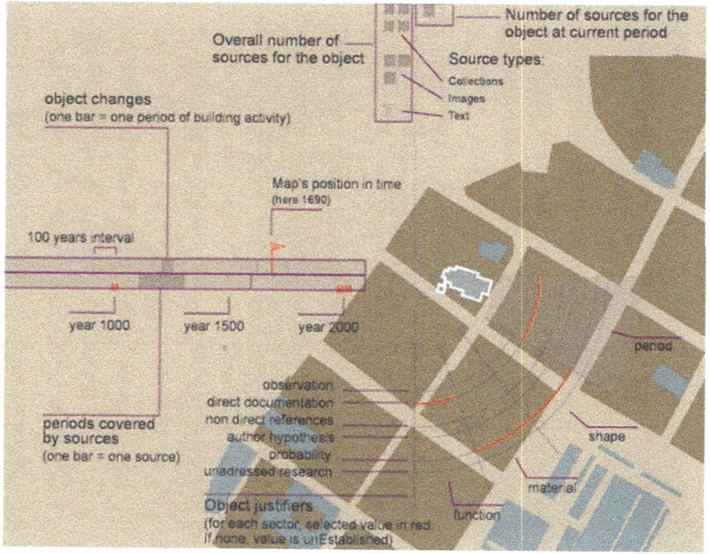


**Figure 4.6** Left: a slider allows user-monitored browsing through time (here on a subselection of components, artifacts only). This slider is also used to show densities of changes over time: each little horizontal bar corresponds to a change in one of the components under scrutiny. Periods of high and low activity are revealed. Right: graphical codification of justifiers and comparative chronology of changes (Dudek and Blaise 2007).
*Credit:* © I. Dudek & J. Y Blaise, UMR CNRS/MCC 3495 MAP.

sources concerning the morphology of the component, the construction material, the function, and its dating. Additionally, a comparative chronology of the transformations of a given component and of historical sources helped in underlining potential anachronisms (i.e., temporal gaps between a change and its report in sources) or factual errors.

The temporal aspects were still underrepresented in the initial VIA development, although time sliders included in the visualization did help with any browsing through time. But this solution only provided "temporal snapshots" of the data—that is, only one period was visible at a time. A comparative chronology of changes and a map of the densities of changes over time (figure 4.6, left; figure 4.7) prefigured solutions we later investigated in order to spot and analyze temporal patterns.

A later development of the VIA platform could target specifically visual sources that could not be unambiguously localized using contemporary methods owing to their artistic nature, their lack of information on the captors, or their inability to correspond to several points of view (e.g., historical panoramas). By contrast, the Locus Imaginis project (Stefani et al. 2013) illustrates the attempt to localize visual sources within the space of a 3D model, but focusing on *well-defined* visual sources. Which approach can the analyst use when dealing with old paintings, hand-drawn sketches, or panoramas? The idea we investigated using the VIA platform was to describe an approximate point of view using a combination of parameters: the viewer's estimated position(s) (*x*, *y*, *z*), an estimated target point(s), a horizontal angular direction(s) of view, a geographical bearing(s), a vertical bearing(s), and the image's ratio and type (figure 4.8). This approach led us to some noteworthy observations, such as when a painting turns out to correspond not to a real view but, rather, to a combination of points of view carefully chosen by the painter to reach a satisfactory artistic result (Dudek and Blaise 2007).

## Understanding Temporal Patterns

Understanding how transformations of a city occur over time, and not only in space, is also a key aspect of historical research. The InfoVis community has been active in the formalization of time as such, introducing models aimed at coping with notions such as cycles, time granularity, density, and determinacy, implemented in real cases. Temporal uncertainty was addressed, for instance, in the PlanningLines experiment (Aigner et al. 2005), whereby time intervals were modeled with "floating" bounds so as to allow assessment of risks of delay in planning chained activities.

A similar mechanism was introduced in the Chronographs project through which we developed a methodological framework to describe life cycles through a diachronic approach (Dudek and Blaise 2010). Here, the focus was

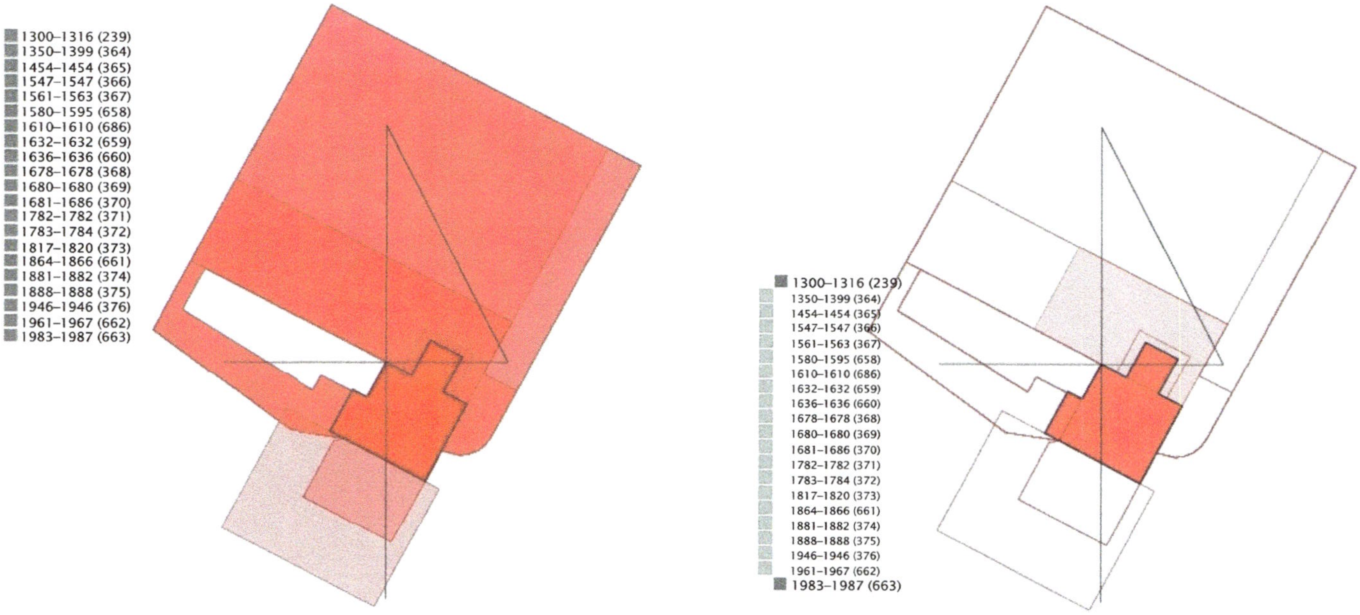


**Figure 4.7** Successive layouts of the town hall are superposed using transparencies, underlining the density of land use over time.
*Credit:* © I. Dudek & J. Y Blaise, UMR CNRS/MCC 3495 MAP.

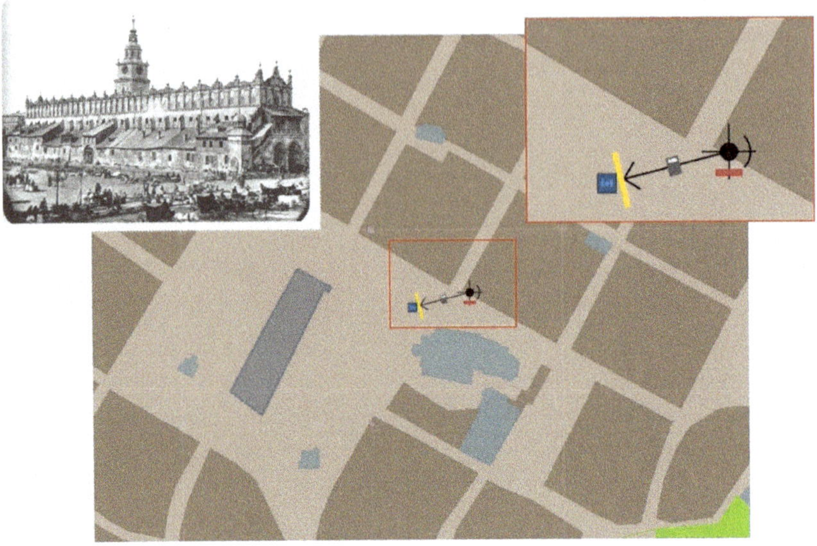

**Figure 4.8** Glyphs used to position in space visual sources qualitatively. Top left: the visual source itself (old photograph). Top right: the glyph, featuring information on qualitative information on the image type, its ratio (proportions), the viewer's estimated position, the angle of view, etc. (Dudek and Blaise, 2007).
*Credit:* © I. Dudek & J. Y Blaise, UMR CNRS/MCC 3495 MAP.

on the way artifacts get transformed, with a grid of notions identifying life cycles as a sum of states and transitions. Thus, the chronographs act as *visual explanations* (a tribute to Tufte 1997) of the artifact's life cycle, underlining the temporal patterns of evolution within a site or across sites, but also pointing to the uncertainty of such temporal bounds. The chronographs combine three elements: diachrograms and variograms that allow a global view of transformations, and feature-variation visual assessments that foster comparisons and enhance the reading of consequences. The whole unit is designed to combine vertical readings—what precisely happens at period *p*—and horizontal readings of how a feature changes with time. In that sense, the disposal matches Bertin's view of graphics as "visual answer to a question" (Bertin 2010, 9).

## Understanding Alternative Temporal Patterns

The chronographs solution has proven quite efficient in performing analysis tasks—it does help in assessing causal relations and orders in time sequences, events, and consequences in a robust manner. But chronographs have the disadvantage of forcing the analyst to choose one *and only one* scenario of evolution for a given urban component; in other words, it does not identify those key historical time slots for which the analyst has difficulty interpreting. In addition, although doubts about dating are duly visualized, there is no means for saying that the very existence/relevance of an event is questionable.

As a solution, we developed a new set of visualizations that capitalize on the user's a priori understanding of the timeline paradigm, but that introduce visual interruptions marking key choices the analyst has to make before adopting an unambiguous scenario (Dudek and Blaise 2012). Three complementary types of graphics were produced, corresponding to different analysis needs:

- *Multi-hypothesis chronology diagrams.* These deliver a quick, synthetic view of where alternatives exist (in time) and what options the analyst faces.
- *Graphs of potential interactions* (figure 4.3). These graphs position alternatives in time, providing a classification of architectural consequences. They help in understanding relative density when compared to well-established indications, and allow the analyst to connect them to the corresponding clues.
- *Visual measures of complexity.* These are diagrams that foster comparisons of the relative complexity of an artifact's evolution across the collection (figure 4.9).

These graphics serve to highlight alternative explanations to available existing data. However, they have their pitfall, as well: they are solutions that put individual cases "side by side." For example, in studying a collection of twenty-seven artifacts, we found Tufte's recommendation to "enforce comparisons within the eyespan" to be a challenge. We opted for a visualization more or less inspired by the "small multiples" principle (a concept attributed to Scheiner's 1626 representation of changes in sunspots over time—see Friendly 2006). Life spans, the number of transformations, and the number of alternatives across the collection were visualized in a single graphic, thereby helping the analyst to read the collection of artifacts as a whole.

### Communicating About Doubts

It is sometimes said that conveying any doubts is irrelevant in the context of a communication effort. In an attempt to contradict what we consider a simplistic view, we introduced an "edutainment"-oriented proof-of-concept prototype called Tangible Chronology (Blaise and Dudek 2013). This prototype combines a master board, with 3D physical models of the artifacts, and with a coding of their position on the board and in ordinal time, as well as a tangible timeline for each artifact (figure 4.10).

The specificity of this model lies in the tangible codifications used for various purposes, from positioning 3D physical models on the master board

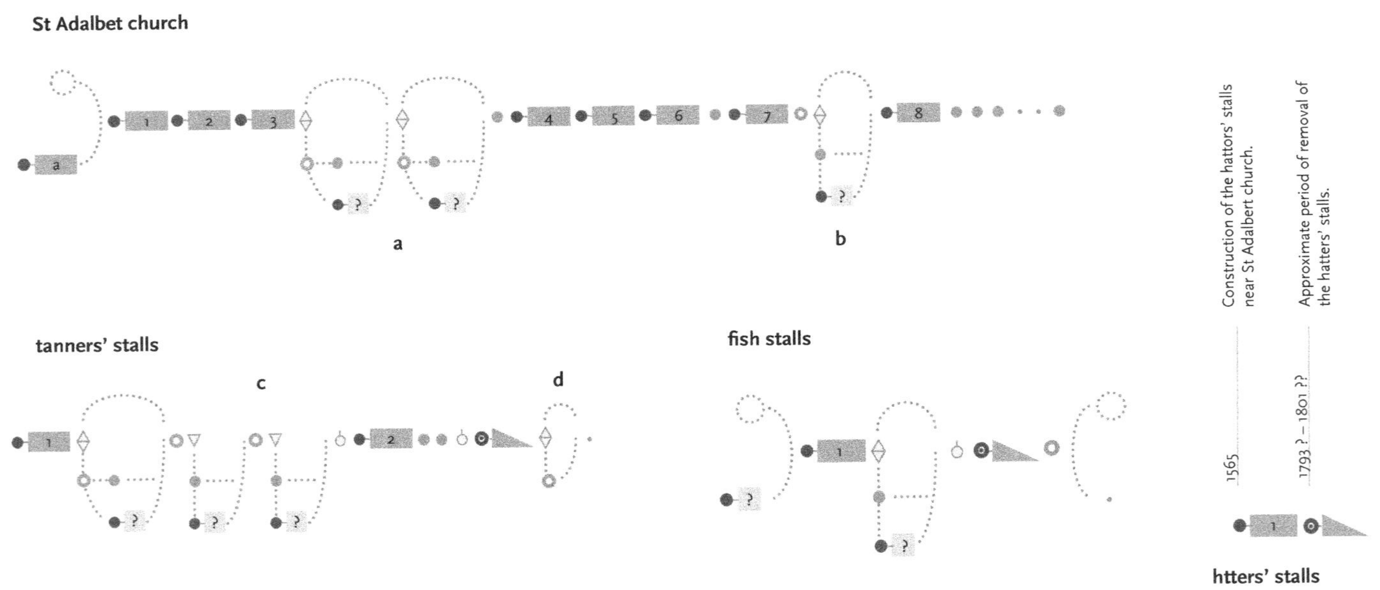


**Figure 4.9** Graphs of potential interactions position changes in ordinal time and allow for the visual assessment of alternative scenarios of evolution. Dark circles and rectangles correspond to known states. Triangles position events that result in temporal "forks" (i.e., moments when alternatives scenarios should be considered). Dotted lines above the triangles mean the event is *not* confirmed (**a**, **b**, **d**). Dotted lines below the triangles correspond to alternative potential consequences. The graphs are read as follows: (**a**), (**b**), and (**d**)—the event may not have occurred at all (triangle upwards, dotted line straight to next step) but if it did occur (triangle downwards) either it has had limited consequences (light gray circle) or it led to a main change (black circle, rectangle with question mark). Outlined circles correspond to destructive changes. For (**c**), confirmed alternative, the event did occur for sure (triangle downwards) but either it has had limited consequences (light gray circle) or it led to a main change (black circle, rectangle with question mark).

Graphs are presented along with explanatory vertical texts, as in the rightmost example.
*Credit:* © I. Dudek & J. Y Blaise, UMR CNRS/MCC 3495 MAP.

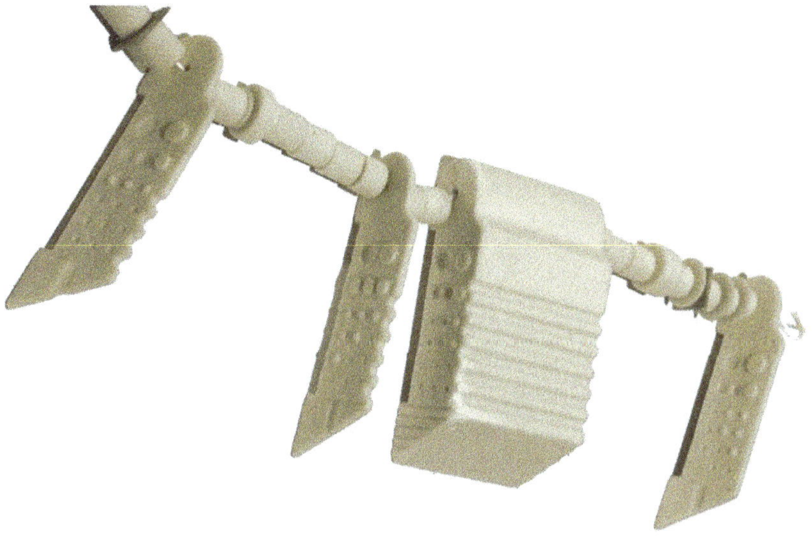

**Figure 4.10** The Tangible Chronology experiment (Blaise and Dudek 2013): a tangible timeline is provided for each artifact, along which occurrences of relevance are positioned in discrete time (chron. 1 year). Tangible codifications are introduced to represent dates, durations, uncertainty in dating, and to differentiate functional transformations from morphological transformations. It is composed of shapes meaning *time* (square—turn of the century; cylinder—decade), and of shapes meaning *events and processes* (thick cylinders—functional transformations; plates—dating of a morphological transformation).
Credit: © I. Dudek & J. Y Blaise, UMR CNRS/MCC 3495 MAP.

to making reliability assessments. The initial concept was to offer users, including ill-seeing users, a means for analyzing the site's urban composition and evolution, for any time slot within the 750 years covered. To do so, users would combine on the master board 3D physical models corresponding to each edifice's evolution for the period chosen. But what the experiment underlined at the end of the day is that consistency with the underlying historical evidence's specificity (temporal aspects, uncertainty) does not undermine scientific communication—rather, it helps it.

### Analysis of Space and Time in a Row

There are a number of explanations for why it is arduous to combine the reading of spatial and temporal patterns in one unique visualization. At the end of the day, whatever scale is employed, visual solutions used to represent spatial features (in 2D or 3D) and temporal features are in most cases simply put side by side—typically a map or 3D model, complemented by a time chart, as in Rød (2003), Stefani et al. (2008), Auer et al. (2011), or Tomaszewski and MacEachren (2012).

The ConcentricTime visualization (Blaise and Dudek 2011) is, by contrast, an attempt at reading in a row the spatial distribution and the changes over time of a collection of artifacts. It capitalizes on basic, well-known, well-understood representations (cartography, timelines), and it introduces a specific, context + focus, graphic combination of spatial features and chronology. A map in the central part of the visualization shows the layout of artifacts. Each artifact is connected with a radiating timeline along which its transformations are localized in time. Moments of change on the timeline, represented by colored rectangles, can be interactively queried: the 2D contours corresponding to the user query are then projected in the central area. Because the contours are drawn with a level of transparency, this mechanism enables user-monitored comparisons of contours, such as neighborhood analysis or transformation analysis (e.g., period of emergence of a given component).

The visualization allows the reading of an individual artifact's lifeline, and it fosters the reading of common behaviors inside the collection of artifacts (impacts of changes of one artifact on others, general patterns of change such as destructions).

ConcentricTime was designed in order to combine, inside a unique information space, particular spatial features and chronology, with constraints such as uncertainty assessment, interface capabilities, and simplicity. As a result, although it is limited by the fact that only one scenario is shown, and the number of radiating timelines is finite, the model does offer efficient support in carrying out analysis of architectural changes on a collection of artifacts, and it presents as fairly generic (Figure 4.11).

### Making the Best of an Ill-Defined Data Set

In a recent experiment (Blaise and Dudek 2014), we focused on poorly documented edifices: wooden stalls known from textual sources to have existed, but for which no visual material is available. These commercial facilities share rather challenging particularities:

- They cannot be precisely localized (such as known only in archival sources by expressions such as "close to the town hall" or "located in the northern sector"). Moreover, they were wooden structures, demolished during the nineteenth century. Accordingly, there is no archaeological indication we could count on.
- There is no direct visual or textual archival material that can help analysts say what was their construction or their spatial layout. In fact, these facilities may even not have been "buildings" in the proper sense of the word; they could even have been rows of movable stalls, not physically

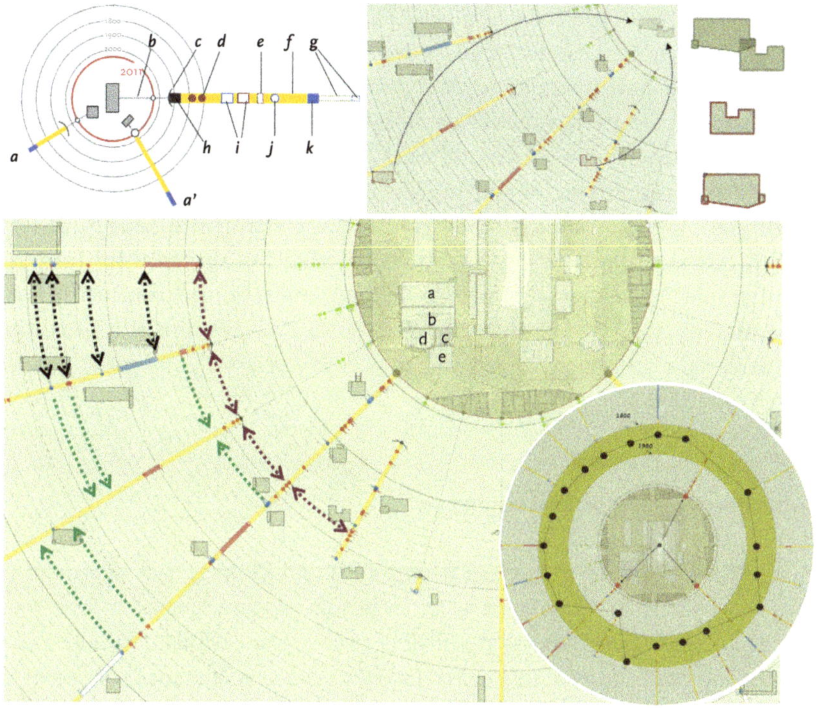


**Figure 4.11** Principle of the ConcentricTime visualization. Top left: artifacts are connected to radial timelines (a, a') that distribute transformations in time. Colors differentiate event types: yellow (periods of stability, no known change, **f**), blue (morphological change, including construction, **k**), red (functional change, **d**), gray (destruction, **c**, meaning here that no visible trace is left above the ground). Shapes differentiate duration and certainty: long-lasting events (duration > year, rectangles—**i**, **e**, **k**), punctual events (duration < a year, circle—**d**, **j**), uncertain dating white filling and outline (**i**, **j**), hypothetical dating dashed outline (**e**, **g**). Top right: contours can be projected inside the central circle on user selection, enabling user-monitored comparisons (neighborhood analysis for instance). Bottom: the visualization can be used in "focus view" so as to allow a fine-grained analysis of temporal patterns across a group of artifacts. Illustrated here is the town hall ensemble: a majority of temporal co-occurrences for artifacts **a** and **b** (granary/town hall, black arrows), a majority of non–co-occurrences for artifact **d** (prison) with neighbors **b** and **d** (town hall, belfry, green arrows), a phase impacting all the elements (purple arrow). Bottom right: the visualization can be used in "context view" so as to spot patterns concerning the whole collection—here, the massive destruction phase on the market square during the nineteenth century (only here artifacts, **a**, remain until now).
*Credit:* © I. Dudek & J. Y Blaise, UMR CNRS/MCC 3495 MAP.

connected to one another, but considered as a consistent group based on their common function (goods/services for sale).

Though very little is known about these commercial facilities (in fact, temporal aspects are better known than spatial aspects, which is not that common), we do have one document that gives useful indications: an eighteenth-century inventory conducted by Kraków's municipal authorities at the time. The challenge we faced was thus: Can we learn anything useful about the layout or organization of these facilities, based on information in only this document? To find out, we reinterpreted two classic eighteenth-century visual formalisms—*poleometric tables* and *candlestick charts* (figure 4.12). Ultimately, we uncovered some noteworthy patterns, such as there being no relation of tax levels to surfaces, no catching-up process in the tax policy, regularities within a consistent group of stalls with few contrasting components, and so on.

This re-vision of classic visualizations revealed the potential for InfoVis-inspired solutions when faced with ill-defined data sets, and particularly how such an approach could facilitate workgroup discussions, reveal unexpected trends in data previously considered as fully investigated, and pose new research questions.

### Collection Reading Visualizations

One of the key services expected from InfoVis solutions is that of providing the analyst with tools to cope with large information spaces and to spot trends or exceptions that are not visible at first glance. This naturally leads to spotting the "outliers" in a collection (whether the "distance" of the item from the rest of the collection is the result of faulty data or a genuine difference), as well as for observing general trends. From a practical point of view, this implies the promotion of visual solutions that implement a well-known mantra in the InfoVis community, expressed by Schneiderman's mantra "Overviews first, details on demand." It also implies, when needed, introducing grouping mechanisms that analysts consider as relevant for this or other analytical purpose.

An example of this approach is the "evolution field" metaphor we introduced in order to analyze the spatial distribution of types of artifacts by "cluster." The objective was to get a quick idea of whether groups that are consistent in space are also consistent in time, in patterns of evolution, in density of sources, and so on. A multidimensional icon summarized various parameters: possible period of construction, overall lifetime, possible period of demolition/dismantling, confirmed or suspected events like fires, possible or definite transformations, and archaeological remains of (or attributed to) the artifact. Icons were combined with regard to a given analysis need. The approach was applied, for instance, to Kraków's market square building ensemble in order to weigh the density and specificity of objects on both sides of Cloth Hall. Wooden commercial facilities on the west side outnumbered by far those on the east side; some emerged earlier in time, their process of decay and demolition standing out as harder to date precisely—in fact, activity as such appears to have been far greater on the west side (figure 4.13).

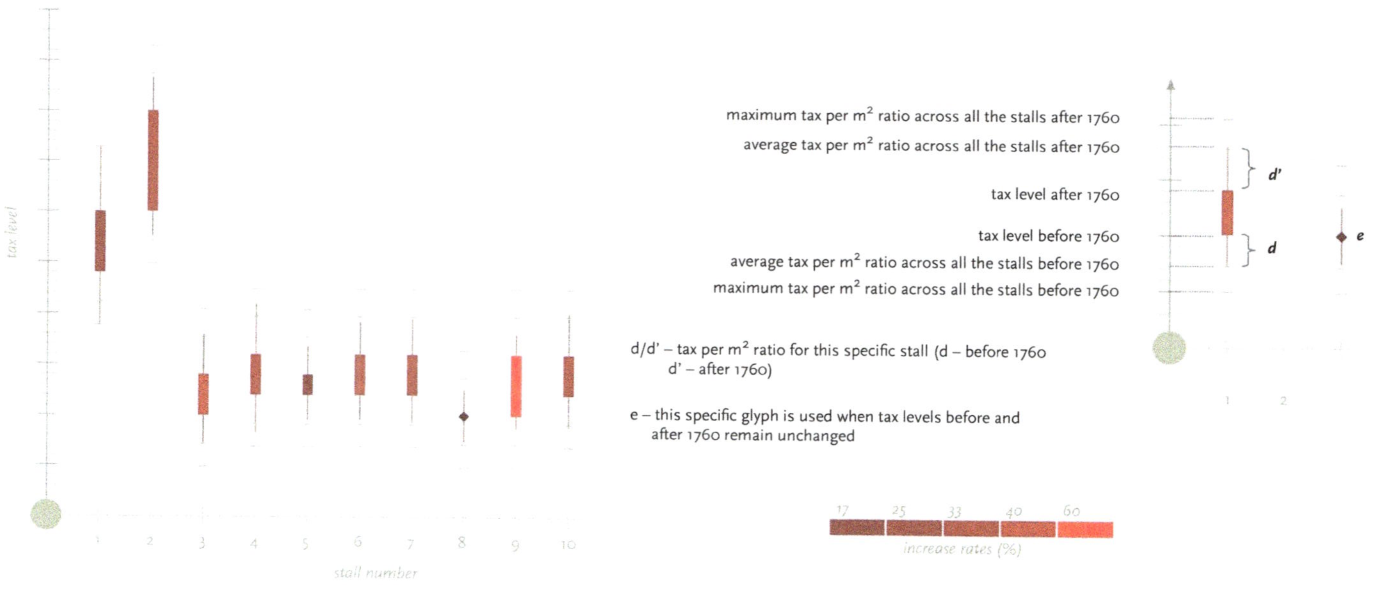


**Figure 4.12** A re-vision of Munehisa Homma's candlestick charts (Blaise and Dudek 2014) is used to analyze tax levels and their variations (as reported in a 1760 inventory) inside one same group of stalls (relation of tax levels to surfaces). "Outliers" are easily spotted (for instance, stalls 1 and 2 appear as more heavily taxed than others, and stall 8 is the only stall for which the tax level remains unchanged).

*Credit:* © I. Dudek & J. Y Blaise, UMR CNRS/MCC 3495 MAP.

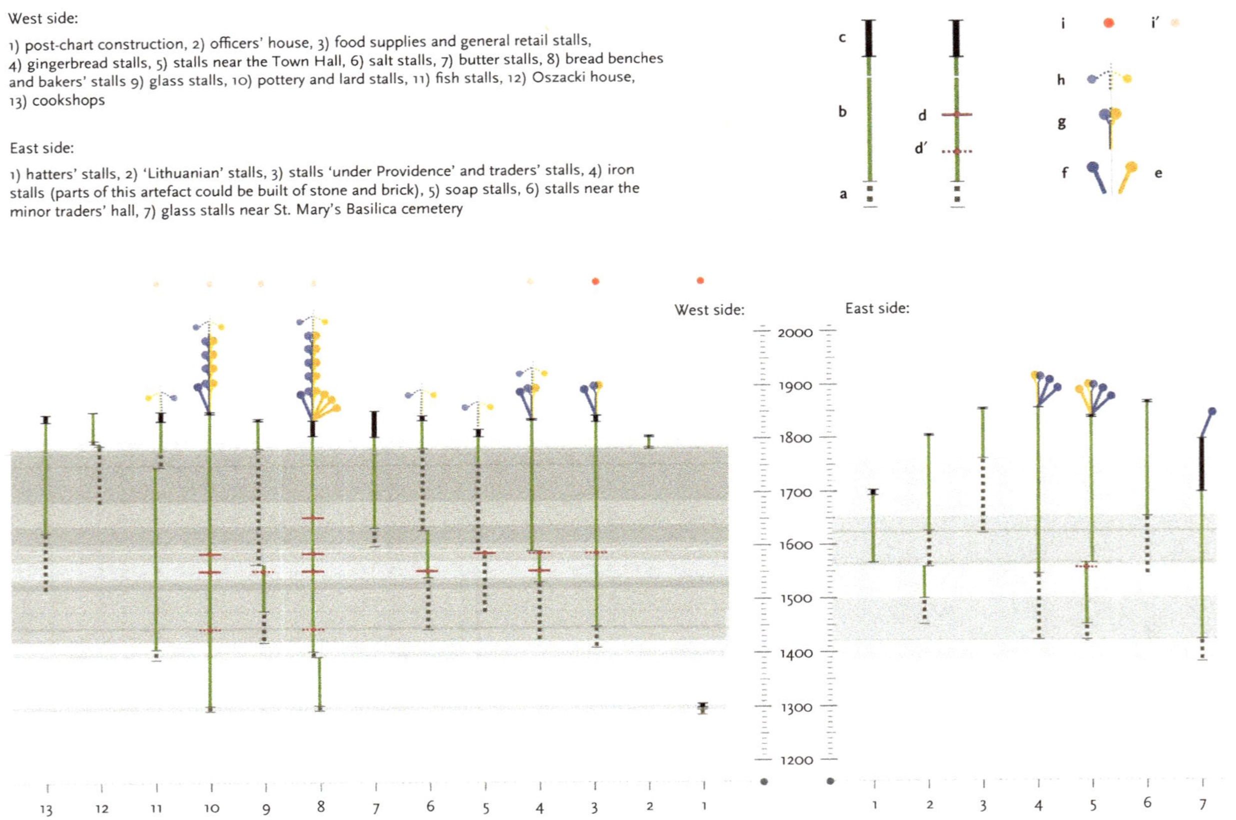


**Figure 4.13** Abstract visualizations designed to facilitate the analysis of collections: the evolution-field metaphor is used to characterize the contrasting behaviors of artifacts located on the west and east sides of the market square. Each vertical line corresponds to an artifact, time is counted from bottom to top. The graphical encoding combines: **a**—possible period of construction; **b**—artifact's overall lifetime; **c**—possible period of demolition/dismantling; **d/d'**—confirmed/suspected fire; **e**—definite morphological transformation; **f**—definite recurrent transformation; **g**—unknown results of a confirmed transformation; **h**—potential transformation; **i**—artifact identified during archaeological; survey; **i'**—archaeological remains not clearly; attributed to the artifact.

*Credit:* © I. Dudek & J. Y Blaise, UMR CNRS/MCC 3495 MAP.

Another example of this approach was the multidimensional transformation pie charts organized in temporal clusters (Dudek and Blaise 2012). Here, multidimensional icons of varying complexity were arranged visually, either as classic time series or as small multiples (figure 4.14). This helped in the observation of the data and its interpretation in a renewed way, typically highlighting source effects like the relation of construction type to the nature, amount, and assertiveness of the information available.

It is important to state that such visualizations show that there is room for comparisons and analysis even when edifices cannot be precisely localized in space—heterogeneity in terms of the level of knowledge naturally remains a concern, but it ceases to be a lock on the process.

### Context Assessment and Temporal Co-occurrences in a Collection

The last experience we mention here was aimed at putting architectural transformations into a context and at highlighting the temporal co-occurrences. Among the uncertainty factors hindering the analyst's reasoning process about a city and its transformations are the potential results of events and facts that cannot be directly linked to this or that urban component (e.g., legislation, occupation by a foreign army, plagues, demographic variations, presence of important people, etc.).

If we are to support the analyst in his or her effort to understand and recount the evolution of a city, we need to go beyond the depiction of consequences—that is, the spatial changes at time *t*, and to correlate those changes with potential causes, or at least place them in a more general context. This is precisely the aim of the "HistorySkyline" visual metaphor. The visualization pulled together in a time-oriented display the data we had on transformations of urban components (along with their uncertainty assessment and with the amount of sources used to document the changes) and a number of contextual indicators (rulers, military events, natural disasters, legal matters, global military context in the country, demography). The design of the visualization was based on the concept of the visual metaphor, described by Kienreich (2006). In practice, it develops the comparison to an urban skyline, with the horizon acting as a timeline (time goes from left to right). The building activity (change) is then represented as "blocks" or "facades," and elements of contexts are superimposed on them (figure 4.15).

At this stage, the HistorySkyline visualization is basically a proof-of-concept prototype needing further development, but it seems promising for identifying temporal co-occurrence patterns, whether expected or not (see Blaise and Dudek 2015).

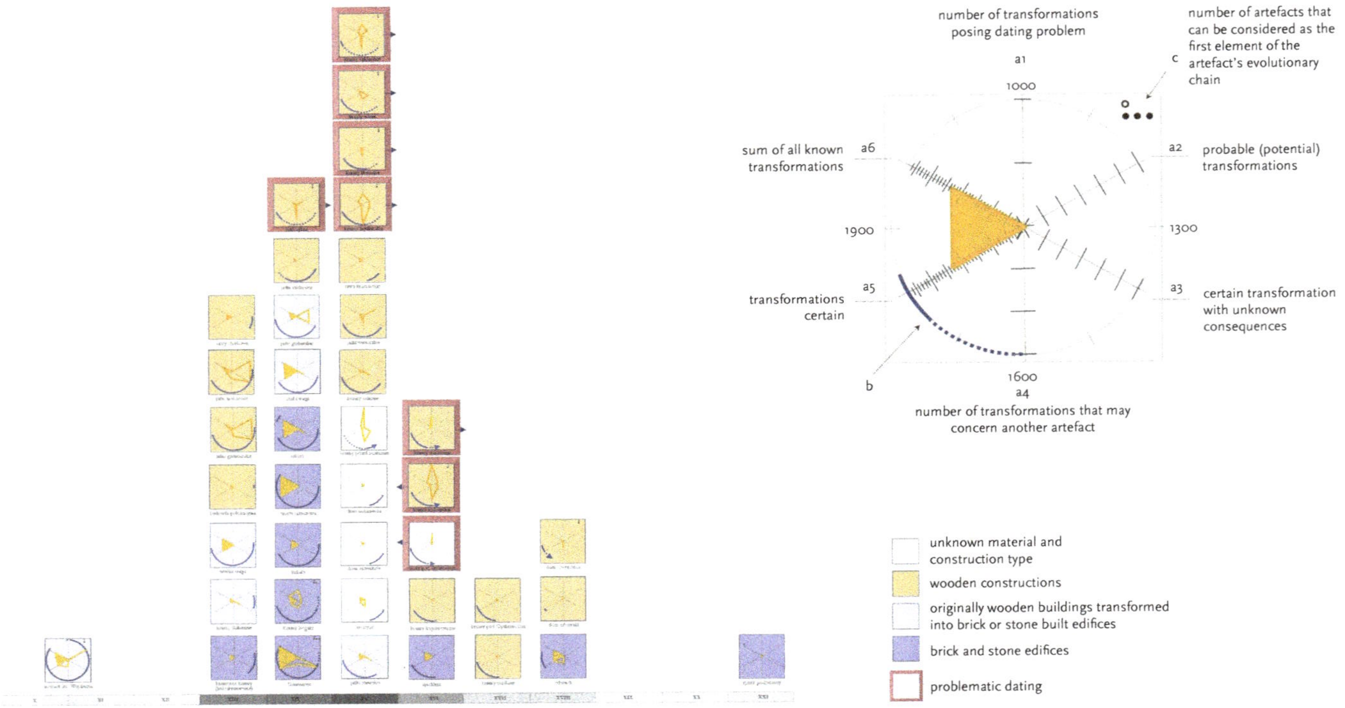


**Figure 4.14** Multidimensional icons summarize characteristics of an artifact (here concerning transformations). They are then grouped into user-defined "temporal clusters" (represented by columns)—a column corresponds to one century. Rhythms and general trends inside and across clusters can be spotted: the most important building activity on the market square in Kraków takes place during the fourteenth century, brick and stone buildings tend to be built earlier, wooden structures appear as harder to date and come in a second row.

*Credit:* © I. Dudek & J. Y Blaise, UMR CNRS/MCC 3495 MAP.

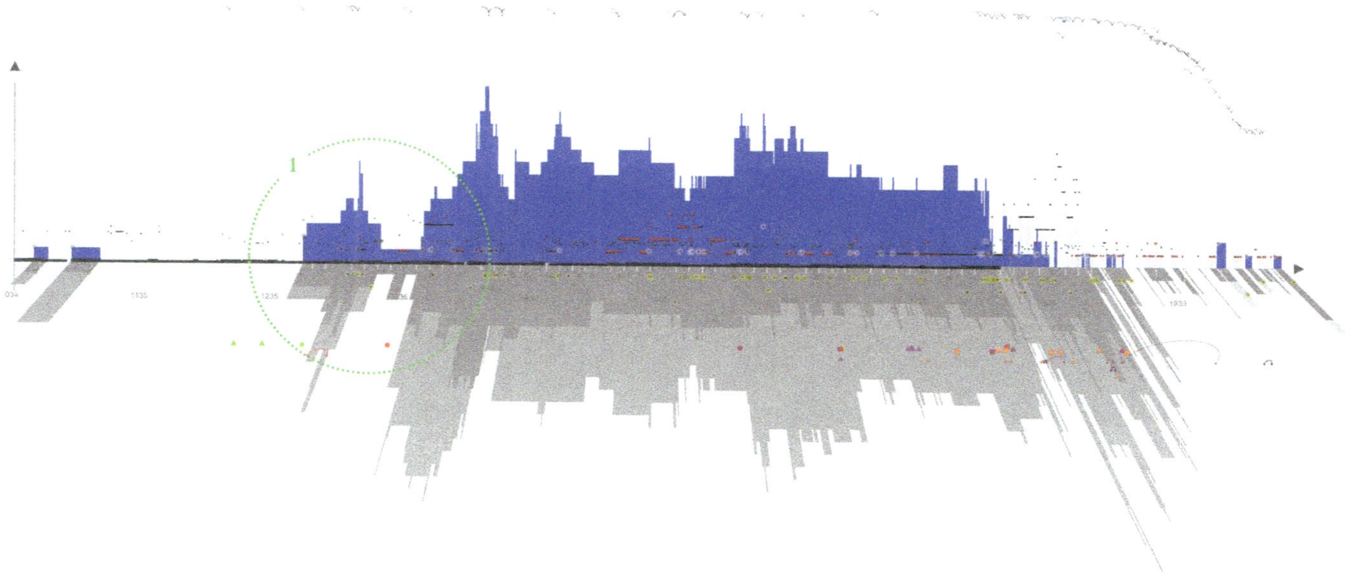


**Figure 4.15** The HistorySkyline metaphor: vertical bars above the horizon line represent the amount of transformations across the whole market square, year by year, since 1035. Blue bars represent "positive" transformations (creation/modification), red lines represent episodic and recurrent changes (maintenance or equivalent), and black lines represent destructive changes. These indications are correlated with the number of documents available (gray "shadow" below the horizon line). In addition, they are also correlated with a variety of contextual information, such as wars, natural disasters, life of important people, new laws. Sieges, wars, invasion and occupations are represented on parallel lines with colors and glyphs representing types of events and actors—in this example, we underline ([1], dotted green circle) a clear peak in the construction activity during the Czech occupation (1291–1306). It has to be said that the visualization does not mention causality: it barely puts facts that occurred at the same period side by side. Actual conclusions remain under the sole responsibility of the analyst.
*Credit:* © I. Dudek & J. Y Blaise, UMR CNRS/MCC 3495 MAP.

## CONCLUSION

Our record of experiences illustrates how we can today complement "conclusive" visual results such as 3D reconstructions with graphics aimed at supporting analysis of the data behind the analysis (e.g., at clarifying choices, considering alternatives, cross-examining the historical evidence, assessing contexts, interrelations, potential causal chains, etc.). That said, 3D reconstructions underline the potential benefits of InfoVis-inspired abstract graphics as a means for exploring ill-defined data sets, observing the relations between the data, and investigating the time parameters. We have in particular discussed four major scientific challenges that the growing use of 3D models in the study of "lost cities" raises: dealing with the ill-known, doing analysis involving temporal data, ensuring the reproducibility of work processes, and enabling the analysis of collections as such.

At the end of the day, our research emphasizes the methodological issues where the information sciences interest with the historical sciences. In that context, modeling and visualizing doubts and recording research processes remain issues to be further investigated. Understanding the time parameters as multidimensional (not only running from past to future but also in terms of densities, rhythms, alternative paths, and so on) is indeed a challenge still to be met. Finally, the quantity and heterogeneity of data sets and of urban components concerned when analyzing entire cities implies a questioning of the way we handle collections.

Our record of experiences regarding the historic city of Kraków shows that exploring visually the known and the ill-known about a city's evolution can benefit from the intersection of these scientific fields, provided the nature of the historical evidence is considered as the cornerstone for research.

## REFERENCES


Aigner, W., S. Miksch, H. Schumann, and C. Tominski. 2011. *Visualization of Time-Oriented Data*. London: Springer.

Aigner, W., S. Miksch, B. Thurnher, and S. Biffl. 2005. "PlanningLines: Novel Glyphs for Representing Temporal Uncertainties and Their Evaluation." *IEEE Proceedings: Information Visualisation 2005*: 457–463. doi:10.1109/IV.2005.97.

Alkhoven, P. 1993. *The Changing Image of the City. A Study of the Transformation of the Townscape Using Computer-Aided Design*. The Hague: CIP-Gegevens Koninklijke Bibliotheek.

Auer, T., A. M. MacEachren, C. McCabe, S. Pezanowski, and M. Stryker. 2011. "HerbariaViz: A Web-based Client-server Interface for Mapping and Exploring Flora Observation Data." *Ecological Informatics* 6: 93–110. http://www.geovista.psu.edu/publications/2010/herbariaviz_prepub_copylock.pdf.

Baik, A., R. Yaagoubi, and J. Boehm. 2015. "Integration of Jeddah Historical BIM and 3D GIS for Documentation and Restoration of Historical Monument." *International Archives of the Photogrammetry, Remote Sensing and Spatial Information Sciences* 40: 29–34. doi:10.5194/isprsarchives-XL-5-W7-29-2015.

Beaude, B., and L. Guillemot. 2012. "Commuting Scales: Cartographie dynamique d'accessibilité temporelle." *M@ppemonde*105. http://mappemonde.mgm.fr/num33/lieux/lieux12102.html.

Bertin, J. 2010. *Semiology of Graphics: Diagrams, Networks, Maps*. Redlands: ESRI Press. [Original edition in French 1967]

Blaise, J. Y., and I. Dudek. 2000. "SOL: Spatial and Historical Web-based Interface for On Line Architectural Documentation of Kraków's Rynek Główny." In *Architectural Computing: From Turing to 2000*, edited by A. Brown et al., 700–707. Liverpool: Ecaade.

Blaise, J. Y., and I. Dudek. 2003. "New Experimentation of a Generic Framework for Architectural Heritage Data Visualisation." *Journal of WSCG* 11(1): 109–117.

Blaise, J. Y., and I. Dudek. 2005. "From Artefact Representation to Information Visualisation: Genesis of Informative Modelling." In *Smart Graphics*, edited by A. Butz, B. Fisher, A. Krüger, and P. Olivier, 230–236. Berlin Heidelberg: Springer.

Blaise, J. Y., and I. Dudek. 2011. "Concentric Time: Enabling Context + Focus Visual Analysis of Architectural Changes." In *Foundations of Intelligent Systems*, edited by M. Kryszkiewics, H. Rybinski, A. Skowron, and Z. W. Raś, 632–641. Lecture Notes in Computer Science. Berlin/Heidelberg: Springer.

Blaise, J. Y., and I. Dudek. 2013. "A Tangible Chronology." In *Archaeology in the Digital Era: Papers from the 40th Annual Conference of Computer Applications and Quantitative Methods in Archaeology (CAA)*, edited by P. Verhagen, 874–887. Amsterdam: Amsterdam University Press.

Blaise, J. Y., and I. Dudek. 2014. "Can Simplicity Help?" In *Proceedings of the 14th International Conference on Knowledge Technologies and Data-driven Business, i-KNOW 2014*, edited by Stefanie Lindstaedt, Michael Granitzer, and Harald Sack, 17:1--17:8. New York: ACM Books.

Blaise, J. Y., and I. Dudek. 2015. "Spotting Temporal Co-occurrence Patterns: The HistorySkyline Visual Metaphor." *Proceedings of 1st International Conference on Metrology for Archaeology* 1: 378–383.

Dima, M., L. Hurcombe, and M. Wright. 2014. "Touching the Past: Haptic Augmented Reality for Museum Artefacts." In *Virtual, Augmented and Mixed Reality. Applications of Virtual and Augmented Reality*, edited by Randall Shumaker and Stephanie Lackey, 3–14, Springer: Cham.

Döllner, J., and H. Buchholz. 2005 "Non-Photorealism in 3D Geovirtual Environments." Proceedings papers from caGIS AutoCarto. http://www.cartogis.org/publications/proceedings.php?year=2005.

Dudek, I., and J. Y. Blaise. 2007. "Visual Tools Decipher Historic Documentation." In *Proceedings of I-Know 2007*, edited by T. Pellegrini and S. Schaffert, 456–463. Graz: JUCS.

Dudek, I., and J. Y. Blaise. 2008. "Profiling Artefact Changes: A Methodological Proposal for the Classification and Visualisation of Architectural Transformations." In *Digital Heritage, Proceedings of VSMM 2008*, edited by M. Ioannides, A. Addison, A. Georgopoulos, and L. Kalisperis, 349–356. Budapest: Archeolingua.

Dudek, I., and J. Y. Blaise. 2010 "Understanding Changes in Historic Architecture: Can We Provide Tools & Methods for Visual Reasoning?" In *Proceedings of IVAPP 2010, International Conference on Information Visualization Theory and Applications*, edited by P. Richard and J. Braz, 91–100. Angers: INSTICC Press. doi:10.5220/0002826000910100.

Dudek, I., and J. Y. Blaise. 2012. "Analyzing Alternative Scenarios of Evolution in Heritage Architecture: Modelling and Visualization Challenges." *Journal of Multimedia Processing and Technologies* 3(1): 29–48.

Dudek, I., J. Y. Blaise, L. De Luca, L. Bergerot, and N. Renaudin. 2015. "How Was This Done? An Attempt at Formalising and Memorising a Digital Asset's Making-of." In *Proceedings of 2nd International Congress on Digital Heritage 2015, Vol. 2, Assessment of Methodologies and Tools in DH*, edited by G.Guidi, R.Scopigno, J.C.Torres, H. Graf, F.Remondino, L.Duranti, P.Brunet, S.Hazan, J.Barcelo, 343–346.

Firlet, E. 2010. Podziemia Rynku Głównego w Krakowie – od eksploatacji do trasy turystycznej, [in] Krzysztofory nr 28, Ed. Elżbieta Firlet, cz. I, MHmK, Kraków, 9–52.

Friendly, M. 2006. "A Brief History of Data Visualization." In *Handbook of Computational Statistics: Data Visualization*, edited by C. Chen, W. Hardle, and A. Unwin, 15–56. Heidelberg: Springer.

Gschwandtner, T., W. Aigner, S. Miksch, J. Gärtner, S. Kriglstein, M. Pohl, and N. Suchy. 2014. "TimeCleanser: A Visual Analytics Approach for Data Cleansing of Time-Oriented Data." In *Proceedings. of the 14th International Conference on Knowledge Technologies and Data-driven Business (i-KNOW 2014)*, 1–8. New York: ACM Press.

Heinonen, A., S. Pulkkinen, and I. Rakkolainen. 2000. "An Information Database for VRML Cities." In *Proceedings of IEEE International Conference on Information Visualization, 2000*, 469–473. London, UK: IEEE. doi: 10.1109/IV.2000.859798.

Johnson, B., and B. Shneiderman. 1991. "Treemaps: A Space-filling Approach to the Visualization of Hierarchical Information Structures." In *IEEE Proceedings of the 2nd International Visualization Conference*, edited by Gregory M. Nielson and Larry Rosenblum, 284–291. Los Alamitos: IEEE Computer Society Press.

Kienreich, W. 2006. "Information and Knowledge Visualization—An Oblique View." *MIA journal* 7–17. http://www.map.archi.fr/mia/journal/articles/vol0/num1/kienreich.pdf.

Lorenzini, M. 2009. "Semantic Approach to 3D Historical Reconstruction." In *International Archives of the Photogrammetry, Remote Sensing and Spatial Information Sciences* 38. (CD Rom): ISPRS.

Marek, Marta. 2011. *Cracovia 3D—Digital Reconstructions of the Historical Built-up Areas of Kraków*. Kraków: MhMK.

Pfeiffer, M., C. Carré, V. Delfosse, P. Hallot, and R. Billen. 2013. "Virtual Leodium: From Aa Historical 3D City Scale Model to Aa Archaeological Information System." In *ISPRS Annals of Photogrammetry, Remote Sensing and Spatial Information Sciences* 2: 241–246. Strasbourg: ISPRS. doi: 10.5194/isprsannals-II-5-W1-241-2013.

Rød, J. K. 2003. "ViewConflicts: Software for Visualising Spatiotemporal Data on Armed Conflicts." *Geography, Conflict and Cooperation ECPR Joint Session*. http://www.geomatikk.ntnu.no/viewConflicts/papers/ECPR_paper.pdf.

Rydzewski, Jacek. 2006. *Materiały Archeologiczne* 306. Kraków: MhMK.

Skeels, Meredith, Lee Bongshin, Greg Smith, and George Robertson. 2010. "Revealing Uncertainty for Information Visualization." *Information Visualization* 9(1): 70–81.

Slingsby, A., J. Dykes, J. Wood, and A. Crooks. 2009. "The Role of Layout and Order in Treemaps for Showing Spatial and Temporal Variation in House Prices." *GeoViz Hamburg Workshop 2009*. http://www.casa.ucl.ac.uk/andrew/research/geovizHamburg-v3-1.pdf.

Spence, R. 2001. Information visualization (1st Edition), Harlow: Addison-Wesley. Collection: ACM Press Books.

Stefani, C., C. Busayarat, J. Lombardo, P. Véron, and L. De Luca. 2013. "A Web Platform for the Consultation of Spatialized and Semantically-Enriched Iconographic Sources on Cultural Heritage Buildings." *Journal on Computing and Cultural Heritage* 6(3): 13:1–13:17. doi: 10.1145/2499931.2499934.

Stefani, C., L. De Luca, P. Véron, and M. Florenzano. 2008. "Reasoning About Space-Time Changes: An Approach for Modeling the Temporal Dimension in Architectural Heritage." In *Proceedings of the IADIS International Conference on Computer Graphics and Visualization*, edited by Yingcai Xiao and Eleonore ten Thij, 287–292. Amsterdam IADIS.

Strothotte, T., M. Masuch, and T. Isenberg. 1999 "Visualizing Knowledge About Virtual Reconstructions of Ancient Architecture." In *Proceedings of IEEE Computer Graphics International*, 36–43. Canmore: IEEE. doi 10.1109/CGI.1999.777901.

Sylaiou, A., and P. Patias. 2004 "Virtual Reconstructions in Archaeology and Some Issues for Consideration." *Imeros Journal for Culture and Technology* 4. http://www.ime.gr/Publications/Print/Imeros/En/04/Article01.Html.

Takase, Y., K. Yano, T. Nakaya, Y. Isoda, T. Kawasumi, K. Matsuoka, T. Seto, D. Kawahara, A. Tsukamoto, M. Inoue, and T. Kirimura. 2008. "Virtual Kyoto: Visualization of Historical City with 4D-GIS, Virtual Reality and Web Technologies." *International Archives of the Photogrammetry, Remote Sensing and Spatial Information Sciences* 307: 975–980.

Thomson, Judi, Elizabeth Hetzler, Alan MacEachren, Mark Gahegan, and Misha Pavel. 2005. "A Typology for Visualizing Uncertainty." *Proceedings of the SPIE-VDA 2005: SPIE Proceedings* 5669. doi: 10.1117/12.587254.

Tomaszewski B., and A. M. MacEachren. 2012. "Geovisual Analytics to Support Crisis Management: Information Foraging for Geo-Historical Context." *Information Visualization* 11(4): 339–359.

Tufte, E. R. 1997. *Visual Explanations: Images and Quantities, Evidence and Narrative*. Cheshire: Graphics Press.

Yabuuchi, A. 2011. "Home to School Surrounds Diagram" In *Informational Diagram Collection*, edited by Yu Fukushi, 215. Tokyo: Pie Books.

Zuk, T., and S. Carpendale. 2006 "Theoretical Analysis of Uncertainty Visualizations." In *Visualization and Data Analysis 2006, Proceedings of SPIE 6060*, edited by Robert F. Erbacher, Johnathan C. Roberts, Matti T. Gröhn, and Katy Börner, 66–79. San Jose: SPIE - The International Society for Optical Engineering. doi: 10.1117/12.643631.

Zuk, T. and S. Carpendale 2007. *Visualization of Uncertainty and Reasoning*, In *Proceedings of SG 2007*, edited by A. Butz et al., 164–177. Berlin and Heidelberg: Springer.

CHAPTER 5

# Experiencing Past, Present, and Future Urban Environments Through Digital Representation, Storytelling, and Simulation

EVA PIETRONI, CNR ISPC, INSTITUTE OF SCIENCES FOR CULTURAL HERITAGE

## INTRODUCTION

The application of digital technologies can result in a significant advance in the research and valorization of the urban built heritage. In the digital domain we can analyze, classify, and represent transformations on an urban scale, integrating information from multiple fields (geography and topography, archaeology, history, architecture, urban planning, art history, social science, literature, art, etc.). Geographic information systems represent a base upon which to build the study of complex, multilayered spatial contexts. Data of different ontologies, otherwise dispersed in multiple archives and in a variety of formats, can be brought together and organized in a unique and coherent 2D or 3D context, with great potential for analysis.

The integration of bottom-up and top-down approaches (Forte 2008) can lead researchers to the 3D representation of a cultural context in its actual state through surveys and digital acquisition, and in its past phases of life through virtual reconstruction based on historical sources and interpretative work. In the past ten years, the London Charter (2009), the Seville Principles (Lopez-Menchero Bendicho 2013), and a widespread debate at the international level have focused attention on the need to make the

Eva Pietroni, *Experiencing Past, Present, and Future Urban Environments Through Digital Representation, Storytelling, and Simula-tion* In: *Digital Cities*. Edited by: Maurizio Forte and Helena Murteira, Oxford University Press (2020). © Oxford University Press.
DOI: 10.1093/oso/9780190498900.003.0006

interpretative process more explicit, and to create repositories where data can be shared and metadata can be opened for verification, criticism, and possible reuse.

Moreover, in the past few years, augmented reality tools and technologies have allowed users to experience an enhanced perception of contexts by overlapping and synchronizing virtual information in the real space. Thus, 3D reconstructions impact not only the way we look at sites and spaces, offering new depth and insight, but also can be useful interpretative tools, supporting researchers in simulating and testing hypotheses about complex contexts or dynamic processes. However, despite advances made in this field, most of the virtual environments dedicated to urban contexts have not realized their full potentialities.

Usually, projects following a historical approach implement visual archives where urban plans, iconographies, historical sources, and 3D renderings are collected and arranged in relation to how these urban environments have been transformed over the centuries. A good example of this is Visualizing Venice, an educational project started in 2009 and promoted by Duke University, the University of Venice (IUAV), and the University of Padova. It is mainly intended for students and architects involved in the creation of 3D digital models and maps of the city of Venice, based on archival and printed sources, maps, documents, cartography, and images (Ferrighi 2013). The goal of this project was to visualize the evolution of urban buildings over time, communicating new knowledge about place and space to the public through portable devices and on the Visualizing Venice website (www.visualizingvenice.org/visu/).

Platforms like this are useful in traditional digital humanities studies, but they are often limited in their ability to represent the "city of stone," shown by means of fixed "pictures" keyed to specific periods. They still lack the means for a multisensorial approach, narrative and intangible contents, and dynamic simulations; in a very few cases, they investigate and communicate the visions and feelings of the inhabitants of the time.

Today's computing capacity can extend research and practices much further. A virtual urban landscape can be conceived as an open digital ecosystem where communities of users can interact in a multimodal way. Recent promising trends in information and communication technologies point to a convergence of research disciplines, AI (artificial intelligence), VR (virtual reality), and AL (artificial life). Even the concept itself of "virtual reality" is evolving, in the sense of expanding virtual environments with artificial creatures. Such evolution of a technology can offer interesting perspectives on the digital heritage domain and the future for the interpretation of historical urban processes.

## 3D MODELING TECHNIQUES FOR ARCHITECTURAL AND URBAN ENVIRONMENTS

Different methods are traditionally employed for capturing in 3D models some architectural elements and monuments, according to the level of detail and the kind of documentation/representation required. For instance, computer-assisted design (CAD) tools are usually considered the best for architectural planning, so as to create precise 2D elaborations based on given dimensions, such as architectural plans, building heights, and area sections. 3D models realized through CAD tools are correct in terms of general measures and typology, but they cannot represent the exact topology of an existing object because they have not been acquired through a process of high resolution measurement.

Aerial and terrestrial photogrammetry and laser scanning are powerful techniques for 3D digitization at different scales, resulting in dense point clouds with a good level of resolution and accuracy. They are usually employed to document and represent in 3D the real topology of the object, its surface, and its conditions of preservation. In both cases, the post-processing work, aimed at creation of a mesh model with the proper level of detail, includes many phases of elaboration and is often long and tedious. One limitation is the fact that only visible parts of the objects can be rebuilt. Portions that cannot be captured by the camera or the laser need, then, to be reconstructed through similarity—where possible. Furthermore, both techniques produce undifferentiated models that represent the entire topological surface of the object, without any "semantic" characterization. That is, there is no distinction between roads, buildings, walls, roofs, windows, doors, gardens, trees—all elements are included in a unique mesh object deriving from the triangulation of the point clouds. Division and characterization have to be made later by hand or with semi-automatic algorithms. Moreover, no further changes to the 3D model's parameters are possible, because 3D models do not come from parametric techniques (as, instead, happens in CAD or procedural modeling); rather, they are derived from detailed digitization of concrete structures.

Procedural techniques represent another powerful solution, guaranteeing nice and faster results for large-scale urban representations. An urban environment includes a huge number of buildings, and in time the extension of urbanized areas are transformed; moreover, each building is composed of varying architectural elements whose style, proportions, and relation between form and function change through the ages, following historic, aesthetic, and statutory rules (Ferdani et al. 2010).

Each architectural element can be, by itself, a complex object, composed of subelements. For this reason, in several cases the reconstruction of holistic and complex urban environments can be successfully done using "procedural modeling" (Parish and Muller 2001). This methodology allows the creation of 3D libraries of architectural typologies, useful for producing repeats of 3D models, which are similar but have some variations, including covering large extensions of space. Typically, as a first step, the basic floor plans of all buildings have to be imported in geographic information system (GIS) format. The files are then converted into units-building ("allotments"). To create a realistic urban model, characterized by a wide range of architectural typologies, these allotments are associated with a specific description (shape grammar) (figure 5.1).

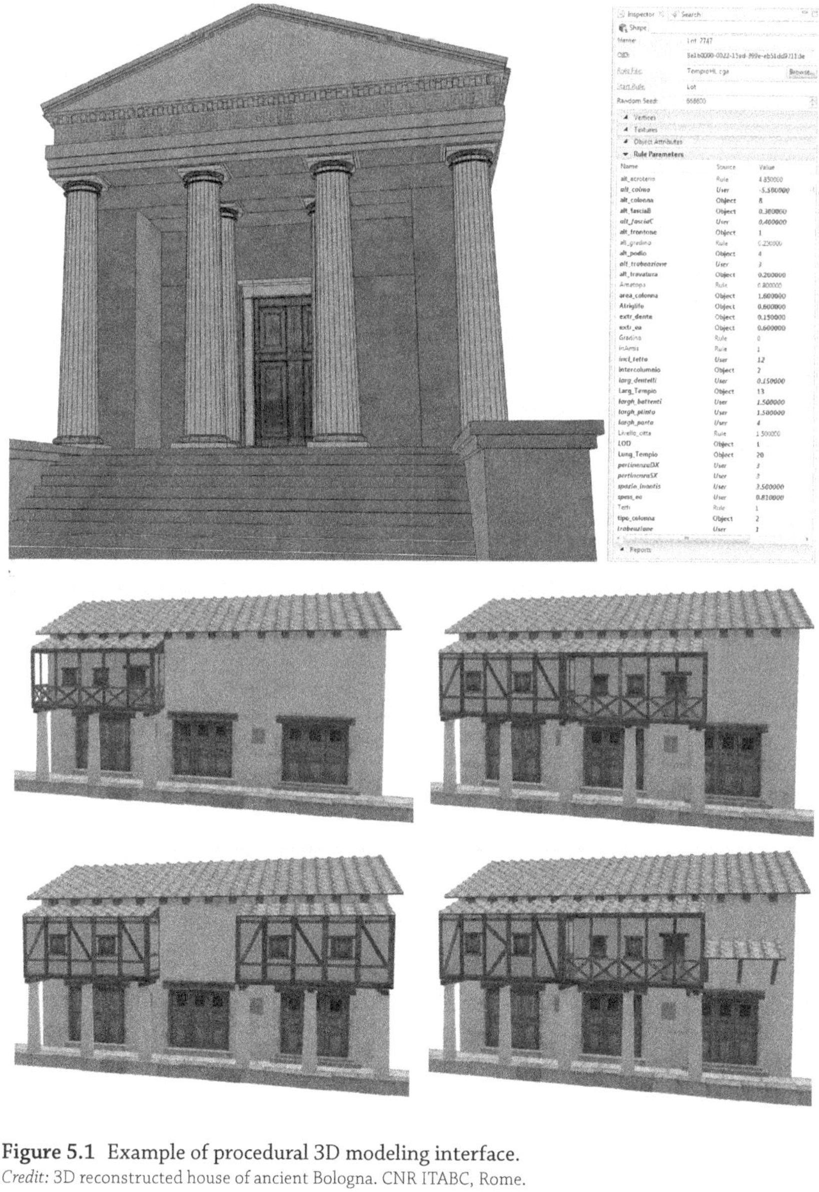


**Figure 5.1** Example of procedural 3D modeling interface.
*Credit:* 3D reconstructed house of ancient Bologna. CNR ITABC, Rome.

This procedure involves a set of rules that define and control appearance, size, and type of building, as well as the hierarchical relationships between the architectural elements and the variables that generate randomness, thus lending realism to the model. With this set of rules, defined by the modeler, the geometry and appearance of the buildings are "constructed" and displayed in the 3D space, using the software. CityEngine is one of the most common types of modeling software (figure 5.2).

To adapt the town to the geo-morphology of the terrain and to place the ground at the correct height above the sea level, an elevation map can be useful, imported in gray scale (Geo-referenced Tagged Image File Format, or GeoTIFF). Finally, a set of textures are created in order to characterize the buildings. A correlation matrix is used to randomly assign the set of textures to each building, taking also into account the size and proportion of each building. 3D models, in this case, do not reproduce the effective topology of

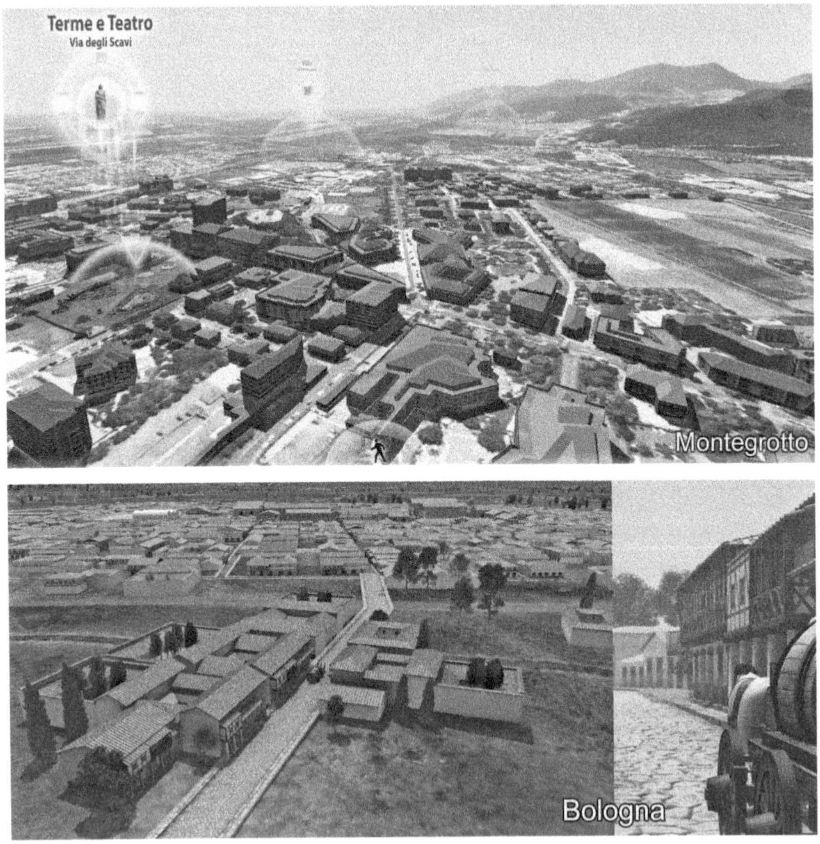


**Figure 5.2** Examples of 3D urban environments constructed using procedural modeling techniques. Above: Montegrotto Terme, in present state (Aquae Patavinae project). Below: Bologna, as reconstructed in Roman age (Apa Discovering Bologna project).
*Credit:* 3D models of Montegrotto Terme and Roman Bologna, CNR ITABC, Rome.

the real buildings because they are obtained following parametric and procedural rules; anyway, they can be very detailed and descriptive.

Procedural modeling techniques can be particularly useful for representing hypothetical 3D reconstructions of past and present urban environments, suggesting their shape and appearance holistically. It is important to know the context scheme so as to place each piece of the puzzle correctly in position and in relation to its neighbors; for this reason, architectural guidelines and rules in old treaties can be helpful (De Luca et al. 2007, Chevrier 2009). Finally, 3D elements/buildings that can be modified through parametric controls can be reused in an unlimited number of practical cases and successive applications, according to user need, with significant simplification and reduction in time (Mueller et al. 2007).

The need to create and share digital libraries, especially online, is heavily debated in Europe. Many parametric objects are available in digital libraries for specific fields (industry, mechanics), but there are few open architectural libraries, especially as scientifically validated. But even if we have many tools for visualizing, streaming, and interacting with 3D models, the tools for coding, extracting, and sharing the semantic content of the 3D media are still lacking. Automatic classification of 3D databases, automatic 3D content annotation, and content-based retrieval represent new lines of research and constitute key topics in the fields of computer graphics and vision research.

In the same way, the knowledge technologies, such as structured metadata and ontologies, can be extremely useful in supporting a stable and standardized approach to content modeling and data sharing. A very interesting project is FOCUS K3D (Foster the Comprehension and Use of Knowledge-intensive 3D media, 2008–2010, FP7-ICT), which promotes the adoption of best practices for the use of semantics in 3D content modeling and processing. The shape of an object is described in a definition of geometric primitive entities, the classification of the reference context, and the meaning associated with it. This new representation can highlight the semantic level at which we code our perception of shape. In fact, through this new modeling paradigm, it is possible to formalize the geometric form of an object (geometric model), together with the set of contexts that could use the model (semantic model). Following this approach, there could be developed new tools for the classification and recognition of shape features and topological structures (Falcidieno 2008).

All 3D modeling techniques described here can be useful for representing architectural and urban environments, depending on the project's goals. In the case of multi-scale representation (from holistic to delimited visions), which usually require several levels of detail (LOD), and if a diachronical approach is followed, the best solution is a combination of these means.

## THE DIGITAL CITY AS A SCENARIO OF SIMULATION

Representing urban ecosystems is a complex activity because it comprises both tangible and intangible heritages—specifically, the city of stone and the city of people.

Reconstructing historical cities is still a problem for digital humanists, notwithstanding the high-quality results of projects such as Visualising Venice, or the virtual museums of cities, such as those of Bologna (Apa alla scoperta di Bologna), Amsterdam (Creative Amsterdam/CREATE project, www.create.humanities.uva.nl) (Oudenampsen 2007), or Lisbon (Lisbon Before 1755 Earthquake) (Murteira and Rodrigues 2016).

Not only planners, city administrators, urban designers, and conservators, but also experts in virtual heritage, information and communications technology (ICT), artificial life simulation, serious games, sociology, anthropology, and smart cities, are all dealing with the same challenging questions: How to investigate urban transformations? How to represent and communicate them?

One of the next most innovative attempts, from a methodological view, integrates traditional disciplines (digital humanities) and simulation algorithms based on artificial life (as proposed by ICT experts). This integration is really crucial if we don't want to limit the work to the 3D representations and visualizations of predefined urban environments (even if at different chronological stages of evolution). On the contrary, our goal is the simulation of these processes of transformation processes so as to reach a deeper and critical comprehension of it.

Artificial creatures are able to self-organize in the territory and modify their behavior (also through learning) to satisfy their needs (to survive and to improve their quality of life). They have sensors that experience the environment and they autonomously establish relations between them and the space (Parisi and Schlesinger 2002, Borghi et al. 2011). The 3D space then becomes the stage for these relations; different properties of the environment will influence the artificial agents' dynamics. The space as well will be influenced and modified according to the evolving behaviors of the agents—for instance, by assuming specific colors or changing volumes. This means that we can interactively and in real time modify some conditions in the 3D environment (changing, for instance, urban plans, roads, buildings, functional properties, sensorial stimulus, and so on), or some properties of the agents (alphabetization, cultural needs, daily needs). Then we can observe the resulting dynamics/behaviors (Annunziato et al. 2008). In so doing, we can better interpret some complex historical processes or the social and ecological aspects of the context, and we can plan with more awareness our future urban developments. Therefore, such an approach could be useful both for the interpretation and comprehension of phenomena and for education and communication.

The most important factors involved in urban transformation must be identified in order to create a conceptual model; additionally, a 3D graphic and interface are needed, which are able to evolve according to the dynamics of the urban environment, which changes in unpredictable way. A promising solution could be the integration of artificial life dynamics and procedural modeling rules. The possibility of operating on parametric values allows creation of a link between simulations, based on the algorithms of artificial intelligence and architectural/urban "mutations." This would allow us to test, visualize, study, and verify, in a photorealistic (and theoretically multisensory) 3D digital environment, the impact of such dynamic processes in spatial urban contexts (Piga 2008, Morello and Ratti 2009).

As explained here, in the procedural approach, the generation of different typologies of buildings is defined through a scripting language, rather than traditional editing tools, using a shape-grammar approach. The simulation of artificial life could influence the parameter that occupies the top of the hierarchy of scenes, modifying all allied entities and automatically changing the virtual geometry. The a-life dynamics developing in the virtual environment could influence the procedural models' parameters and thus determine evolving shapes, volumes, and aggregations. In this way, a realistic graphic interface would be added to traditional artificial-life applications related to urban environments, adding a third dimension.

## THE DIGITAL CITY AS A SPACE OF SOCIAL INCLUSION

Today we are living in a transitional phase that urgently needs new tools to refine our digital cities into places of learning, enjoyment, and social inclusion. The incredibly fast progress of the visual and ICT technologies has brought totally new ways of social interaction and cultural engagement that a few years ago might have been unthinkable.

Digital cities have a twofold potential today: (1) to innovate the way people approach, understand, and learn about their environments and cultural heritage; and (2) to enrich the cultural contents and provide additional information to the visitors, This is in comparison to traditional supports (guides, websites, etc.). The new challenge, for cultural heritage professionals, is to transform the traditional and static one-to-many knowledge-transfer model, used until now in classical and academic communication channels, into a more dynamic and interactive many-to-many communication model that integrates the social media assets. The social media are a new domain compared to the classical communication paradigms, with a participative kind of knowledge sharing. These social platforms enable whole communities to interact, share, co-create, and comment on their cultural heritage and urban environment, thus reusing and spreading cultural contents, in the process preserving the cultural identity but also promoting cultural diversity.

The digital cities of the future will bridge the real and virtual worlds, the tangible and intangible cultural heritage, to link past, present, and future, while embedded in a scientific and trusted environment based on personal profiling. In this way, the representations of digital cities will offer additional and improved experiences for visitors, previously difficult to achieve with traditional methods.

The key concept is the connections linking stories, urban structures/spaces, and users. This interconnection will lead to the creation of many potential scenarios reflecting local, global, individual, and collective dimensions and perspectives that will help users to feel included in their personalized cultural experience. By these means, the role of the audience little by little gains importance—no longer passive viewers or observers of what is shown in the digital domain but, rather, uploaders of data and contents able to enrich the digital cultural ecosystem.

In order for planners to explore, "reconstruct," and communicate not only the physical space of the city but also the "mindscape," the digital city needs to integrate varying levels of visualization, multilayered contents, metadata, storytelling, tools of interaction, and structured digital libraries and repositories. To avoid confusion, however, these digital cities—consisting in virtual museums and social platforms for cultural heritage—should be addressed to different groups of stakeholders: (1) tourists, (2) cultural heritage professionals, (3) researchers and students, (4) general community and inhabitants. For each group, the contents should be structured differently. Knowledge, languages, and communication should follow diverse approaches in terms of concept and design, so as to fit the needs and the expectations of the different targets. Thus, the digital city becomes a space of experience, work, enjoyment, and personal engagement. It becomes also a space of storytelling and "dramatization," even of embodiment and performance in case of a digital city in 3D.

Today, social platforms are mainly in 2D because they are conceived for fast and immediate exchange among users. However, the potential for 3D web communities is much greater in terms of the embodiment and cognitive impact, even if at the actual state of the art they are less immediate to use by the general public. In an online 3D environment, users not only can meet and exchange information but they can also perform actions, touch objects, and transform the surrounding space: knowledge is built essentially on sensory motor skills and direct experience in the 3D cyberspace (Mellet-D'Huart 2006, Varela et al. 1991).

The first example of such an environment is Second Life (http://secondlife.com), a virtual reality platform launched in 2003, which enables its users, called "residents," to build personalized scenarios, cities, museums, and gardens and to interact with each other through avatars. They can explore, dialogue, organize, or attend individual and group activities, can create and trade abilities or services, can invent new jobs, and even can exchange money (Gerosa 2007). Different from other platforms, Second Life is a virtual world completely built by its users. In these "islands," the residents can integrate multimedia, audio and video, images from libraries and collections, and the like. By 2013. Second Life had approximately 1 million regular users, according to Linden Lab, which owns the platform. Although use of Second Life has been decreasing in recet years, this experience points to how the general public's image of the "ideal" city or virtual museum played the role of protagonist.

Another example of 3D performative space comes from the Integrated Technologies of Robotics and Virtual Environment in Archaeology project (2006–2009), financed by the Italian Ministry of the University and Scientific Research. In this framework, CNR ITABC pioneered a multi-user VR environment addressed not to the general public but to the community of archaeologists. This collaborative environment was conceived as a space for simulations and education, an open laboratory allowing experts to share and exchange cultural contents and interpretations. Users could load a 3D scenario, related to an archaeological site, and meet, chat, and interact in real time, as well as share data. The scene could be edited by the user; that is, it could be disassembled and recomposed again, models could be added or hidden or moved, or otherwise modified according to a different hypothesis. It was possible to add links to metadata, save new versions of a scene, then share it and discuss it within the community. In essence, users became creators of ideas and contents, establishing a strong relationship with other users (Forte and Pietroni 2009).

Similarly, the Virtual Rome project (2007–2009), developed by CNR ITABC, was an open-source VR project based on geo-specific data, with front-end (VR WebGIS) and back-end (VR WebLAB) solutions, for the interpretation, reconstruction, and 3D exploration of the archaeological and potential landscape of ancient Rome (Pescarin et al. 2008). The purpose was to create a 3D environment embedded in a web browser. The back-end framework was conceived as a collaborative platform; through a 3D content management system, different professionals could be involved in the complex, open, and never-ending activity of landscape reconstruction.

Imago Bononiae offers a different example of performative urban space. This VR application, realized in 2013 by CNR ITABC in collaboration with CINECA, allows the user to explore a large-scale 3D reconstruction of the city of Bologna in Roman times (first century CE). To interact in real time in the virtual scenario, the visitor uses his or her body movement, performing a set of gestures in front of the projection (Fanini and Pagano 2015) (figure 5.3).

The urban environment is populated by a virtual crowd, controlled by artificial life algorithms. The crowd's behaviors, in fact, take place along ancient roads and in ancient architectures, envisioning hypothetical paths according to urban "chronotopoi" and guiding the real users toward different visions of the future urban layer of Bologna (Fanini and Calori 2014). Gesture-based interactions allow the enhancement of the user inside the VR environment, constituting a new frontier in the communication and learning processes (Featherstone and Burrows 1995, Pietroni 2013).

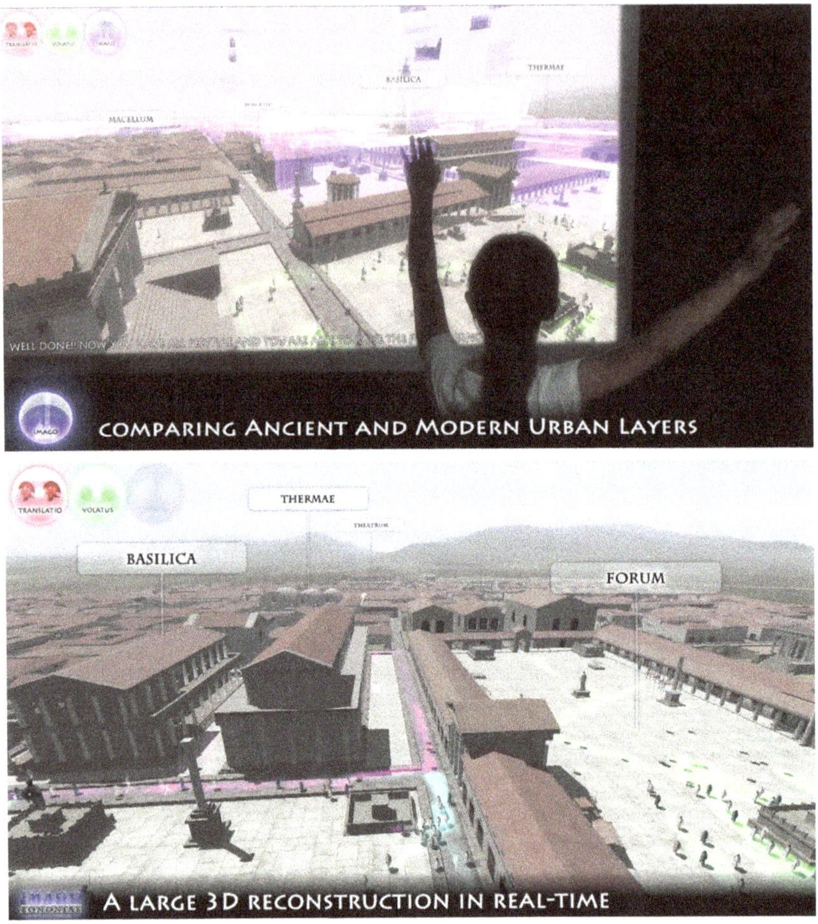


**Figure 5.3** Imago Bononiae project, VR application with gesture-based interaction and artificial life dynamics.
*Credit:* 3D model of Roman Bologna, CNR ITABC, Rome, in collaboration with Cineca, Bologna.

## THE DIGITAL CITY AS A SPACE OF NARRATION

In the virtual museums of an urban context, storytelling and powerful visualization play key roles, together with the multiplication of communicative registers and perspectives. The main goal is to stimulate the curiosity, the interest, and the affection of the public toward our habitat and cultural heritage, and thus to convey spontaneous actions of fruition, self-identification, and personal involvement. For this purpose, narration and interaction are powerful means.

Google Earth and Street View are visual platforms with powerful tools especially designed to satisfy punctual needs: to find a specific information, to visualize, or to create an itinerary. Despite their great potential, they are not designed for deeper educational purposes nor for narration; in fact, they adopt quite "flat" communication modalities.

Storytelling, on the contrary, establishes the need for a precise "direction," as is also the case for interactive applications and virtual reality environments. To structure this "direction," novel communicative approaches are needed, mixing techniques that come from virtual reality, cinema, augmented reality, and games. Besides words and traditional themes, stories are told by managing colors, light, and soundscapes, with a visual mood that produces another layer of communication. This subliminal layer, in particular based on emotions and perceptions, represents the immersive part of the experience.

In the virtual museum of a city, different communicative styles need to be adopted: educative, emotional, explicit, subliminal, intimate, collective. Storytelling can be supported by the integration of real and virtual elements—for instance, a real actor integrated into the computer graphic context, or the overlapping and alternation of real space and its virtual reconstruction, implemented through camera tracking techniques.

A typical element in storytelling is the creation of a story/fiction within a wider cultural context. This is the case of the Apa Discovering Bologna project, an animation in stereoscopy carried out by CINECA for the new Museum of the City, Genus Bononiae. The animation tells the story of the city through an imaginary character, Apa the Etruscan, who jumped out of an ancient vase in the museum, and it combines philological accuracy and innovative computer graphics technologies. The 3D reconstructions of the potential urban landscape of Bologna in Etruscan and Roman times have been carried out by CNR ITABC, through procedural modeling techniques, as shown in figure 5.2. This project evolved from a previous one dedicated to the city of Bologna, called N.u.M.E.; both brought to the creation a set of open 3D data sets available for a wide range of different applications, from real-time crowd simulations within large urban contexts (as the case of Imago Bononiae) to serious gaming environments (Apa Game), allowing the visitor to interact with 3D urban reconstructions and understand their evolutions (Guidazzoli et al. 2011).

## THE DIGITAL CITY AS A SPACE OF MULTIPLE PERSPECTIVES: MATERA TALES OF THE CITY

The Matera Tales of a City project was carried out in 2009–2011 by CNR ITABC, and was supported by the Regional Promotion Agency (APT) and the Basilicata Regional Government. Matera (Basilicata region, southern Italy) has been a UNESCO World Heritage Site since 1993, and was designated European Capital of Culture for 2019. This city is modeled as in the calcareous rock of a deep canyon called Gravina, where thousands of natural caves have been used for protection and living, from 12,000 years ago until now (Fonseca et al. 1999). The territory in fact has plenty of rupestrian sites: prehistorical settlements (Paleolithic and Neolithic age), rupestrian churches and monasteries, caves used as houses, pastoral villages excavated into the calcareous rock, and medieval manor farms.

The most famous areas of the city are called "Sassi" (Sasso Barisano and Sasso Caveoso), where the fronts of the houses are built with calcareous bricks obtained by excavating the inner caverns. In most parts of the Sassi, the streets run atop the rooftops of other houses (Laureano 1993), which presents such a peculiar spatial configuration that it has inspired famous writers, film directors, and artists.

For many centuries, families lived in caves together with animals; they aggregated around open courtyards, sharing everyday experiences. This situation lasted until the 1950s, when the Italian government forcefully relocated most of the people living in very bad conditions to apartment buildings outside the Sassi neighborhoods. Here, the people started living in much better conditions, but they lost their ancestral status based on a communal life. After a period of neglect, at the end of the 1980s the rupestrian center of Matera was "rediscovered"; and in 1993, it was the first site in southern Italy to be declared a World Heritage Site. Restoration and systematic re-occupation of Sassi areas was started again for commercial, tourist, and craft activities, but also for living. Today, many people occupy the same houses as owned by their ancestors thousands of years ago.

The aim of the project, thus, is to provide tourists visiting Matera and the surrounding park of rupestrian churches with cultural content related to the place, the history, and its heritage in order to enhance their cultural experience and encourage them to spend a longer time in this urban context. A digital platform was created to support the public before and during a visit to Matera, through access to cultural content on the website (www.matercittanarrata.it) and in specific applications for mobile devices. The project included 3D reconstructions of the city and of its landscape in eight chronological phases, as well as movies, virtual reality environments, and audio guides adapted to different formats and for different audiences and devices.

From the beginning, we tried to plan a visualization that would not compete with local guides (many people of whom base their living on this activity) and that could be perceived as adding value to what was already existing in terms of cultural offering. We decided, for instance, to reconstruct the landscape and urban evolution in eight different chronological periods, from 2 million years ago until the present day. This was the first time such work has been attempted for Matera and presented to the public.

This evolution is described and shown through video clips based on a 3D rendering that shows the context at different stages: the emergence of land from the sea more than 2 million years ago, the terrain and Gravina formation, the birth of Neolithic entrenched settlements (5000–4000 BCE), the development of life in the caves, the construction of the "built" and fortified space in the Middle Ages, and the expansion from the Civitas to Sassi suburbs, and from Sassi to the planar area where the modern and contemporary city developed. The 3D rendering is supported by historical documentation, iconography, and literary citations. A critical info-graphic apparatus was also developed to highlight, on the actual images of the town, the interpretation process that was followed and to show the correspondences between the 3D reconstruction and what is still visible today.

The 3D reconstruction also was used to create a timeline, enabling an "augmented" perception during the visit. From six fixed panoramic points, the user can explore the landscape and switch among the different ages. In each panorama, the actual image and the 3D reconstruction are perfectly overlapping, so it is possible to perceive the area's evolution. A cursor allows the user to modify the transparency of the active level so as to have more comparisons with the actual landscape (figure 5.4). Other comparative views consist in historical photos taken inside the Sassi areas. Every couple of images shows, from the same point of view, the life and the places before and after the depopulation of 1950s or their successive restoration (figure 5.5).

The main purpose of the 3D reconstruction in eight chronological phases is to show the general transformation of the urban context, rather than focusing on specific buildings. Because of the rupestrian nature of the city, none of the

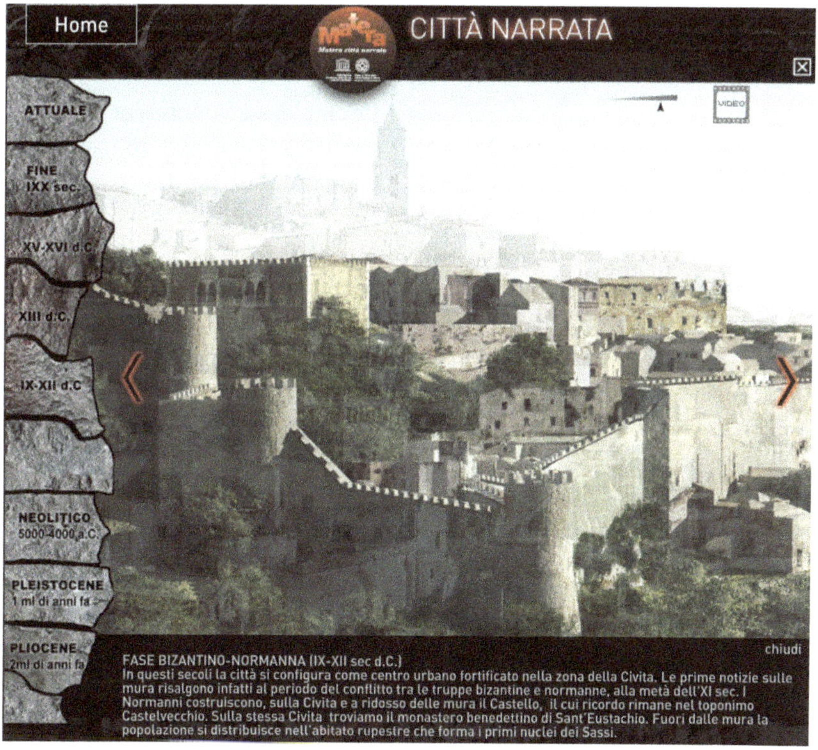


**Figure 5.4** Matera Tales of a City project, with interactive timeline.
*Credit:* Matera 3D reconstructions and panoramas, CNR ITABC, Rome.

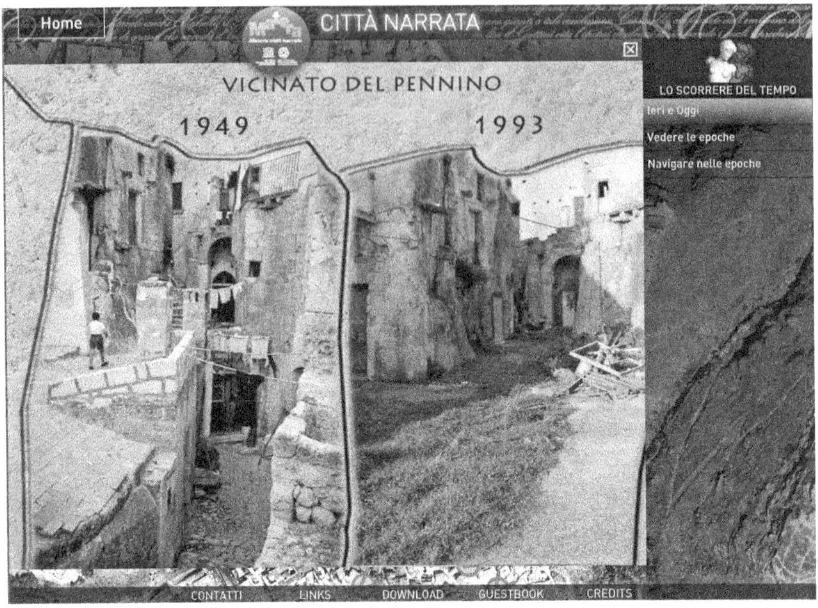


**Figure 5.5** Matera Tales of a City project, with the same place shown in different ages through historical photos.
*Credit:* Photos purchased by CNR ITABC, Rome, from Archivio Buonsanti, Matera.

3D modeling techniques could effectively be employed to represent the urban context on a large scale. Matera is sculpted in the rock, and no parametric or procedural rule could be helpful. Laser scanning or photogrammetric techniques would have been too long and complex, and not consistent with the main purpose of representing the whole city along its entire evolution. Therefore, the 3D model was executed in computer graphic techniques, using 3D Studio Max and Vue (a 3D landscape generator), and based on an actual city plan, topographical data (bottom-up approach), and iconographical and historical sources (top-down approach) (Pietroni 2012).

A greater scale and, in order to grab a wider portion of landscape including the Rupestrian Churches Park, a VR environment has been implemented. Moving and tapping the iPad, the user can guide the icon of a Lesser Kestrel (a typical little hawk living at Basilicata) to fly all over the territory. This bird is the user's avatar. The 3D reconstruction of context is based on a GIS, where all the data are geo-referred and integrated. It is a "holistic" vision of the context, where the user is able to visualize and locate the main typologies of cultural sites present in the area. Some of these sites are interactive, and the user can access more detailed stories about them, told by local experts, artists, and citizens.

The *Matera Città Narrata* guide is not a descriptive traditional guide inspecting monuments and mentioning who made them and when. Instead, it tells the stories that took place in those ancient sites, presenting real fragments of life, myths, events, characters, and memories. Local people, inhabitants, children from primary schools, or artists were involved, asked to give their particular and original vision, their interpretation of Matera's story and "soul," through their own memories, experiences, and expectations. This narrative dimension, carried on by inhabitants, is maybe the most important aspect of the project, also taking into consideration that this generation is the last one with a direct connection to life in caves. Moreover, some itineraries and excursions through the urban environment or the surrounding landscape are suggested.

The contents are accessible, for free, from a variety of mobile technologies. The main access possibilities implemented by the system are:

1. *Database access while moving through the city:* The visitor stops in front of a monument and finds a sign with its name and ID. He can access the website and, keying in the ID, obtains a story about the monument. He can also download a stand-alone application if he owns a tablet or a smartphone (iOS, Android, Symbian, Java), in order to avoid reconnecting to the internet during successive steps in the visit. If no connection to the internet is available, he can call the toll-free number indicated on the sign. In this case, he will reach an audio guide.
2. *Guide access from a fixed position*: The user can access the website from home, school, or other static base. He finds cultural contents organized in thematic itineraries; he can explore the contents or download them in different formats (*.mp3, *.pdf, stand-alone applications for smartphones) in order to access them in a second moment, while the real visit to Matera is in progress.
3. *Guide access while moving through the site*: The user is in front of the monument and he has just received the information about it. After this and in relation with this, the system will suggest what else the visitor can enjoy: thematic itineraries, witnesses, video clips with personal stories or memories told by local people related to that monument.

The project is still today quite in its pioneering state. Its main goal was to create something new not only, and not necessarily, from a technological point of view but, especially, in the general infrastructure, in the design of the applications, in its robustness and possibilities to be maintained by the local administrations, and of course, in the choice of cultural contents, as of the proper communicative formats. In recent years, many projects intended to provide tourists with cultural guides on mobile devices ceased after a few months. Sometimes this was due to the extremely sophisticated technological solutions that have been employed, requiring for instance the combination of inter-dependent services (wifi, GPS, Bluetooth) that make the general infrastructure delicate and fragile. We wanted to build something more robust, that could work even in the absence of a GPS signal or a wifi connection (we provide these systems but visitors are not totally dependent on them). In fact, the urban spaces of Matera are often very narrow and places are located one above the other, so GPS cannot always be precise.

Thus, we tried to fit the project to meet the needs of the people visiting Matera. A wifi infrastructure has also been provided, to cover the most attended places in the Sassi neighborhoods. The project, coordinated by ITABC and realized also through the collaboration of public institutions and private companies, was made available to the public in 2012. Efforts, during the digital platform design, were oriented toward keeping the costs of maintenance as low as possible. After six months of experimentation and economical support by CNR ITABC, maintenance is now handled by the Municipality of Matera.

## CONCLUSIONS

It is commonly accepted that 3D can greatly enhance our understanding of the outside world and of its phenomena, because it multiplies the possibilities of analysis and of interaction between us and the digital ecosystem. Up till now, projects approaching digital cities from a historical point of view have been focused on the creation of visual archives mostly related to the "city of stone," where topographical data, historical documentation, iconography, and 3D reconstructions are used to present predefined interpretations of chronological phases in the city's evolution. The next generation of digital cities could go a step forward, exploring the possibility to represent 3D urban environments and their transformations as digital living ecosystems, exploring how and why spaces, architectures, human behaviors, sensorial perception, and social phenomena change over time. A digital city, reconstructed in 3D, could become the scene for artificial simulations (Ferrauto et al. 2013) set to investigate, or verify, the dynamic processes from which transformations originate.

The economic and environmental crisis, and the migrations currently taking place worldwide, are changing the needs, perspectives, spaces, and countries involved. Places destroyed by war and natural disasters may be lost forever, together with their heritage and identity. Preserving and transmitting the past and present image of a city, not only the city of stone but also the city of people, is a fundamental step in guaranteeing our cultural awareness and social sustainability (Porfyriou 2010, Porfyriou and Sepe 2012). Artificial simulations could constitute a suitable tool to study how to face the change and how to encourage adaptability and resilience to crises, as well as how to guide future urban transformations and projects. Such an innovative approach requires the convergence of disciplines: planning, architecture, archaeology, art, conservation and valorization, social sciences, anthropology, smart cities, ICT, a-life, robotics.

Furthermore, nowadays it becomes urgent to merge research efforts in considering the role of the end user and interactions among visitors. The European Cultural Heritage sector has witnessed the emergence of new technological models leading to the creation of digital social platforms able to encourage active participation by stakeholders with diverse backgrounds, and by citizens as well, in the cultural processes. The Matera Tales of the City project has been presented here as an example of such an approach. Apart from the 3D reconstruction of the city along the main steps of its transformations, which follows a traditional descriptive style, the narration of the actual city, and of its traditions and intangible heritage, is told by its local people, reflecting multiple perspectives. An improved integration of stories, urban structures/spaces and users is needed to enhance awareness, responsibility, and participation. Layered narratives, embodiment, and novel integration of media sources are fundamental in the creation and transmission of cultural values for future digital cities. They are directly connected to emotions that can motivate understanding, conceptual appropriation, and self-identification, contributing to a higher cognitive process of learning.

## REFERENCES


Annunziato M., E. Bonini, P. Pierucci, and E. Pietroni. 2008. "Cultural Mirrors: An Epistemological Approach to Artificial Life for Cultural Heritage Communication." In *Digital Media and Its Applications in Cultural Heritage*, edited by J. Al-Qawasmi, M. A. Chiuni, and S.El- Hakim, 219–135. Amman: CSAAR Press.

Borghi A. M., A. Di Ferdinando, and D. Parisi. 2011. "Objects, Spatial Compatibility, and Affordances: A Connectionist Study." *Cognitive Systems Research* 12: 33–34.

Chevrier, C., and J. P. Perrin. 2009. "Generation of Architectural Parametric Components: Cultural Heritage 3D Modelling." In *Joining Languages, Cultures and Visions. Proceedings of the 13th International CAAD Futures Conference,* edited by T. Tidafi and T. Dorta, 105–118. Montréal: Les Presses de l'Université de Montréal.

De Luca L., P. Véron, and M. Florenzano, M. 2007. "A Generic Formalism for the Semantic Modeling and Representation of Architectural Elements." *Visual Computer* 23(3): 181–205.

Falcidieno, B. 2008. "From Geometric to Semantic 3D Content: The FOCUS K3D Initiative." In *Digital Heritage: Proceedings of the 14th International Conference on Virtual Systems and Multimedia* (VSMM), 20–25 October 2008, Limassol. Cyprus, edited by M. Ioannides, A. Addison, A. Georgopoulos and L. Kalisperis. Milano: Archaeolingua.

Fanini, B., and L. Calori 2014. "3D Interactive Visualization of Crowd Simulations at Urban Scale." In *Virtual City and Territory 2013 Proceedings*, a cura di M. Cerasoli, 276–284. Rome: Università degli Studi Roma Tre, ArcAdiA-Open University Archive, ISBN 9788897524090.

Fanini, B., and A. Pagano. 2015. "Interface Design for Serious Game Visual Strategies: The Case Study of Imago Bononiae." In *Proceedings of Digital Heritage International Congress 2015*. Granada, 28 September-2 October 2015, edited by G. Guidi, J. C. Torres, R. Scopigno, H. Graf, Volume 2, 623–626. European Association for Computer Graphics—EUROGRAPHICS; Institute of Electrical and Electronics Engineers: IEEE. IEEE Catalog Number CFP1508W-USB.

Featherstone, M., and R. Burrows. 1995. "Cultures of Technological Embodiment: An Introduction." In *Cyberspace, Cyberbodies, Cyberpunk: Cultures of Technological Embodiment*, edited by M. Featherstone and R. Burrows, 1–20. Thousand Oaks, CA: Sage Publications. doi: http://dx.doi.org/10.4135/9781446250198.n1

Ferdani, D., S. Pescarin, and E. Pietroni. 2010. "A Procedural Approach to the Modeling of Urban Historical Contexts." In *Fusion of Cultures. Proceedings of the 38th Annual Conference on Computer Applications and Quantitative Methods in Archaeology*, Granada, Spain, April 2010, edited by Francisco Contreras, Mercedes Farjas, and Francisco Javier Melero. BAR International Series 2494, 63–70. Oxford: Archaeopress.

Ferrauto, T., D. Parisi, G. Di Stefano, and G. Baldassarre. 2013. "Different Genetic Algorithms and the Evolution of Specialisation: A Study with Groups of Simulated Neural Robots." *Artificial Life* 19(2): 221–253.

Ferrighi, A. 2013. "Visualizing Venice: A Series of Case Studies and a Museum on the Arsenale's Virtual History." In *Built City, Designed City, Virtual City: The Museum of the City*, edited by Donatella Calabi, 137–151. Rome: Centre for Studies of Rome (CROMA), Roma Tre University.

Fonseca, C. D., R. Demetrio, and G. Guadagno. 1999. *Matera*. Bari: Editori Laterza.

Forte, M., ed. 2008. *La Villa di Livia, un percorso di ricerca di archeologia virtuale*. Rome: Erma di Bretschneider.

Forte, M., and E. Pietroni. 2009. "3D Collaborative Environments in Archaeology: Experiencing the Reconstruction of the Past." *International Journal of Architectural Computing* 7(1): 57–76.

Gerosa, M. 2007. *Second Life*. Milano: Edizioni Meltemi.

Guidazzoli, A., L. Calori, F. Delli Ponti, T. Diamanti, S. Imboden, A. Mauri, A. Negri, G. Boetto Cohen, S. Pescarin, and M. C. Liguori. "Apa the Etruscan and 2700 Years of 3D Bologna History." In *Proceedings SA'11 of SIGGRAPH Asia 2011 Posters*, edited by Zhigeng Pan, Hong Kong, China, December 12–15, 2011, Article n.2. NY, USA: ACM New York. doi 10.1145/2073304.2073307

Laureano, P. 1993. *Giardini di pietra: i Sassi di Matera e la civiltà mediterranea*. Torin: Bollati Beringhieri.

London Charter. 2009. http://www.londoncharter.org.

Lopez-Menchero Bendicho, V. M. 2013. "International Guidelines for Virtual Archaeology: The Seville Principles." In *Good Practices in Archaeological Diagnostic, Natural Science in Archaeology*, edited by C. Corsi et al., 269–283.: Springer International, Charm, Switzerland.

Mellet-d'Huart, D. 2006. "A Model of (En)Action to Approach Embodiment: A Cornerstone for the Design of Virtual Environments for Learning." *Journal of Virtual Reality* 10 (3/4): 253–269.

Morello, E., and C. Ratti. 2009. "A Digital Image of the City: 3-D Isovists in Lynch's Urban Analysis." *Environment and Planning B: Planning and Design* 36: 837–853.

Mueller, P., G. Zeng, P. Wonka, and I. Van Gool. 2007. "Image-based Procedural Modeling of Facades." *ACM Transactions on Graphics* 26(. 3): 1–9.

Murteira, H., and Paulo Simões Rodrigues. 2016. "Lost Cities in the Digital Era: The Case of Pre-Earthquake Lisbon." In *Space and Time Visualisation*, edited by Maria Bostenaru Dan and Cerasella Craciun, 151–172. Cham: Springer. doi:10.1007/978-3-319-24942-1.

Oudenampsen, M. 2007. *Back to the Future of the Creative City: Amsterdam's Creative Redevelopment and the Art of Deception*, Mute Books.
Parish Y. I. H., and P. Muller 2001. "Procedural Modeling of Cities." In *SIGGRAPH'01 Proceedings of the 28th annual conference on computer graphics and interactive techniques*, 301–308. ACM New York, NY, USA.
Parisi, D., and M. Schlesinger. 2002. "Artificial Life and Piaget." *Cognitive Development* 17: 1301–1321.
Pescarin, S., L. Calori, C. Camporesi, M. Di Ioia, M. Forte, G. Galeazzi, S. Imboden, A. Moro, A. Palombini, V. Vassallo, and L. Vico. 2008. "Back to 2nd AD A VR On-line Experience with Virtual Rome Project." In *VAST'08 Proceedings of the 9th International conference on Virtual Reality, Archaeology and Cultural Heritage*, 109–116,edited by Michael Ashley, Sorin Hermon,Alberto Proença, Karina Rodriguez-Echavarria. Aire-la-Ville, Switzerland: Eurographics Association.
Pietroni, E. 2012. "An Augmented Experiences in Cultural Heritage Through Mobile Devices: 'Matera Tales of a City' Project." In *Proceedings VSMM 2012*, 117–124, edited by Gabrielle Guidi and Alonzo C. Addison Conference Chairs, Piscataway, NJ 08854IEEExplore Digital Library. doi: 10.1109/VSMM.2012.6365915.
Pietroni, E. 2013. "Natural Interaction in VR Environments for Cultural Heritage: The Virtual Reconstruction of the Regolini Galassi Tomb in Cerveteri." *Archeologia e Calcolatori* 24., pp. 231–247
Piga, B. 2008. "Attualità dei laboratori di simulazione urbana. La lezione dell'Environmental Simulation Laboratory di Berkeley a Times Square, New York." *Territorio* 45: , p. 63. Franco Angeli, Milano.
Porfyriou, H. 2010. "Spazio urbano come luogo. Camillo Sitte e il ruolo della visione nella Modernità." In *Spazio fisico- Spazio vissuto*, edited by M. Di Monte and M. Rotili, 261–278. Milan: Sensibilia.
Porfyriou, H., and M. Sepe. 2012. "Conservation of Urban Heritage and Monitoring Tourist Impact: An Integrated Approach." In *Measuring Heritage Conservation Performance*, edited by S. M. Zancheti and K. Similä, 155–164. Rome: ICCROM.
Varela, F., E. Thompson, and E. Rosch. 1991. *The Embodied Mind: Cognitive Science and Human Experience*. Cambridge, MA: MIT Press.

CHAPTER 6

# Simplified Crowd Simulation in Virtual Heritage Sites

LUÍS MIGUEL SEQUEIRA

Traditionally, reconstructions of heritage sites using computer graphics, especially within three-dimensional environments, have focused on replicating the architecture and artifacts of heritage sites (Styliani et al. 2009), creating "virtual museums" that can be "virtually visited" both by researchers and the public.* The digital capture of such heritage sites could furthermore be used by the movie industry—both for pure entertainment and as imagery for historical documentaries (Sequeira and Morgado 2013, Frischer et al. 2002)—but also by the games industry for the purposes of entertainment, as well as for education involving use of "serious games" (Anderson et al. 2010). Thus, one of the interesting aspects of virtual heritage projects is that the produced content has several possible applications and is not limited to pure research in archaeology and history.

However, such exhibits of heritage sites, be those on location (through *augmented reality* technologies; see Anderson et al. 2010, 15), in a museum, or recreated digitally, tend to show architecture and artifacts only—rather than the actual people who inhabited those spaces—mostly for the simple reason that physical exhibits require actors to be present, while digital re-creations simulate such actors. Therefore, for purely pragmatic and practical reasons, as well as for financial constraints, visitors usually view heritage exhibits that are "empty" of human presence. It is left to the movie industry, producing

* The author wishes to thank the Portuguese Science and Technology Foundation (FCT) for the PhD grant that allowed the development of this research project.

Luís Miguel Sequeira, *Simplified Crowd Simulation in Virtual Heritage Sites* In: *Digital Cities*. Edited by: Maurizio Forte and Helena Murteira, Oxford University Press (2020). © Oxford University Press. DOI: 10.1093/oso/9780190498900.003.0007 documentaries or entertainment films, as well as to the games industry, to "populate" such exhibits with actors—either filming on location, or recreating the sites in a studio, or recreating it digitally.

The three-dimensional re-creation of heritage sites ("virtual heritage sites") usually employs multidisciplinary teams: archaeologists to capture the relevant information from existing sites, historians to frame the background and context of the exhibit, and graphic artists to create the 3D models from existing data and documentation (pictures, engravings, textual descriptions, and so forth). Historically, in the 1990s, when advanced computer graphics became available on common desktop personal computers, the first projects were mostly a display of technical prowess, and archaeologists and historians were only asked to validate the created models at the end of the project—and sometimes their opinions were not even taken into account (Frischer et al. 2002). Such models might have looked photorealistic, but they lacked historical accuracy, especially in such cases where the original heritage site was seriously degraded, in ruins, or nonexistent (that is, its digital reconstruction was made based on historical documentation and/or pictures or engravings, not on actual location data).

With the beginning of the twenty-first century, the lack of a methodology to create historically accurate models of heritage sites prompted several groups of historians to propose new methods of planning and implementing such projects, such as the London Charter (Beacham et al. 2006) and the Seville Charter (Lopez-Menchero and Grande 2011).

These methodologies propose a different organization: research teams are led by archaeologists and/or historians, and their focus becomes that of achieving historical accuracy through the acquisition and documentation of relevant data to sustain the formulated hypothesis when constructing the 3D model. The actual model development, done by computer experts, is subsidiary to the entire project and not its primary goal; however, the technical team is expected to adapt the methodology to implement the models with rigor and accuracy (Gea et al. 2013). As a result, virtual archaeology projects are susceptible to being transparently evaluated by independent researchers as to their historical accuracy, since all relevant information to address the formulated hypothesis is captured by the process, as well as the issues and decisions made when implementing the 3D models.

## INTERACTIVITY AND ITS CHALLENGES

One hurdle that all virtual archaeology projects developed under the London or Seville Charters have to face is the communication problem between historians/archaeologists and computer experts on the team, since each group is focused on overcoming completely different obstacles. In traditional 3D modeling, designers create 3D objects and submit them to a rendering stage to create photorealistic imagery. Such rendering may take a long time (which naturally depends on the amount of data, the complexity of images to be created, the level of realism desired, and so forth). Effectively, by the time the historians or archaeologists leading the team are able to see the "final" images of the project, several months might have passed. If these images do not accurately represent what the historians had in mind—in other words, when the models fail to be validated by the historians and archaeologists—this requires a stepping back, changing the original models and running the rendering stage once more. As a consequence, virtual archaeology projects may take a long time to be developed iteratively—but methodologies like the London Charter actually advocate such an iterative approach to virtual archaeology projects.

Traditionally, the final 3D models were mostly used either for displaying static pictures (either in an exhibit or perhaps in an academic publication or presented at a conference), possibly made into an animation (which requires additional human resources, such as a director with a storyboard and digital video editing know-how) or an "interactive" 3D model, which could be set up in a museum (or even on a web page), allowing visitors to navigate around the virtual environment. The 3D models of the Rome Reborn project, for instance, were converted to displays inside the Google Earth website in 2008 (Wells et al. 2009), a process that took seven months of intensive development (working from existing models for ancient Rome, created by the team led by Frischer (Dylla et al. 2009)), in order to allow the public to navigate around the models.

Models that are to be viewed by the public also tend to be interactive; at the very least, they incorporate useful information about the items on display. But such models can allow more kinds of interaction: doors that may open; lights that can be turned on and show the effect of using old methods of illuminating interiors (Frischer 2004, 70; Kennedy et al. 2012); and, of course, offer changes to engage in conversation with virtual agents, or non-player characters (NPC), as they are known in the gaming industry (Bogdanovych et al. 2010).

Such interactivity requires an additional set of skills from the research team— namely virtual agent programming. While many graphic designers and 3D modelers might be acquainted with some programming languages appropriate for adding simple interactive elements, the realm of automating behavior of virtual agents is a different field of research (Bogdanovych et al. 2009, 2012; Maïm et al. 2007; Trescak et al. 2014; Bogdanovych et al. 2010; Ijaz et al. 2011; Slater and Burden 2009; Forte 2009).

Therefore, as a research project, the new challenge that the team faces is how to add interactive elements—especially interaction with simulated crowds of virtual agents—in order to enhance the user's immersion in the virtual environment without, however, sacrificing historical accuracy and the ability to systematically evaluate the results of the choices made when programming such interactivity. The usual approach is to separate the teams further: one team of graphic designers will model not only the architecture and eventually the virtual representation of simulated humans (also known as *avatars*), while a different team, acquainted with artificial intelligence—more specifically crowd simulation and intelligent agents—will provide the programming. Both teams are coordinated, as per the London or Seville Charters, by the historians and/or archaeologists leading the project.

Evaluating the actual interactive elements, especially in cases of human interaction with virtual agents, has its own pitfalls. Morgan advises researchers to resist the temptation of using virtual agents solely for the purpose of adding "information" (which can be done more effectively with pop-up boxes on the display area or audio recordings) or, worse, as "decoration" (treating historical characters as props in the virtual environment) (Morgan 2009). Such ill-advised attempts to incorporate virtual agents in the project will reduce its credibility as a research and educational tool.

Therefore, traditionally, such research projects were usually done in closed environments, either developed from scratch or by using existing commercial tools, for which the required skills are usually available (from the movie/gaming industry, for example). Programming interactivity adds another layer of complexity to the project, not only in terms of additional human resources with different skills (for the programming) but also requiring separate testing, documenting, and evaluation.

Extending such projects is also not very easy. Once the project is completed, the necessary resources for maintaining the project may not be available; funding might not be enough to keep the project alive for a long period of time; and new technology or new paradigms might not be easy to incorporate in such projects later on. Thus, the projects will be subjected to *bit rot* (Cerf 2011), or the inability for contemporary technology to run applications developed a decade or two prior. Such considerations about the future should be part of any virtual archaeology project (see principle 6 of the Seville Charter [Lopez-Menchero and Grande 2011]).

## VIRTUAL WORLDS

By the turn of the twenty-first century, programmers, game designers, and researchers were developing what was termed *virtual worlds*—persistent computer-generated environments where participants experience others as being there with them (Schroeder 2008). In particular, researchers in many different areas started studying how to best employ virtual-world technologies as aids in their fields of research. One reason for using virtual-world technologies is that they allow real-time communication with others.

Morgado et al. (2010) explain the key difference between "simulations" and "virtual worlds." While visually both might look similar and require equivalent skill sets to develop and maintain, simulations can address only a limited set of possible explorations of the virtual space (as well as have limited interaction with NPCs, since they are only programmed to work within the constraints set by the simulation). And the programming of the simulation requires a certain overhead until a final result is produced: changes to that result, as mentioned, will demand a new project. Virtual worlds, by contrast, allow human participants to interact directly within the environment. As a teaching tool, virtual worlds allow teachers to answer questions from the students while they are simultaneously exploring the environment; similarly, virtual museum tour guides can be explored by many different visitors at the same time, and can be guided by the researchers (or staff trained for this purpose) to adapt the tour according to the wishes of the visitors.

Virtual worlds, therefore, present a distinct advantage in terms of reusing content: the same environment can be used for tours, teaching classes, even performances or re-creation of certain events using the existing content as background, and so forth; such uses are only limited by the imaginations of those researchers willing to work within the context of virtual worlds.

Morgado et al. (2010) give many examples of such uses of virtual worlds in several different areas of research. Some projects specifically address heritage and archaeology. One such example is the Theatron project (Childs 2008), created by the Centre for Computing in the Humanities (CCH) at King's College, London; the team for this reconstructed several historical European theatres, from ancient Greece to the twentieth century. While such reconstructions could be freely visited and could include interactive information about each theatre, the project's main goal was to allow students and researchers to use the existing content in order to enact drama performances and learn about scenography. Such projects were actually done by different universities collaboratively; once the content is developed in a virtual world, it can be used by anyone (with the proper permission set), and even further extended and developed by different teams.

Among the several virtual worlds that have been employed to develop virtual heritage sites, one popular choice has been Second Life—a commercial product developed by Linden Lab and launched in 2003. It is a mature technology that allows user-generated content in a visually contiguous landscape. Access to Second Life is free, but deploying the content requires paying a monthly fee to Linden Lab: the more "virtual land" you own, the more objects you can place in it for others to see and explore, and the more you pay.

Second Life, now entering its seventeenth year of operation, still boasts about a million regular users. Therefore, any project that plans to have visitors or to be used for training purposes can benefit from a considerably large audience—even though, for technical reasons, not more than a hundred visitors can be in the same spot simultaneously. Second Life users are also a good pool of human resources with skills in 3D modeling and interactive programming.

An alternative to Second Life is the free and open-source virtual-world OpenSimulator. It has been in continuous development since 2007, and it uses the same viewer application as Second Life—its communication protocol has been reverse-engineered and independently developed by a team of volunteers, which at some point in time included programmers by Intel, IBM, and even Microsoft. From the perspective of the end user, OpenSimulator is not visually different from Second Life—the same viewer application can display content from either platform in exactly the same way. In terms of actual features, OpenSimulator emulates over 90% of all the functionality of Second Life, and adds some functionality of its own. The software can run on any server (or even on a laptop) running Linux, Windows, or Mac OS X; the only costs of operation are server maintenance and bandwidth, no matter how large the virtual world is, thereby making OpenSimulator a perfect solution for low-budget projects. And, of course, the content created and developed in OpenSimulator can be relatively easily ported to Second Life and vice versa.

Besides Second Life and OpenSimulator, there have been two major developments in this field. One is Linden Lab's next platform for virtual worlds, code-named Project Sansar (Linden Lab 2015). The other one has been in development for a few years, by Linden Lab's former owner and CEO, Philip Rosedale, who created a new company to develop its own virtual world, High Fidelity, which at the time of this writing is still in its early stages of development. Many virtual worlds for specific projects, however, are often created from scratch using one of the popular virtual-world development platforms, like Unity 3D (see, for example, Forte 2014 or Trescak et al. 2014). With Facebook's acquisition of the company Oculus, which has been developing a new haptic device allowing personal computers to experience immersion via a (relatively low-cost) virtual reality headset, the media have been speculating about a forthcoming virtual world launched by Facebook's founder Mark Zuckerberg (Poladian 2016), which, however, has not released further information so far. Therefore, there are expectations of a renewal of mainstream interest in virtual reality technology. Such a platform has not been released yet; on the other hand, Microsoft's HoloLens technology (a direct competitor to Oculus) might also encourage developers to create "mixed reality" environments, which also use virtual-world technologies, as seems to be the case with a few selected companies and organizations, at least according to Microsoft's own blog (Bardeen 2017).

## THE VIRTUAL LABORATORY OF ARCHAEOLOGY

Virtual worlds render content in real time—that is, there is no "rendering stage" requiring days or weeks to produce a final image. Once the content is "dropped" into the virtual world (or, rather, uploaded into it), it becomes immediately visible to all participants. In virtual worlds with user-generated content, objects can usually be manipulated directly by the human participants. Second Life and OpenSimulator even incorporate a set of tools in their viewer applications allowing users to build content inside the virtual world. Such activity can be performed collaboratively; and as the users' avatars deploy shapes and drag them across the virtual landscape, others will "see" the actual content building taking place before their very eyes.

This allows the teams developing a virtual-archaeology project to adopt a new methodology of work. Instead of completely separating the research (produced by historians and archaeologists) and the content design (created by the 3D modelers and interactivity programmers), both teams can meet inside the virtual environment and evaluate the progress of the development. Even more than that, historians can effectively manipulate the created objects themselves, or annotate them and mark them for correction and so forth, during in-world[1] meetings. This blurs the border between the teams; historians also become, in a sense, participants in the creation and design process, thanks to a relatively user-friendly interface that allows simple manipulation of objects without requiring many computer skills.

This particular characteristic of virtual worlds with user-generated content creates what Murteira and Rodrigues (2016, 151) call the *virtual laboratory of archaeology*: "an interactive and immersive model and in a collaborative real-time environment, so as to give the project an innovative *laboratory* dimension."

How is the laboratory used in practice? Besides being a "virtual conference room," its most important ability is to allow historians to validate documental sources. That is, the recreated 3D models can be compared to existing paintings, engravings, or pictures (if existing) of the actual historical site, or even with textual descriptions (when no images persist); additionally, hypotheses formulated by the historians regarding the layout of architectural elements (for example) can be visually checked—and validated or not. These abilities are, of course, even more important on those historical sites that do not exist any longer: Murteira and Rodrigues (2016) refer as an example to the Baroque city of Lisbon, which was destroyed in 1755 and was completely rebuilt; only maps, drawings, paintings, engravings, and textual description exist to show the previous architecture. A similar case can be made for the Rome Reborn project: even though some monuments of ancient Rome are still extant today, most of the city was rebuilt over the centuries; documentation from the period, listing descriptions of the buildings—termed "regionary catalogues"—have lasted to our days (Dylla et al. 2009, 62).

In those cases where the original site does not exist any longer, historians typically attempt a "reconstruction" (be it textually or with diagrams/maps) based on certain sources that are assumed to be authoritative for the period, while discarding the remaining evidence. Such a process is aided by extensive research in the field, correlating information from reputable sources. But it's not always easy to choose the "best" sources—especially when new sources are unexpectedly made public (a new painting of the historical site that was discovered in a private collection; a set of maps or of blueprints previously unknown; a collection of letters from the period which describe certain buildings or streets, and so on). In the case of the engravings of the city of Lisbon made during the earlier part of the eighteenth century, it was not uncommon that the artists for many of those were commissioned without ever having visited the city, and were merely given textual descriptions of it; in other cases, the artist might actually have visited the city, but drew it from memory, without any previous sketching at the location. Therefore, many such paintings and engravings, typical of the period, depict the city from "impossible" viewpoints (i.e., from the middle of the river, or from far above the sea level, as if the painter were flying over the city!). Even if the painter or engraver had been on location, their viewpoints often distort the perspective or fail to show certain important details of the buildings. And of course there is also artistic licensing; not all artists were interested in "realistically" painting a panorama view of the city, as they did not feel certain details were necessary for conveying the "feel" of the city.

Through the virtual laboratory of archaeology, it is possible to evaluate such sources by modeling them in 3D and positioning one's avatar at the viewpoint presented in certain engravings or paintings—and then compare what the model shows to what the artists have depicted. Similarly, a textual description of a certain building, describing in detail what the rooms looked like, and the impression of size/dimension given by the author of a letter, can be easily simulated: the modelers can create such buildings based on hypothetical formulations on how they ought to look (based on similar, still-existing buildings designed by the same architect). Then the historian makes a "walk-through" using his or her avatar, and contrasts the building with the textual description. In the Lisbon 1755 pre-earthquake project, many sources have thus been validated, while others have been discarded, based on a "walk through" and/or by subtly manipulating the buildings, dragging them around the virtual landscape until they "match" certain engravings, maps, or paintings. Sometimes certain approaches had to be abandoned (rejecting a particularly detailed painting because it simply does not depict the architecture accurately); sometimes, interpolation was necessary, as in the case of precise blueprints which, however, use unknown measurements (but by assuming certain "standard" sizes—like the width of doors or windows, comparing that blueprint with others from still-existing buildings of the same architect).

It is important to note that in this kind of virtual laboratory, done inside a virtual world, it is the historians themselves who can do the checking of sources and the validation (sometimes on their own and sometimes helped by architects). This not only speeds up the process of project development—historians will not need to wait for the technical teams to deliver all content, and instead can evaluate it as soon as each object gets produced and deployed in the virtual world—but also keeps historians (and archaeologists) in control of the overall project, as per the recommendations of the London and Seville Charters. Historians become the first "beta testers" of the developed environment—including testing the eventual interactive elements—and can provide useful input from the user perspective well before the intended audience visits the project.

## UPGRADING THE SIMULATION: CROWDS

Early virtual-archaeology projects were mainly focused on capturing data from existing archaeological sites and modeling the data in 3D with accuracy; or in the case that the original artifacts did not exist any longer, they suggested virtual re-creations of those artifacts based on documental evidence describing them. In the case of architectural objects, or even sections of cities (or whole cities!), the virtual-archaeology project was deemed "complete" once the buildings were accurately modeled and validated. In the Google version of the Rome Reborn project, the experience of visiting ancient Rome, navigating it using the interface provided by Google Earth, is a solitary affair: it is like walking through a museum exhibit that is empty of people.

Because virtual worlds allow users to have a virtual presence—or avatar—in the simulated environment, this adds a new level of realism to a virtual-archaeology project: not only can the buildings and the artifacts be observed but visitors can also interact with each other (or even with the architecture—that is, by opening doors or clicking on information panels to learn more about the historical site). Project developers can also give "virtual tours" of the historical site by entering with their own avatars while visitors are traveling through the space, and promptly answer any questions or point out important features.

Events inside the recreated site can also be held by hiring human actors and asking them to perform a staged event according to a script. Projects like Theatron (Childs 2008) allows stage performances to be done inside a re-creation of the actual theatre where the play was originally performed. Such performances are not much different from similar performances held on physical historical sites, often even by historical reenactment societies, university students, or museum staff. In fact, some organizations hold mixed events, where human actors are superimposed on virtual models, visited by users connecting to the virtual world where the models are deployed (Kennedy et al. 2012); such events were found to increase the awareness of visitors not only to the architecture of the past but also to how such spaces were used.

The main problem with employing human actors is that visitors in a virtual world can log in from any part of the world, at any hour. Although virtual worlds like Second Life and OpenSimulator offer a plethora of events for regular users, held at specific times, there is often a need to attend to the "casual" visitor, who might not be willing to wait until the next staged performance and would prefer to visit on his or her own. Ideally, virtual-archaeology projects ought to be attended by staff around the clock, 24 hours per day and 7 days per week; in practice, however, few projects are able to sustain the costs of such a substantial staff—all of whom require training in the specific context of the epoch depicted in the virtual world.

An alternative, of course, is to simulate the crowds of the past by using *virtual agents*, "autonomous computer programs that are represented by human-like 3-dimensional figures (called avatars) that move around the reconstructed environment and simulate ancient citizens of the reconstructed site" (Trescak et al. 2014). Virtual agents rely on a well-established field of research around the autonomous agents, crossing robotics with artificial intelligence; in effect, the developments that make programming autonomous robots possible can be successfully employed for simulating humans in a virtual environment.

Modern computer games, of course, rely on virtual agents to model the interactions between the player and a vast array of other characters (as mentioned earlier, known as non-playing characters, or NPCs) to distinguish them from the "playing characters," or the humans who play the game). However, as Trescak et al. (2014) warn us, the development of such realistically behaving NPCs is expensive—in general, it's far more expensive than modeling the buildings in the virtual environment and would easily deplete the budget of an academic project. Fortunately, there are many alternatives to convey the same effect for a fraction of the cost; nevertheless, one should be aware of the trade-offs, remembering the critical advice from Morgan (2009) that such virtual agents ought not to be employed as "decorative elements," adding little to the project and even in certain cases discrediting its seriousness.

A popular approach (also evaluated by Trescak et al. 2014) is to employ crowd-simulation techniques. Crowd simulation is a parallel development in the area of autonomous agents, where the purpose is to model a large number of agents, programming them with simple rules. It is an evolution of the original flock-generating algorithms developed by C. W. Reynolds (1987) and is extensively used by the movie industry to provide backgrounds with large groups of simulated people, as opposed to casting them with thousands of human actors. The area of crowd simulation continues to be developed (Liu and Passino 2000) and has spun off into new areas of research, where the original assumptions—very simple rules to create flocks of "artificial animals" (or *boids*, as Reynolds called them) that can converge toward a goal in what is perceived as "realistic," natural movement—are applied to finding solutions of complex mathematical problems. There are, therefore, several approaches to solving such problems, as well as many methods to evaluate the algorithms according to their performance and accuracy.

Virtual-archaeology projects employing crowds of virtual agents populating the reconstruction of the historical site can at least convey to visitors an impression that the space is "inhabited" (Maïm et al. 2007). Crowd simulation exhibits interesting visual possibilities, since the simulated crowd appears to have a "purpose" (convergence toward a goal) and moves in a realistic way, avoiding each other (even avoiding the avatars of human visitors), as happens with a "real" crowd. In certain cases, simulated crowds can do more than merely walk from one point to another; they can also stop to pick certain objects and bring them to the goal. With a little creativity and some resources, such crowds can even feature different animations (Foni et al. 2003) for different kinds of characters in the crowd (different crowds can be deployed in the same environment, since the crowd-simulation algorithms automatically avoid obstacles, including virtual agents belonging to a different crowd), therefore conveying a broader sense of realism.

Mixed approaches certainly exist as well, where "crowds" are in essence virtual agents with a certain degree of independence and which can learn not only from the environment but also from interactions with humans (Bogdanovych et al. 2010). Such projects, from the perspective of historians, allow them for further experimentation, questioning if certain aspects of the heritage site did have any influence on the humans inhabiting the space. For example, a typical question could be: "Was this street wide enough for the traffic in the area?" This could be answered by populating the virtual environment with virtual agents having a certain degree of autonomous behavior and learning abilities, and see what happens—will the virtual agents, when going from A to B, passing through that specific street, get too crowded so that traffic will be effectively jammed, or will such agents use alternative routes to reach their goal?

Therefore, the development of virtual agents goes beyond a purely "realistic" effect—populating a historical site for added immersion; it becomes, by itself, a further object of study in the context of the virtual-archaeology project. In a similar way to how the reconstructed architecture was validated by looking at images or descriptions of the historical site, historians can now evaluate the crowds depicted in certain images by tuning certain parameters in the programming of the virtual agents and visually observing the resulting crowd simulation, contrasting it with the images of the period. Furthermore, whereas images, paintings, and engravings can show only a static experience of the past—a snapshot—crowd simulation with virtual agents conveys the dynamic nature of this very past.

The tradeoffs in this case are many, of course. The more sophisticated the algorithms are, the more time they require to be programmed, and that implies a higher cost for the overall project; at the same time, in general, the more sophisticated the development is, the less likely historians are able to interact with the crowd simulation by themselves and will require talking with the human programmers each time a parameter has to be changed—which implies more time until the overall project can be validated as to its accuracy. Some projects include simplified programming tools to allow historians, with proper training, to configure the simulation by themselves (usually, such configurations are done outside the scope of the virtual-world environment where they run); but in general, such programming will always require some know-how. And of course the more time is spent in developing a user-friendly interface for the crowd simulation, so that it becomes simpler to configure it, the costlier the project will be.

## A PROPOSED MODEL FOR SIMPLIFIED CROWD SIMULATION

Computer-game designers also have to deal with the interface problem: How can the complexity of a game increase dramatically (for instance, by deploying more and more sophisticated artificial intelligence in the simulated environment), while at the same time keeping the computer game easy to use? These days, in fact, most computer games do not even include a manual (as was common in the 1980s); instead, designers aim to develop the simplest, most intuitive interface possible—one that is immediately obvious as to how it works, even if it actually drives a complex system, so that it can appeal to a vast number of users, including those with little or no technical expertise.

In the case of simulation games (those that model aspects of reality; for a more complete definition and discussion, see Galvao 2000), certain activities, like configuring parameters of the simulation, are described by abstractions. Consider the scenario of a strategy game like *Command & Conquer*, which simulates a battlefield at the tactical level. Players fight against the artificial intelligence commanding the enemy's forces, and tactical commands to move, attack, or retreat (as an example) are given by merely clicking on representations of the forces and their intended destination. The game does not actually simulate the complexities of transmitting orders to several units in a modern battlefield; instead, it relies on the abstraction of clicking on the units displayed on the screen and clicking on the destination to simplify the actual configuration of the simulation. In other words, by changing the position of one unit in the battlefield, the player has actually changed the simulation parameters, which now have to take into account how the new position affects the overall battle (and react accordingly). One might argue that this is a form of visual programming—changing the parameters of a complex algorithm via simple visual tools that are highly abstract—even though, from the perspective of the player in such a game, he or she does not experience the interaction as "programming."

Such abstractions have become more popular in the past two decades, during which computer performance and video graphics have constantly increased to the point that a large number of the most popular games in the industry include certain simulations driven by AI routines with which the player interacts using abstractions. Several ideas and concepts about how to develop such interfaces to make them more useful have been explored by active research in the field, as well as, naturally, from the game-development companies (see a review of past research done in this area of usability and human-computer interaction in games by Jørgensen 2004).

It seems reasonable to assume therefore that a good strategy is to borrow concepts from computer games in order to let users drive complex simulations. The model presented here assumes that historians are mostly interested in simulating movement across a city—establishing where people would gather to work, to eat, to rest, or to join (or decide not to join) a specific event (church mass that has just started, a religious procession on the street, etc.). This leads to devising a means of representation of the simulation parameters by using concepts not unlike what has been used in popular social-simulation games such as *The Sims* by Electronic Arts, created by Will Wright.

Those games represent AI-driven autonomous agents (NPCs) by a set of skills or abilities, such as the level of energy/vitality they have and requirements they need to fulfill (as, for instance, needing to go to work or to sleep). By placing certain objects in the environment of the game, the player can subtly control what the NPCs will do. However, they are not "mindless robots"; in those types of games, the NPCs will perform a series of actions, according to their skills, abilities, or characteristics, independent of the human player. Here's a typical example: The NPC might require a period of rest. If there is no place to sleep, it will sleep on the floor, which will not restore its "life energy" to maximum; the player, however, can buy a bed and place it for the NPC to sleep—it will detect the presence of the bed and prefer to sleep there instead of doing it on the floor. By simple clicking and dragging-and-dropping, the player can therefore affect the simulation, or even significantly alter its course—without needing to learn a programming language to change the NPC's behavior.

To capture the essence of the required simulation parameters for the historians, a model was developed whereby each virtual agent has an *energy level*. Moving around the virtual environment will make the agent lose energy—so, at some point, the agent has to find an energy source. This model is also used in robotics, where physical robots require locating a source of power to replenish their batteries (Mondada and Floreano 1995): genetic algorithms (one approach of artificial intelligence) are used to determine the "best" path, given the current variables of the simulation (obstacles, speed of the agent, current energy level and expected reduction of energy required for moving toward the energy source, etc.).

The next concept used is *money as motivation*. Agents will require working a certain amount of time in order to fulfill a daily quota that allows them to survive. This means that agents have to evaluate these two factors: working more will earn them more money, but it will also deplete their energy levels; working farther away from "home" might raise the chances of a better job, but it also means expending more energy in the commute.

Finally, working is hard and boring, and to simulate that, agents also have a *value for happiness*. If it decreases below a tolerable level, agents will leave their workplace in search of entertainment. But, of course, moving toward the entertainment area also requires expending energy. Therefore, the genetic algorithm that deals with path planning requires balancing all these variables, depending on the current levels of those three attributes.

How do historians interact with—and modify—the simulation? Essentially, "energy sources," "money sources," and "entertainment/happiness sources" are represented abstractly by colored cubes, which the historian, while logged in to the virtual environment, drags and drops on the ground (figure 6.1). The historian can set different levels for those sources—for instance, putting very high energy replenishment at the "home" location—and this is also accomplished easily by dragging the mouse over a slider on one of the faces of the cube (the cubes can also be made invisible and do not count as "obstacles" for the purpose of the simulation).

By strategically placing such cubes in key locations in the virtual environment, historians can therefore manipulate the simulation, changing it as appropriate—in real time. For example, they can suddenly drop a huge source of "happiness" to simulate a special event in a location, and then they can see how the population will react. Some will probably continue to work; some might wish to attend the event but have not enough energy to move to its location; and others still might feel attracted to the event and start moving toward it.

Historians might also be able to simulate panic during historical catastrophes—fires, earthquakes, revolution—simply by removing all cubes of the simulation and dropping them all at the end of an escape route. They can furthermore increase the number of obstacles in the way by dropping appropriate "props" in the environment (fallen debris, carts, and so forth). As the energy levels—as well as the happiness levels—of the agents start to diminish, the genetic algorithm will recalculate the best path (the "route of escape") toward the place where all sources of energy and happiness are located.

Variations on the theme, which might be further explored, include moving sources—for example, a priest conferring blessings might simply be an avatar of the historian, picking up a happiness cube and walking around the

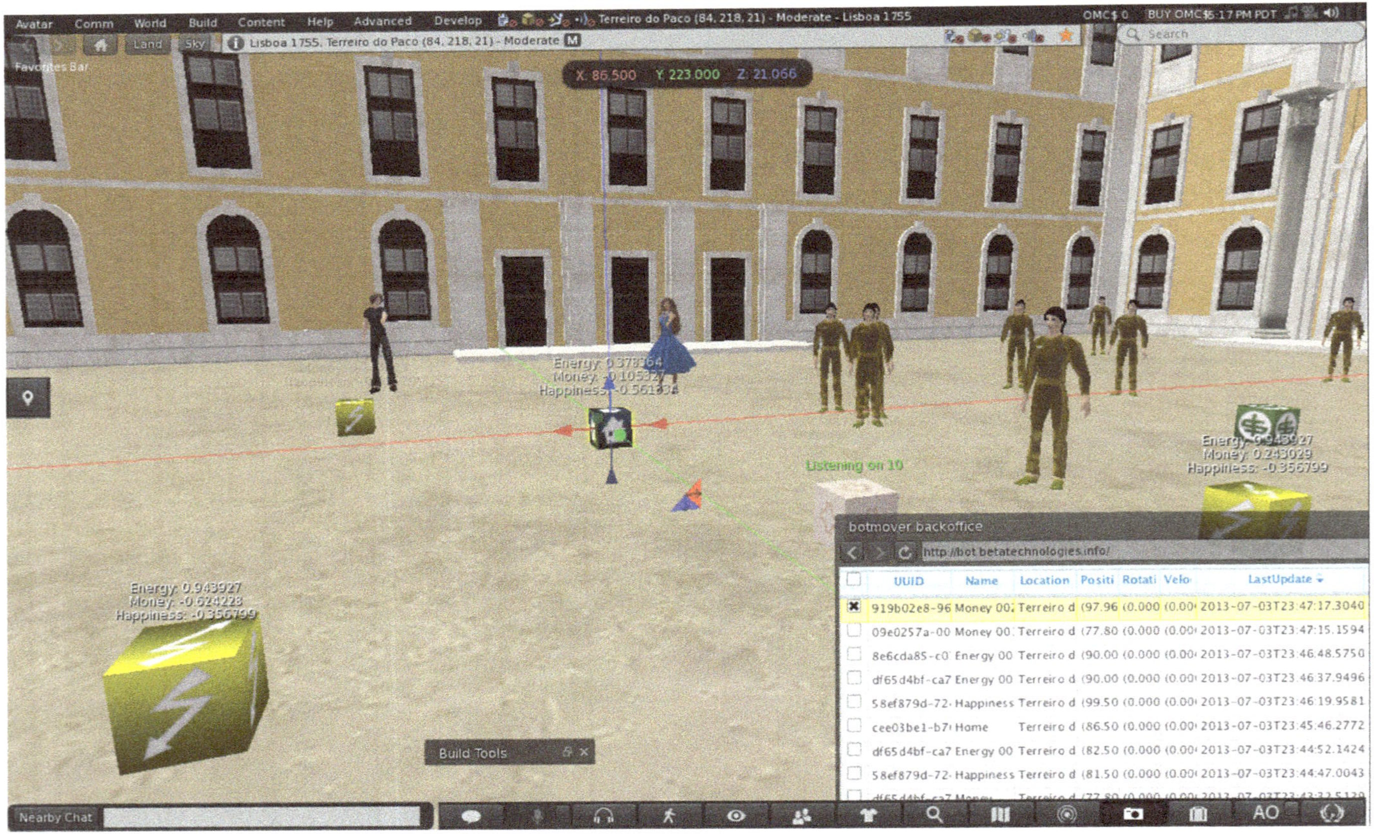


**Figure 6.1** Staged preview of two historians (left and center avatars) setting up cubes in the environment. NPCs (right, in golden uniforms) wait for the simulation to start. A web page, inside the 3D viewer application, shows positioning data for the deployed cubes.

*Credit:* Luís Miguel Sequeira.

simulation, thus attracting nearby agents which require a "happiness boost," and this will simulate a procession led by the historian.

By allowing historians to creatively affect the simulation, merely shuffling cubes around the virtual environment and using the simple sliders to set abstract levels of "energy," "money," and "happiness," they will be able to experiment with crowds in a re-creation of a historical site and see which configuration best represents a realistic, or historically accurate, representation of the crowd behavior.

For the Lisbon 1755 project, the requirements specified an additional level of simulation: the rules of interaction between different social classes. Typically, peasants would avoid noblemen or clerics when walking around the city; by contrast, noblemen might expect peasants to move out of their way and not even change their paths—collisions might have produced an angry response (including violence).

To simulate this aspect, the simulation includes an *interaction matrix*, which is basically a table listing the types of interactions between different agent "classes" (priest, nobleman, merchant, foreigner, soldier, peasant, beggar, and so forth) that are simple friend/foe/neutral relationships. "Friend" means mostly that the simulation will not avoid a collision between agents of those classes, while "foe" means that the collision should be avoided (in other words, the simulation will calculate the path toward a goal by treating a "foe" as an obstacle). Because agents dynamically recalculate their paths, and everything happens in real time, it is conceivable that sometimes (just like in reality), a "foe" cannot be avoided; in that case, there will be a collision between agents and an interaction between them (which can be shown using a special animation, like, for example, showing a nobleman hitting a peasant for crossing his path).

Currently, the different classes (or profiles) are fixed, as well as the animations for the interactions, but historians have access to a web page where they can define their own relationships (again, in real time) and see how they affect the simulation; in the near future, the relationship between two classes might become an analogue value instead, ranging from –1.0 ("hate/contempt") to +1.0 ("love/awe"), and the web-based interface might act as a template: each virtual agent will get initial values for its relationship with other classes, and those values might be affected by the simulation itself as the agent encounters other agents from different classes and adjusts the values depending on how successful the interaction had been with previous elements from that class. This would allow simulating situations where certain agents of the peasant class might never have been "hit" by a nobleman, and therefore might feel predisposed to engage in conversation with them—which, in turn, might create a violent response from the nobleman, affecting both agents' view of the interaction between their respective classes.

Such complex interactions, however, are left for a future stage of development; note, though, that the interface permitting all of the above does not require any fundamental change—it remains a simple web-page listing a table for all possible interactions between all classes. Future development might even include the possibility of adding additional classes; the current restriction to a limited set is mostly because each class requires its own avatar, with a selection of clothing, with specific animations for the interactions, and so on. Producing all that content is, at the moment, not easy for historians; therefore, each additional class still requires some content development by the technical team before being added to the overall simulation.

A more thorough discussion on possible future developments, as well as a more detailed technical implementation of the crowd simulation solution presented here, is found on Sequeira et al. (2014).

## CONCLUSIONS AND FURTHER WORK

Virtual representations of archaeological sites in 3D computer-generated environments, especially those that allow collaboration and user-generated content (termed *social virtual worlds*), allow historians and archaeologists to formulate hypotheses and test them themselves by going inside the virtual environment. Until recently, such laboratories of virtual archaeology were mostly restricted to architecture and artifacts. However, those historical sites have been populated by humans. Simulating crowds inside a virtual environment, although technically possible and an area of active research with quite mature solutions and technology, is not always readily accessible by historians and/or archaeologists, who usually do not have the required programming skills to develop complex AI solutions for crowd simulation.

This chapter proposes a model—currently in active development—that borrows ideas from popular simulation computer games, where players also have to interact with sophisticated AIs driving the NPCs in the game. The computer gaming industry, desiring to reach a vast audience in order to get the necessary return on investment (since such high-end games have production budgets similar to the movie industry), therefore had to develop solutions to allow simple manipulation and interaction with the game simulation.

The model described here is being researched using the design science paradigm (Hevner et al. 2004), which is appropriate for developing a technical solution for a problem, by creating several prototypes iteratively, analyzing the results at the end of each iteration, and applying the acquired knowledge to produce further iterations, until the developed prototype conforms to the specifications—in this case, until validation is provided by a group of historians researching the city of Lisbon before the 1755 earthquake. Ultimately, historians without technical intervention should be able to configure, manipulate, and interact with the crowd simulation, allowing them to explore themselves the hypothesis formulated regarding the human usage of the historical site.

The proposed solution, however, might not be appropriate for large-scale simulations (for instance, simulating a crowd of 50,000 individual spectators in the Roman Colosseum); this is the case for several reasons, starting with the technical limitation of Second Life and OpenSimulator as restricted to a maximum of 100 avatars per area[2]. Configuring a large-scale simulation would also be a tedious process, requiring many colored cubes all over the virtual environment, which might be hard to track. There are also performance criteria to take into account, both on the side of the simulation routines and in communication with the virtual world to give the NPC avatars commands for movement and animation; the practical consequences of a large-scale simulation have not been tested.

Further research might add the possibility of linking the source cubes together, thereby allowing scenarios like the representation of a church, which gives NPCs some happiness if they are located there, but the happiness levels would increase further if a priest is also present (the priest, for this purpose, would be attracted to a "money" cube, which only allows priests to be attracted).

Because the simulation interface is based on certain classes of computer games, it is also conceivable that the simulation could effectively be turned into a real, educational game by which participants learn about a historical site and its population while attempting to compete with other players, each controlling a set of NPC avatars by placing cubes in the environment to modify the simulation and scoring points if they achieve certain levels of satisfaction, money, and so on.

For this particular project, there is no way for potential visitors to interact with the NPC avatars using text chat, as Bogdanovych et al. (2012) have done for the city of Uruk reconstruction. Such a module could conceivably be added and linked to the main simulation at the point where characters collide (or come very near to each other) in order to interact. Thus, humans might simply be a "special" class that will stop the NPC avatar in its tracks when it encounters a human, and that triggers a new kind of interaction: processing natural-language requests, allowing visitors to know not only general aspects of the historical site (e.g., answering questions like "What is that building behind you?") but also the internal status of the NPC avatars. Imagine asking "Where are you going now?" and having the NPC answer: "I'm going home because I have earned enough money for today, but now I'm tired," corresponding to the current goal assigned by the simulation based on the current levels of "energy," "money" and "happiness."

As most of the actual simulation aspects run off-world on a separate server (currently developed in the Go programming language), it ought to be possible to configure future versions to run in other environments as well, if and when those become available, by changing just the interface with the virtual environment. In effect, most of the code specifically developed for Second Life and OpenSimulator is small, consisting of routines that make external calls to web services and that reply to similar calls from the server where the main simulation core is running. Any virtual environment which allows objects (and avatars) to be remotely commanded by fast web calls ought, within limits, to be supported with a relatively small number of changes; in effect, the development of the simulation itself—creating extra features, dealing with more virtual agents, optimizing the algorithms, adding further modules and configuration settings, and so forth—can be entirely separate from the actual interface to the virtual world itself.

## REFERENCES


Anderson, Eike Falk, Leigh McLoughlin, Fotis Liarokapis, Christopher Peters, Panagiotis Petridis, and Sara Freitas. 2010. "Developing Serious Games for Cultural Heritage: A State-of-the-Art Review." *Virtual Reality* 14: 255–275. doi:10.1007/s10055-010-0177-3.

Bardeen, Lorraine. 2017. "Mixed Reality Momentum Continues in the Modern Workplace and Microsoft HoloLens Expands to 29 New Markets." *Windows Blogs*. https://blogs.windows.com/devices/2017/11/01/mixed-reality-momentum-continues-modern-workplace-microsoft-hololens-expands-29-new-markets/.

Beacham, Richard, Hugh Denard, and Franco Niccolucci. 2006. "An Introduction to the London Charter." *The E-Volution of ICTechnology in Cultural Heritage, CIPA/VAST/EG/EuroMed*, 263–269. Kings Visualisation Lab. http://www.londoncharter.org/.

Bogdanovych, Anton, Kiran Ijaz, and Simeon Simoff. 2012. "The City of Uruk: Teaching Ancient History in a Virtual World." In *Intelligent Virtual Agents—12th International Conference*, edited by Yukiko Nakano, Michael Neff, Ana Paiva, and Marilyn Walker, 28–35. Santa Cruz, CA, and Berlin: Springer. doi:10.1007/978-3-642-33197-8-3.

Bogdanovych, A., L. Papaleo, M. Ancona, V. Mascardi, G. Quercini, S. Simoff, A. Cohen, and A. Traverso. 2009. "Integrating Agents and Virtual Institutions for Sharing Cultural Heritage on the Web." *Workshop on Intelligent Cultural Heritage*, Reggio Emilia, Italy, pp. 1–10. www-staff.it.uts.edu.au/~anton/Publications/Heritage_Workshop_09.pdf.

Bogdanovych, A., J. A. Rodriguez-Aguilar, S. Simoff, and A. Cohen. 2010. "Authentic Interactive Reenactment of Cultural Heritage with 3D Virtual Worlds and Artificial Intelligence." *Proceedings of the 9th International Conference on Intelligent Virtual Agents* 24(6): 617–647. doi:10.1080/08839514.2010.492172.

Cerf, Vinton G. 2011. "Avoiding 'Bit Rot': Long-Term Preservation of Digital Information [Point of View]." *Proceedings of the IEEE* 99(6): 915–916. doi:10.1109/JPROC.2011.2124190.

Childs, Mark. 2008. "Using a Mediated Environments Reference Model to Evaluate Learners' Experiences of Second Life." *Sixth International Conference on Networked Learning*, Halkidiki, Greece, pp. 38–45. www.nlc.ell.aau.dk/past/nlc2008/abstracts/PDFs/Childs_38-45.pdf.

Dylla, Kimberly, Bernard Frischer, Pascal Mueller, Andreas Ulmer, and Simon Haegler. 2009. "Rome Reborn 2.0: A Case Study of Virtual City Reconstruction Using

Procedural Modeling Techniques." *Making History Interactive: Proceedings of the 37th CAA Conference*, pp. 62–66. www.romereborn.virginia.edu/rome_reborn_2_documents/papers/Dylla2_Frischer_Rome_Reborn.pdf.

Foni, A., G. Papagiannakisa, and N. Magnenat-Thalmann. 2003. "A Virtual Heritage Case Study: A Modern Approach to the Revival of Ancient Historical or Archeological Sites Through Application of 3D Real-Time Computer Graphics." *CIPA 19th International Symposium*, Antalya, Turkey, pp. 235–240. http://www.miralab.ch/repository/papers../477.pdf.

Forte, Maurizio. 2009. "Participatory Research in Cyber Archaeology." In *Proceedings of the International Conference: Making History Interactive. Computer Applications and Quantitative Methods in Archaeology*, edited by Bernard Frischer, Jane Webb Crawford, and David Koller, 22–26. Williamsburg, VA, and Oxford: Archaeopress. http://proceedings.caaconference.org/files/2009/11_Forte_CAA2009.pdf.

Forte, Maurizio. 2014. "Virtual Reality, Cyberarchaeology, Teleimmersive Archaeology." In *3D Recording and Modelling in Archaeology and Cultural Heritage: Theory and Best Practices*, edited by Fabio Remondino and Stefano Campana, 113–27. BAR International Series. Oxford: Archaeopress. www.academia.edu/9542464/Virtual_Reality_Cyberaerchaeology_Teleimmersive_Archaeology.

Frischer, Bernard. 2004. "Mission and Recent Projects of the UCLA Cultural Virtual Reality Laboratory." In *Actes Du Colloque Virtual Retrospect 2003*, edited by R. Vergnieux and C. Delevoie, 65–74. Biarritz: Editions Ausonius. http://archeovision.cnrs.fr/pdf/vr03_pdf/03_frischer.pdf.

Frischer, Bernard, F. Niccolucci, N. S. Ryan, and J. A. Barceló. 2002. "From CVR to CVRO: The Past, Present and Future of Cultural Virtual Reality." In *VAST 2000*, edited by F Niccolucci, 7–18. Oxford: Archaeopress. http://kar.kent.ac.uk/21942/1/From_CVR_to_CVRO_The_Past_Present_and_Future_of_Cultural_Virtual_Reality.pdf.

Galvao, J. R., P. G. Martins, and M. R. Gomes. 2000. "Modeling Reality with Simulation Games for a Cooperative Learning." *Simulation Conference, 2000: Proceedings. Winter* 2 (January 2000): 1692–1698. doi:10.1109/WSC.2000.899158.

Gea, Juan M. Carrillo, Ambrosio Toval, José L. Fernández Alemán, Joaquín Nicolás, and Mariano Flores. 2013. "The London Charter and the Seville Principles as Sources of Requirements for E-Archaeology Systems Development Purposes." *Virtual Archaeology Review* 4(9): 205–211.

Hevner, A. R., S. T. March, and Jinsoo Park. 2004. "Design Science in Information Systems Research." *MIS Quarterly* 28(1): 75–105. doi:10.1.1.103.1725.

Ijaz, Kiran, Anton Bogdanovych, and Simeon Simoff. 2011. "Enhancing the Believability of Embodied Conversational Agents through Environment-, Self- and Interaction-Awareness." In *Thirty-Fourth Australasian Computer Science Conference (ACSC 2011)*, edited by Mark Reynolds, 107–116. Perth: Australian Computer Society. http://citeseerx.ist.psu.edu/viewdoc/download?doi=10.1.1.185.4578&rep=rep1&type=pdf.

Jørgensen, Anker Helms. 2004. "Marrying HCI/Usability and Computer Games." *Proceedings of the Third Nordic Conference on Human-Computer Interaction—NordiCHI*, pp. 393–396. doi:10.1145/1028014.1028078.

Kennedy, Sarah, Lisa Dow, Iain Angus Oliver, Rebecca Jane Sweetman, Alan Henry David Miller, Anne Campbell, Christopher John Davies, et al. 2012. "Living History with Open Virtual Worlds: Reconstructing St Andrews Cathedral as a Stage for Historic Narrative." In *Proceedings of the 2nd European Immersive Education Summit: EiED 2012*, edited by Michael Gardner, François Garnier, and Carlos Delgado Kloos, 146–160. Paris: Universidad Carlos III de Madrid, Departamento de Ingeniería Telemática. http://hdl.handle.net/10023/3332.

Linden Lab. 2015. "Linden Lab Invites First Virtual Experience Creators to Project Sansar Testing." www.lindenlab.com/releases/linden-lab-invites-first-virtual-experience-creators-to-project-sansar-testing.

Liu, Yang, and Kevin M. Passino. 2000. "Swarm Intelligence: Literature Overview." *Measurement and Control* 614: 9. doi: http://dx.doi.org/10.1.1.135.8765.

Lopez-Menchero, Victor Manuel, and Alfredo Grande. 2011. "The Principles of the Seville Charter." In *Proceedings XXIII International CIPA Symposium*, Prague, Czech Republic, pp. 12–16. http://congressgate.com/cipa/proceedings/pdfs/B-2 Seville charter/Lopez Menchero.pdf.

Maïm, Jonathan, Simon Haegler, Barbara Yersin, Pascal Mueller, Daniel Thalmann, and Luc Van Gool. 2007. "Populating Ancient Pompeii with Crowds of Virtual Romans." *Eurographics* VAST'07 Proceedings of the 8th International conference on Virtual Reality, Archaeology and Intelligent Cultural Heritage: 109–116. doi:10.1.1.93.4719

Mondada, Francesco, and Dario Floreano. 1995. "Evolution of Neural Control Structures: Some Experiments on Mobile Robots." *Robotics and Autonomous Systems* 16(2–4): 183–195. doi:10.1016/0921-8890(96)81008-6.

Morgado, Leonel, João Varajão, Dalila Coelho, Clara Rodrigues, Chiara Sancin, and Valentina Castello. 2010. "The Attributes and Advantages of Virtual Worlds for Real World Training." *Journal of Virtual Worlds and Education* 1(1): 15–36. http://home.utad.pt/~leonelm/papers/jvwe-volume-one-number-one.pdf.

Morgan, Colleen. 2009. "(Re)Building Çatalhöyük: Changing Virtual Reality in Archaeology." *Archaeologies—Journal of the World Archaeological Congress* 5(3): 468–487. doi:10.1007/s11759-009-9113-0.

Murteira, Helena, and Paulo Simões Rodrigues. 2016. "Lost Cities in the Digital Era: The Case of Pre-Earthquake Lisbon." In *Space and Time Visualisation*, edited by Maria Boştenaru Dan and Cerasella Crăciun, 151–171. Cham: Springer International. doi:10.1007/978-3-319-24942-1-10.

Poladian, Charles. 2016. "Facebook's Oculus Rift Is Tops with VR Game Developers As Virtual Reality Gets Real In 2016." *International Business Times*. January 20, 2016 http://www.ibtimes.com/facebooks-oculus-rift-tops-vr-game-developers-virtual-reality-gets-real-2016-2272856.

Reynolds, Craig W. 1987. "Flocks, Herds, and Schools: A Distributed Behavioral Model, in Computer Graphics." In *SIGGRAPH '87*, edited by Maureen C Stone, 25–34. New York: ACM Press. doi:10.1145/37401.

Schroeder, Ralph. 2008. "Defining Virtual Worlds and Virtual Environments." *Journal of Virtual Worlds Research* 1(1): 1–3. doi:10.4101/jvwr.v1i1.294.

Sequeira, Luís Miguel, and Leonel Morgado. 2013. "Virtual Archaeology in Second Life and OpenSimulator." *Journal of Virtual Worlds Research* 6(1). doi:10.4101/jvwr.v6i1.7047.

Sequeira, Luís Miguel, Leonel Morgado, and E. J. Solteiro Pires. 2014. "Simplifying Crowd Automation in the Virtual Laboratory of Archaeology." *Procedia Technology* 13: 56–65. doi:10.1016/j.protcy.2014.02.009.

Slater, Stuart, and David Burden. 2009. "Emotionally Responsive Robotic Avatars as Characters in Virtual Worlds." *2009 Conference in Games and Virtual Worlds for Serious Applications*, Coventry, UK, pp, 12–19. doi:10.1109/VS-GAMES.2009.7

Styliani, Sylaiou, Liarokapis Fotis, Kotsakis Kostas, and Patias Petros. 2009. "Virtual Museums: A Survey and Some Issues for Consideration." *Journal of Cultural Heritage* 10(4): 520–528. doi:10.1016/j.culher.2009.03.003.

Trescak, Tomas, Anton Bogdanovych, and Simeon Simoff. 2014. "Populating Virtual Cities with Diverse Physiology Driven Crowds of Intelligent Agents." *Proceedings of the 2014 Social Simulation Conference*, Barcelona. https://staff.scm.uws.edu.au/~anton/Publications/SSC_2014.pdf.

Wells, Sarah, Bernard Frischer, Doug Ross, and Chad Keller. 2009. "Rome Reborn in Google Earth." *Making History Interactive: 37th Proceedings of the CAA Conference*, pp. 373–379. http://romereborn.frischerconsulting.com/rome_reborn_2_documents/papers/Wells2_Frischer_Rome_Reborn.pdf.

PART TWO

# *Conservation, Requalification, and Communication*

CHAPTER 7

# At-Risk World Heritage and Virtual Reality Visualization for Cyber-Archaeology

## *The Mar Saba Test Case*

THOMAS E. LEVY, CONNOR SMITH, KRISTIN AGCAOILI, ANISH KANNAN, AVNER GOREN, JÜRGEN P. SCHULZE, AND GLENN YAGO

### INTRODUCTION

For years, humans have been fascinated with the preservation of material culture, from small artifacts to entire landscapes, and have seen them as having significant cultural value, as tangible links between them and their history.* However, though the ages, people have used the destruction of a cultural heritage as military and ideological tools to dominate others. In the Middle East, this has a long and well-documented history. For example, Jerusalem, the capital of ancient Israel, was systematically destroyed twice in antiquity,

* We are grateful to Father Ioannes and Father Vasilias of the Mar Saba Monastery for their enthusiastic support of and collaboration on this project and allowing us to carry out the field work at their home. Thanks also to Craig Smitheram for his help in the field. We also thank Matt Howland for his critical comments on this chapter and help with fieldwork; thanks to Stephen Savage for producing the Kidron Valley watershed map. The laboratory work for this research was carried out with funding support from the University of California Office of the President Research Catalyst Awards CA-16-376911, with Thomas E. Levy serving as principal investigator. Special thanks to Dr. Andrew Viterbi (La Jolla, California) for his generous support of this work; see –Fox 2015).

Thomas E. Levy, Connor Smith, Kristin Agcaoili, Anish Kannan, Avner Goren, Jürgen P. Schulze, and Glenn Yago, *At-Risk World Heritage and Virtual Reality Visualization for Cyber-Archaeology* In: *Digital Cities*. Edited by: Maurizio Forte and Helena Murteira, Oxford University Press (2020).

DOI: 10.1093/oso/9780190498900.003.0008

first by the Babylonians in 586 BCE (Magness 2012) and then by the Romans in 70 CE (Goodman 2007), with the aim of erasing the cultural identity and memory of a people. By the mid-twentieth century, during the 1948 Arab–Israeli War, systematic destruction of the Jewish Quarter of the Old City of Jerusalem took place again. The Jordanian commander Abdullah al-Tal, who was in charge of the Jordanian assault, justified the destruction of the Jewish Quarter, saying that had he not destroyed the homes, he would have lost half his men and that "the systematic demolition inflicted merciless terror in the hearts of the Jews, killing both fighters and civilians" (Fernea and Hocking 1992, 53). During the Jordanian occupation, thirty-four of thirty-five synagogues in the Jewish Quarter were destroyed (Jewish Virtual Library 2016). During this period, Palestinian refugees were relocated to the Jewish Quarter; following Israel's conquest of the Old City during the Six-Day War, the Jewish Quarter Development Company was established under the auspices of the Construction and Housing Ministry of the State of Israel to rebuild the Jewish Quarter. Following the Six-Day War, approximately 4,000 Palestinian Arabs left the area, were forced to leave, or were prevented from returning to the Jewish Quarter where they had lived during the Jordanian occupation (Central Bureau of Statistics 1968, 6; Dumper 2002). Unfortunately, in the Middle East today, with the birth of extremist ideologies typified by the Taliban and Islamic State (ISIS), the decision to consciously destroy cultural heritage sites with powerful explosives has become common.

Perhaps the most vulnerable places in the Middle East today for the destruction of cultural heritage are Iraq and Syria, where the on-going conflict and wars have led to museums, libraries, and other cultural repositories to be burned and looted, and UNESCO World Heritage sites destroyed. Many of these heritage and archaeological sites are at risk of destruction. Recently, in January 2016, examination of satellite imagery showed that Iraq's oldest Christian monastery was completely obliterated by ISIS, resulting in the loss of a major cultural icon for the Christian community in that country, as well as for the world (Smith and Deaton 2016). However, damage is not always so extreme, and many locations are being polluted, looted (Kersel et al. 2008), and/or generally neglected. Though there is no clear solution to this problem, public awareness and engagement are important first steps, with possible solutions now coming from cyber-archaeology. In this chapter, we describe the development of virtual reality (VR) models for the second oldest Greek Orthodox monastery in the Middle East—Mar Saba, located in the Judean Desert (figure 7.1). While the site is not in immediate danger of destruction, the adjacent polluted stream is damaging the local environment and cultural heritage features around the monastery.

Mar Saba in Israel/the Palestinian territories is not in imminent danger of destruction from human-induced conflict, like the Syrian and Iraqi examples noted. That said, before the Arab Spring in 2011, few thought that sites in

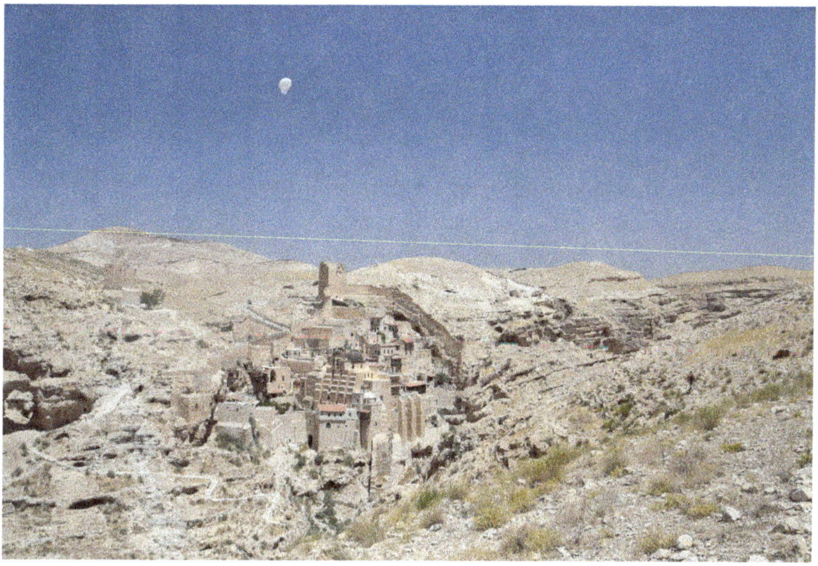

**Figure 7.1** Overview of the Mar Saba Monastery in the Kidron Valley, in the Judean Desert of Israel/Palestine, with the helium balloon used for photogrammetry in this project.
*Credit:* Photo T. E. Levy, UC San Diego Center for Cyber-Archaeology and Sustainability.

Syria and remote parts of Iraq were in danger of destruction. In the Middle East, though, one should always expect the unexpected. Thus, recording all major heritage sites in the Middle East in the most effective ways possible should be a priority. In addition to the negative cultural forces that impact site preservation, there is a wide range of natural forces, such as earthquakes, floods, fires, erosion, and other factors, that can negatively impact the preservation of cultural heritage sites. Cyber-archaeology and virtual reality are two tools that archaeologists and cultural heritage professionals can use today to preserve sites digitally, as well as engage the public in heritage preservation, thereby mitigating the risks that cultural heritage faces in today's world.

Cyber-archeology is a relatively recent field that involves collecting and storing archaeological data in digital form, such that it can easily be accessed and visualized when necessary following a workflow best characterized as: digital data capture, curation, analyses, and dissemination (Levy 2013). If displayed to the right audiences, and through the right medium, the data can transport individuals across the world; more importantly, it can provide narrative regarding at-risk cultural heritage sites. As VR is a real-time interactive computer-generated experience for the viewer in a simulated environment, it can transport people to the heritage site, where they can experience the magnificence of the site and the environmental forces negatively impacting the site, such as pollution and erosion. This experience can influence viewers, both officials and the public, to appreciate the value of the heritage site and to support conservation and preservation. The challenge is finding effective ways to present the digital archaeological data to the general public in an interesting and engaging form.

Virtual reality, an emerging technology and innovative digital medium, has been slowly growing over the past years, only recently becoming widely acknowledged and expanded. The term "virtual reality," for purposes of this research, refers to several different systems that can be considered virtual environments. One major solution to virtual reality is a CAVE, an acronym for "cave automatic virtual environment," which consists of several screens and often multiple computers streaming content around a user (Knabb et al. 2014). A second and more recent form of virtual reality is the head-mounted display, appearing in the form of consumer-ready hardware such as the Oculus Rift or even the Google Cardboard. These devices contain two screens, one for each eye, which display an image and track head rotation when worn. Both systems offer an incredibly immersive experience, and can convey information in a way that involves deep engagement and interactivity.

With the continuing evolution of cyber-archaeology and the arrival of virtual reality as a viable medium for personal data visualization, the present goal is to display archaeological information inside such a virtual environment. This is a task that involves integrating technology that has been largely unexplored, with information already catalogued and stored in a database. The resulting virtual content should be stimulating, informative, and appealing enough to be part of a public display, and should be implemented across both CAVE and head-mounted display platforms to ensure a spectrum of availability. As explained, these VR CAVE platforms and head-mounted displays can provide important experiential learning experiences for policymakers, researchers, and the public, who can influence national and international heritage-preservation efforts.

## SETTING: MAR SABA MONASTERY AND THE KIDRON VALLEY

The risks faced by world cultural heritage, especially in the Middle East, make it crucial to develop innovative models for making cultural heritage into assets, such as archaeological and historical sites, that are of greater economic value when preserved than when destroyed. The Milken Institute, Jerusalem's Financial Innovations Lab, has proposed creating "heritage districts" to conserve such archaeological sites (Milken Financial Innovations Lab 2011, 6). This would be a great tool for heritage site preservation and would benefit local communities at the same time. That is, the potential financial return from assigning a greater value to archaeological heritage is vast and untapped. It includes archaeo-diplomacy, archaeo-tourism, infrastructure growth, and educational outreach; it can provide a unique opportunity to enhance community development by generating job creation, rescuing fragile cultural assets, and funding exploration. Taken together, these goals help mitigate the standard risks to cultural heritage. As shown here, by constructing what we call the Cultural Heritage Asset District model of finance and management for the Kidron Valley watershed, we can strive to meet the model's needs as a sustainable, revolving partnership of multiple stakeholders. The result is conservation of cultural heritage resources.

Cultural heritage tourism is the fastest-growing sector of global tourism. Indeed, cultural tourism is the major element in international tourism consumption, accounting for over 39% of tourism arrivals, and it has become a driver for economic development in developing regions (Richards 2018). Emerging bi-modal distribution of age structure (characterized by peak millennials and baby boomers) are demographic drivers of demand for this cultural heritage tourism. This includes cultural consumption, cultural motivations, heritage conservation, and the relationship between those elements of cultural and the creative economy. This relationship between tangible and intangible heritage and attention to indigenous and other minority groups and geographical expansion all energize the need for conservation of cultural heritage for sustainable development.

Over the past twenty years, there has been an increase in the identification and preservation of cultural heritage sites. Twenty-six new inscribed properties were added to the UNESCO World Heritage list in 2016, making a total of 1,052 current sites. Global emphasis on cultural heritage preservation was evident in the 1995 proclamation by the UN World Commission on Culture and Development, which proclaimed "the explicit role of culture, such as built heritage, as a strategy against poverty in the third world countries, since the concepts of culture and of development are inextricably intertwined."[1] Continuing in this vein, there were agreements at the 1998 and 1999 UNESCO meetings in Stockholm and Florence, respectively, on "the vital impact of culture for economic development in future policy making."

Beyond sharing expertise from the public, private, and nonprofit sectors, effective financing and management structures for these districts must be self-sustaining, revolving instruments (MacDonald 2011), and must digitize to prevent degradation of and educate about cultural heritage (Skarin 2011). As mentioned, cultural heritage boosts job creation and household income. A European assessment of these efforts found that historic rehabilitation and conservation of cultural heritage sites created 16% more jobs than new construction, and that every direct job in the cultural heritage sector creates 26.7 indirect jobs (Rypkema 2008, Baycan and Girard 2011).

The Cultural Heritage Asset District model of finance and management proposed for the Kidron Valley meets these needs as a sustainable, revolving partnership of multiple sectors, the dual mandate of which is to increase local economic activity and to preserve the cultural heritage assets (Dimitriyadis et al. 2012). The model is based on similar tenets as the archeological development bonds and tourism improvement district model (Milken Institute 2008, Milken Financial Innovation Center 2011), which builds a sustainable public–private partnerships (PPPs) and ensures a revolving financial mechanism based on tax revenues, bond issuances, and other tools to stimulate local tourism and preserve local heritage assets (European Investment Bank Institute 2013). The model brings financially and culturally sustainable answers to meet the needs for innovation in the field of financing and management that can benefit cultural heritage conservation.

## DATA AND PLANNING FOR A CULTURAL HERITAGE ASSET DISTRICT

A Cultural Heritage Asset District invests in a variety of assets capable of bringing visitors to the district along with the opportunity to share costs of infrastructure and revenues. Financing is based on capturing and leveraging incremental value that is created from strengthening the cultural assets and using the shared infrastructure (both physical and IT) so as to increase tourism to these sites, both virtually and through actual visits. This process begins with a needs assessment of infrastructure improvements, reuse and historical restoration possibilities, antiquity site maintenance, education and training, business development, and site operations management. Possible sources of financing would include user and entry fees, co-marketing revenues from sites generating income to support PPPs, and grants from government and private foundations. In the past, the economists involved in our Mar Saba work helped develop asset financing and identified a basic financial partnership model (see figure 7.2) that included a database for inputs to these financial and development sites. This information would then be applied to assessing the types of tourists (recreational, heritage, ecotourism) and destinations. By combining groups of sites into higher-value districts, we would create for each target the clusters, themes, and infrastructures to combine historical, environmental, and recreational assets in those districts. In essence, sites and projects would be combined into a pooled portfolio of district projects.

Therefore, the essential features of a Cultural Heritage Asset District are:

- High-value attractions capable of bring visitors to the district
- Sharing of costs and revenues throughout district to support these attractions
- Financing based on incremental value created by strengthening the attractions

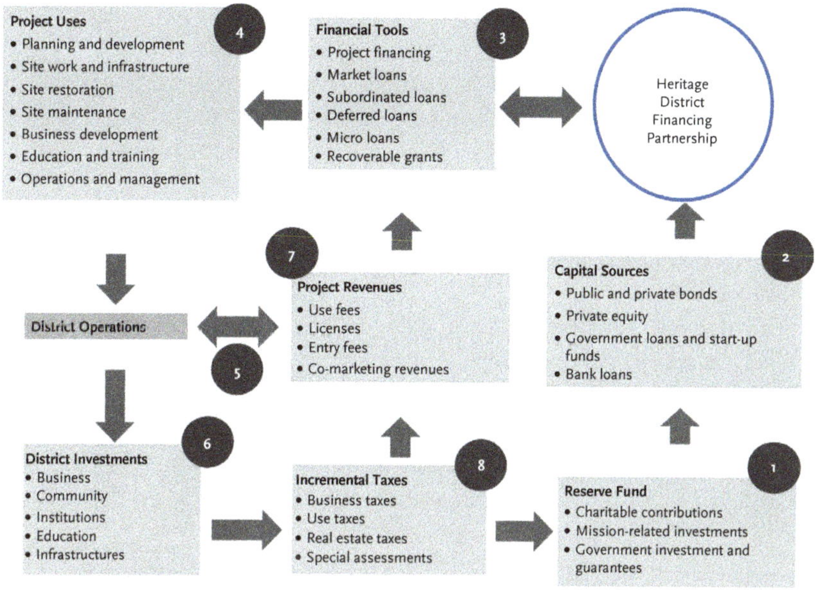


**Figure 7.2** The Heritage Asset Financing model.

This model builds on the way classic economics considers generating value from an underutilized asset and recycling that value back into the asset or ecosystem surrounding it. Using the conceptual approach depicted in figure 7.2, our approach involved the following steps or stages for heritage districts to be developed and financed:

1. A combination of government funds and philanthropic contributions and investments are used to create a reserve fund.
2. Using these sources of capital, a combination of public and private funds, government loans, and bank loans are invested in a public-private partnership for heritage district financing.
3. The heritage district financing partnership uses a series of financing tools, including recoverable grants, micro loans for small businesses, market-rate loans and bonds, and project financing.
4. The financing tools are used to finance a series of activities needed to build the heritage district, including sites, projects, business development, and hard and soft infrastructures, and including physical and virtual access.
5. The activities allow the district to be open, offering a full range of tourism, education, recreation, and cultural experiences to onsite users and off-site virtual visitors (see Girardeau-Montaut, Daniel. 2011). These activities create direct revenues and indirect revenues.

6. The direct revenues, such as use fees, licenses, and entry fees, are used to pay for district operating expenses and repayment of debt.
7. Indirect revenues are used to provide credit and collateral support for meeting financial obligations and to provide supplemental capital to fund the reserve funding and reduce the exposure of the initial reserve funders.

This model is used to evaluate options for how to approach each asset, including preservation, restoration, adaptive reuse, or deployment of addition technologies.

The integration of cyber-archaeology in the project planning is integral, as well. It consists of building Cultural Heritage Asset Districts and decision support tools at the local level. Cyber-archaeology provides the way for archaeologists, government organizations, local societies, and other interested stakeholders to promote the Cultural Heritage District Asset over the internet, in media outlets, and in the different VR platforms (see figure 7.3).

## THE CASE OF THE KIDRON–WADI EL-NAR WATERSHED

Our Cultural Heritage Asset District was formed around a natural watershed that served as the basis for settlement in the ancient world. Water is key to survival, so humans have organized their settlements and settlement strategies based on access to water sources. Thus, the watershed is the basic subsistence system that unites the region; it can be easily identified and mapped.

Consider that a watershed is a region of landscape that drains all the water falling upon it and running through it into a larger body of water. Watersheds combine with other watersheds to make a network of seasonal or and/or perennial rivers and streams that eventually drain into larger water bodies. As watersheds constitute distinct environmental zones, for preindustrial societies, these landscape units often reflected unified cultural systems, making them clear organizational and analytical units.

This chapter presents our preliminary effort at developing an economic model for the Kidron Valley, an important watershed that borders the holy city of Jerusalem (figure 7.3). The Kidron Valley—from Hebrew: נחל קדרון, *Naḥal Qidron*; also Qidron Valley; Arabic: وادي الجوز, and *Wadi al-Joz* for the upper segment near the Temple Mount and *Wadi an-Nar* for the remainder—drains into the western shore of the Dead Sea. The beginning of the Kidron Valley separates the east side of the Old City of Jerusalem's Temple Mount, where Solomon is purported to have constructed the First Temple, from the Mount of Olives. This is the holiest site in Judaism. As the Kidron River flows eastward through the West Bank's Judean Desert, it drops approximately 1,220 meters in elevation along its approximately 32 kilometer length. Accordingly, the Kidron Valley is exceptionally rich in history and archaeology. In addition

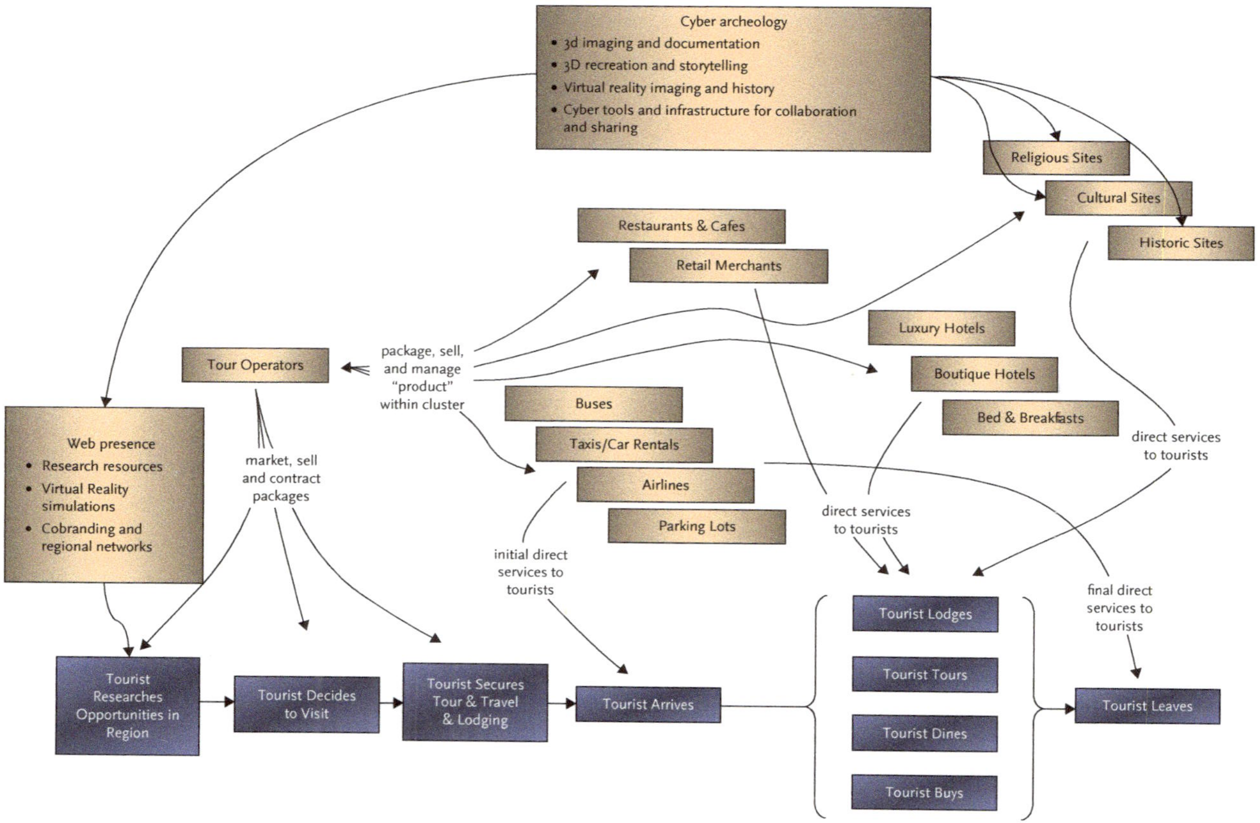


**Figure 7.3** Model for Building Cultural Heritage Asset Districts.

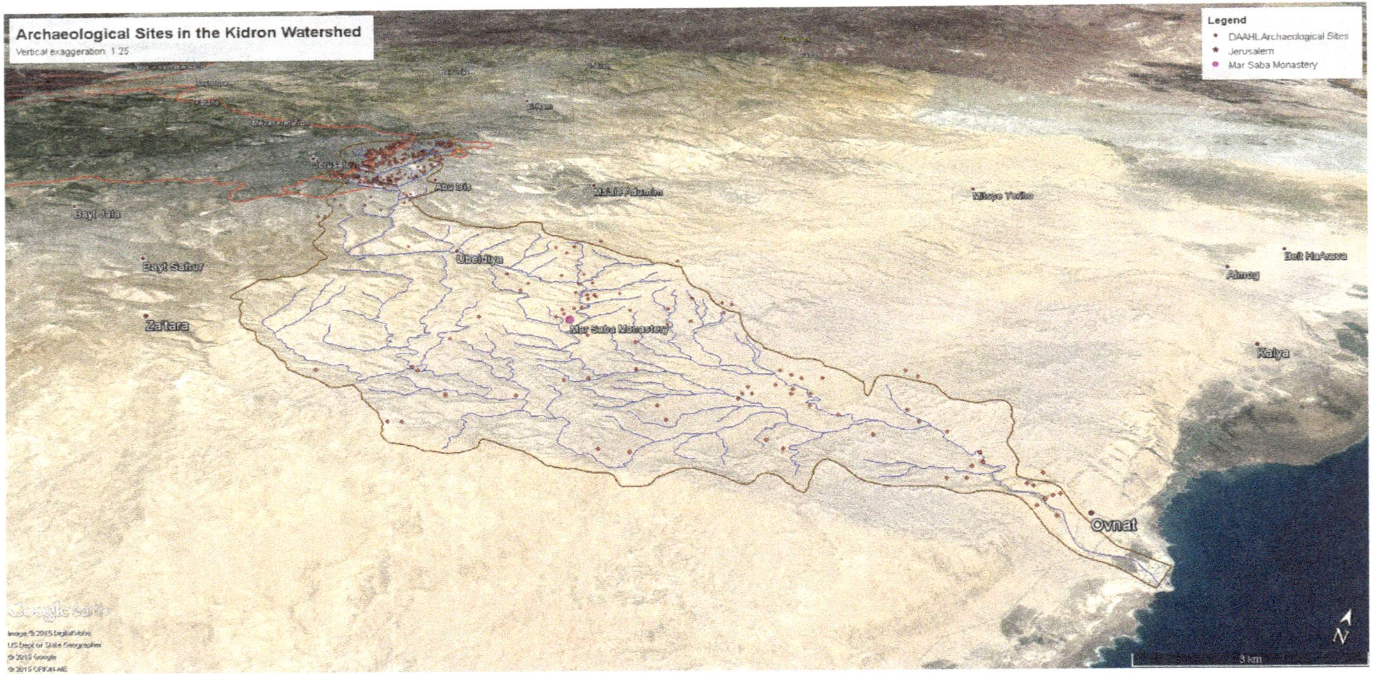


Figure 7.4 View of the Kidron Valley and Wadi El-Nar Catchment Basin.

to the Old City's Temple Mount (in Islamic tradition it is known as Haram esh-Sharif [Arabic: الحرم الشريف, *al-Ḥaram al-Šarīf*, "the Noble Sanctuary," or الحرم القدسي الشريف, *al-Ḥaram al-Qudsī al-Šarīf*, "the Noble Sanctuary of Jerusalem"]), the Kidron's drainage is at the confluence of Jerusalem's richest concentration of rock-hewn tombs dating from the First Temple period (ninth to seventh centuries BCE) and Second Temple period (c. 530 BCE–70 CE; Rahmani 1981).

Approximately halfway down the Kidron Valley, in the Judean Desert, is the Greek Orthodox monastery called Mar Saba (area = 90 m by 170 m, or c. 15,300 sq. m), situated on the south bank of the Kidron and constructed around 483 CE. It was founded by 'Sabbas the Sanctified of Mutalaska, Cappadocia, and today houses around twenty monks; it is the second oldest continuously occupied Greek Orthodox monastery, after Saint Catherine's Monastery in the southern Sinai Peninsula (Ganon 2014; Hirschfeld 1995, 5). Unfortunately, wastewater from eastern Jerusalem flows through the Judean Desert to the Dead Sea. The sewage that flows though the Kidron Valley is damaging for both the natural environment and cultural heritage resources such as the Mar Saba Monastery, as well as the hundreds of ancient sites in its vicinity. Indeed, approximately 12 million cubic meters of sewage flows through the Kidron Valley every year. This is damaging to the natural and cultural environments, as well as the ecosystem as a whole. But as the drainage flows through both Israeli- and Palestinian-controlled lands, finding a solution to this situation is a huge challenge. A number of possibilities have been proposed, addressing the complexities of this asymmetrical political setting where Israelis, Palestinians, and the Greek Orthodox Church are the primary stakeholders (Dombrowsky et al. 2010). The VR project described, thus, is an initial attempt to apply the methods of cyber-archaeology (Levy 2013), including data capture, curation, and dissemination within the context outlined here.

The Cultural Heritage Asset District model was predicated on the environmental cleanup of the river through construction of a wastewater treatment plant. The Kidron Sewage Treatment project (approved by the Joint Israel Palestinian Water Commission in 2017 and currently under construction) will enable further implementation of heritage site projects within the watershed. As explained earlier, the model combines the public and private sectors to create a business model for the watershed in terms of both direct and indirect economic activity, in both Israeli and Palestinian territories.

### Project Financing

As illustrated in figure 7.5, a special purpose vehicle (SPV, or 1) is created as a single-purpose corporate entity. The SPV is owned by partners, including

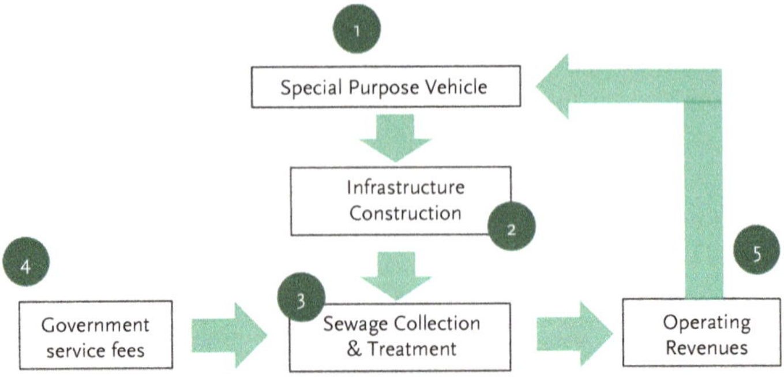


**Figure 7.5** Project Model, showing direct revenues.

limited partnerships such as cash equity investors, a professional team, and contractors and operators. It is managed by a general partner that has the responsibility for all operations. The SPV hires all professional services, including the contractor for the design and (2) construction of the infrastructure and treatment plant itself. The SPV then (3) hires the operator for the collection, treatment, and distribution of services.

The SPV signs a long-term contract with the Israel Water Authority and the Palestinian Water Authority to provide the sewage collection and treatment services. In turn, the governments return a variety of tap-in, flushing, drainage, and water treatment fees (4) to the operator.

In addition to the direct government fees, the SPV is able to sell a variety of products, including gray water, compost, and energy sales on the market and sign long-term contracts to strengthen the (5) operating revenues (and underlying credit) of the project.

To strengthen the direct revenue, the project financing model includes several indirect revenue sources, as well. As part of the project plan, the infrastructure will support the growth of tourism and tourism-related revenues in the greater Jerusalem region. As shown in figure 7.6, new businesses and the expansion of existing businesses (6) will provide incremental value-added taxes, real estate taxes, improvement taxes, use fees, and license fees (7), which can make the case for a partial government guarantee for the project financing (8). Additionally, the incremental value created in the region will permit the use of revenues from events, services, and even a portion of the incremental taxes themselves from the expanded hospitability and tourism activities in the region, such as hotel rooms (9). These indirect revenues are based on the increases in tax revenues from the new infrastructure, which can provide additional reserve funding (10) for the project. All these direct and indirect sources support the plan for a project financing (11) for the SPV (Milken Innovation Center-Jerusalem Institute 2013).

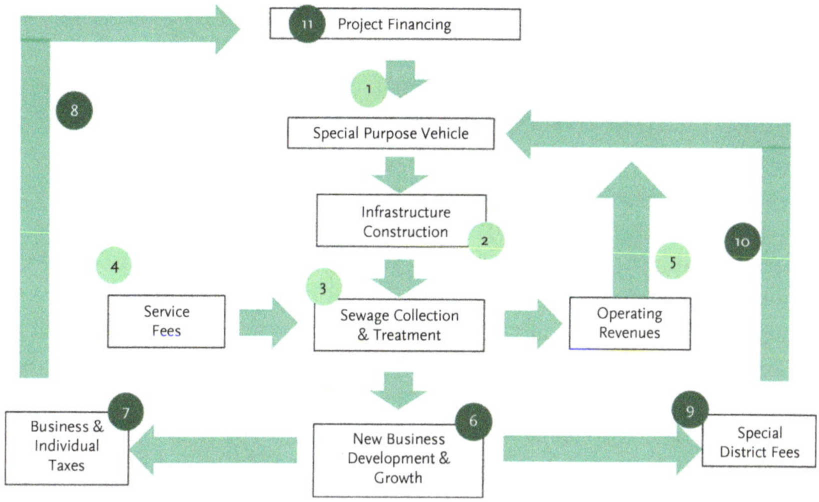


**Figure 7.6** Project Model, showing indirect revenues.

## Previous Research

There have been a variety of projects exploring the use of data visualization and simulation software for archaeological analysis. Since early 2008, there has been a gradual shift from text-based descriptions to 2D photography and then to 3D presentation, owing to improvements in the capability of perceiving data. The impact of 3D models is seen as revolutionary in terms of "registration, documentation, dissemination, presentation, and ultimately the preservation of cultural heritage." According to Mikropoulos et al. (2008), the process of recording archaeological information has five steps: digitization in 3D, processing and storage of 3D data, archiving and management of 3D data, visualization and dissemination of 3D data, replication and reproduction of 3D data. Our research followed this process and attempted to process and visualize the 3D data through the medium of virtual reality. Michael Bawaya (2010:140-141) described how archaeologist Sam Paley was able to use VR to simulate lighting conditions in a virtual throne room based on photos and eventually concluded that "the torches could have been fueled by several types of fish oil and positioned to enhance the art." Another research group utilized photogrammetry on rock cairns in the Tongass National Forest, in the United States, as a means to obtain new information on their functions and highlight their importance. That study attempted to encourage communication with the public on the preservation of archaeological and cultural heritage features, and found that PhotoScan Pro by Agisoft and digital photogrammetry were ideal for their situation (Chodoronek 2015).

Previous researchers have primarily focused their 3D visualizations on computer screens. But in order to immerse users in the archaeological sites and settings, more studies must be done in the field of virtual reality (Knabb et al. 2014, Levy et al. 2008). Indeed, recent advancements in the field of virtual reality have been made in the technology industry, especially in its capabilities for visualizing the past for public audience (Bittel 2013). As Pujol (2004) points out, virtual reality "defines virtual reconstructions as a vehicle for preservation, access and economic development at the service of archaeological remains valued for their artistic qualities." He believes that virtual reality has immense capability for presenting archaeological data to the public because its multimedia environment allows "different formats for presenting information [which] can be adapted to individuals" skills' depending how the audience would like to absorb the information (Pujol 2004). In the application presented here, our research highlighted the strong potential for virtual reality in our mission to preserve and digitize cultural data.

## Data Capture and 3D Dissemination

In order to photograph the Mar Saba watershed, team members developed a custom balloon-based, low-altitude aerial photography (LAAP) system. This system consisted of a large Kingfisher Aerostat balloon (c. 2.29 m by 1.56 m, volume about 4.23 sq. m and lift about 3.39 kg) tethered to and manipulated by a ground-based operator. We used a Brooxes Gent-X Picavet cross-frame capable of holding a high-resolution (15.1 megapixel) Canon EOS 50D digital single-lens reflex (DSLR) camera equipped with an 18–135 mm lens. The camera was triggered by an intervalometer set to an interval of 10 seconds. This setup is well suited to yield the hundreds of high-resolution images appropriate for SfM processing.

As shown in figure 7.1, the monastery is located in a narrow valley, making it difficult to navigate the tethered balloon. The first day we tried to fly the balloon, winds blew so strong that we had to postpone the fieldwork for several days. While we were ultimately successful in using the balloon, there were a number of problems, including the very high cost of helium, the difficulty in moving the heavy tanks around the research area, and the narrow setting that made it difficult to maneuver the balloon into the correct location for image capture. Although the flight time for a balloon is almost indefinite, in the future we would use drones, in spite of their short flight time owing to battery charge. Our lab now uses both systems (Howland et al. 2014, Smith et al. 2014).

Since virtual reality can be interpreted as the "cyber" in cyber-archaeology, the main goal was to use existing data captured by archaeologists in the field at Mar Saba so as to create a visual experience for public display. Our team has prior experience with the Unity game engine, so the first step was to determine how large the amounts of data that could be imported and visualized in Unity—a program used most often to create 2D and 3D games. Furthermore, initial development was targeted for the Oculus Rift head-mounted display, as the CAVE system was still being built. The point cloud captured at the Mar Saba Monastery by photogrammetric processing of images taken by the balloon system consisted of roughly 85 million points, with a size of 2.2 gigabytes (Figure 7.7). Since many VR displays, especially the newest head-mounts, require a great deal of computational power, the starting goal was to display this point cloud, in some form, inside Virtual Reality without latency. To the team, this meant achieving a refresh rate of at least ninety frames per second, as is the expectation with VR experiences.

After initial testing, it was clear that the entire point cloud could not be displayed at once in VR. In fact, it was difficult just to import the data into Unity. Therefore, the next step was to reduce the point cloud using a program called CloudCompare, cropping it to a size that would allow for easy testing and viewing. At the conclusion of this process, a custom point cloud destructor was written, capable of reducing different levels of points in a point cloud such that the data can easily be viewed inside virtual reality. In order to achieve this, an algorithm was designed that skips over a certain number of specified points in a point cloud file, and then saves the new data into a new

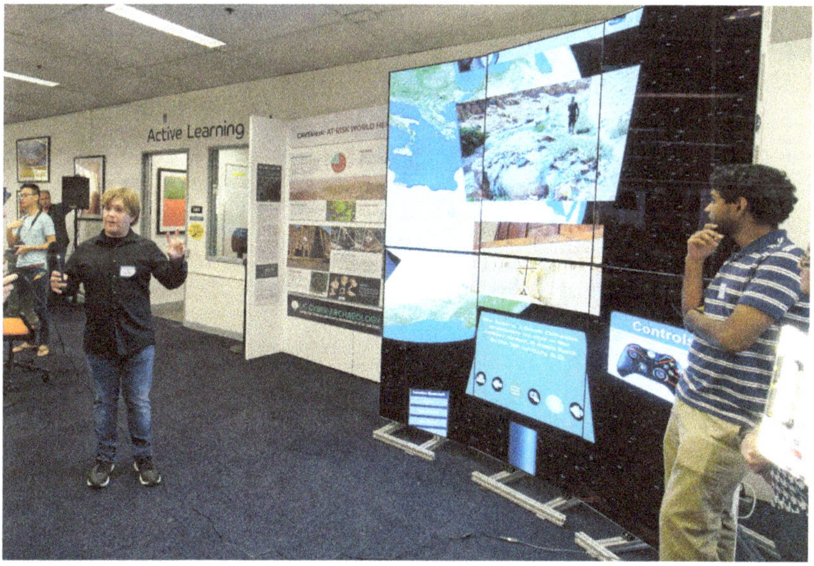


**Figure 7.7** CAVEkiosk 3D visualization platform, developed for the UC Office of the President, Catalyst Grant for At-Risk World Heritage and the Digital Humanities in the UC San Diego Geisel Library.
*Credit:* Courtesy T. E. Levy, Center for Cyber-Archaeology and Sustainability, UC San Diego.

file for quick loading next time. The correctness of this approach is based on the assumption that nearby points are grouped together in the data file, such that deleting neighbor points still allows for an accurate representation of the original site. Once ready, this system was integrated into the Unity application, such that any point cloud with a.off extension can be reduced and loaded automatically.

After we were capable of loading, displaying, and visualizing the point clouds, we attempted to create a mesh (Cignoni and Ranzuglia 2014), defined as a solid geometrical surface, out of the points. It seemed the points contained information about surface normals, but it was still difficult to make a mesh look realistic enough to be usable. The entire purpose of a mesh was to reconstruct the site from points, but for this particular site, the data weren't complete enough to create a fully coherent or complete Mar Saba. Additionally, these meshes were demanding in terms of performance and the frame rate often dropped below sixty frames per second, which is unacceptable for our goals. Following several attempts using Mesh Lab to create a connected version of the data, we decided to move forward and revisit the possibility for a mesh later in the project.

The next system for this project was designed to load 2D images, such as those taken from a generic camera, in order to support different forms of media for archaeological sites. Written entirely in the C# programming language (Microsoft), the photo loader can display up to twenty-four images at one time in three rows, with each picture acting as a point on an octagon. The images slowly rotate around the user, giving a 3D appeal to the initially 2D data. When the pictures are rotating, the user has the option to select a photo that will then emerge from the rotating rows and sit in front of the user for examination. Once the user is finished looking at the photo, he can send it back and it will take its correct place in the corresponding row. The loader is also capable of loading any number of images below twenty-four. It calculates how many images should be on the top and bottom rows by dividing the starting number of images by 3, and then all remaining images are placed into the center row. If the number of images in a single row is less than four, then that row will not rotate but, rather, just display in front of the user. Otherwise, the pictures place themselves geometrically around the user in their respective rows, such that a row with six images will have the layout of a hexagon, seven images a heptagon, and eight images an octagon. After eight images, the photos begin to get cramped, so the maximum number was placed at twenty-four to ensure at most three rows of eight images are loaded. This system can also result in cases that have only some moving rows—for example, with a total of ten pictures, there will be three images on the top on bottom rows and four in the middle. The four image row will rotate, but the ones with three images will not, based on the rule described earlier. This allows for some interesting and unique combinations with different amounts of images. To actually load these images, all a user has to do is type in a directory or folder where their images are located, and the system will always load the first twenty-four. Upon completion of the photo loading system, a very similar movie loader was designed to load movie files in Unity, and an audio loading system that can play specified sound clips whenever necessary. All of these were written inside Unity using C#, and are integrated with the VR system.

At this point, many of the pieces to create a full VR experience were in place. The technical portion was mostly over, and the largest challenge that remained was to connect all systems into an appealing, interactive, and enjoyable experience that can provide interesting narrative and full immersion. Since the goal was to load data for multiple different archaeological sites, the experience begins with a model of the Earth, and the user is standing upon a virtual "podium" with a screen in the center. This podium was an important consideration for the problem of motion sickness. When designing for virtual reality, it's vital that the user is always grounded and motion is intuitive in the experience; otherwise, the experience will be both ineffective and nauseating. By placing a user on the platform, there is no longer a feeling of floating, and motion can be justified as motion from the podium rather than the user, making it an effective method of transportation inside the experience.

On the three-dimensional Earth, there are several "points of interests," represented by large red arrows (https://daahl.ucsd.edu/DAAHL/). Each arrow is a different site that's been loaded into the system, and these points of interest can contain information ranging from point clouds and panoramas to two-dimensional images and audio clips. The designer of the site can also include brief text to be displayed when the POI is selected, which shows on the podium screen and is visible to the user. For actual selection and interactivity, the user interface uses Razer Hydra controllers, which track position and rotation in 3D. The hydras work together with a paradigm of pointing at objects to select them, using a line to indicate the direction of selection. Selecting a POI with this system causes the virtual monitor to display text information and buttons to access the other information.

With the podium comes a movement system designed to work with 3D controllers. There is a sphere in the center of the podium which, when selected, starts a movement mode. When moving, the speed of the podium matches the distance from the starting point of selection, allowing the user to move her controllers further for faster speeds. For example, if the user wants to go up, she will select and move her hand up. Rotations work in a similar way by rotating the controller. This allows users to move around the earth model and each individual archaeological site, but also select points of interest and move to them automatically if manual movement isn't desired.

At this point, most of the technical programming work was completed. This is the first time the details of the VR program are published in depth. As Mar Saba is in a contested area, it is not easy to visit. The programming described here enables visitors to the CAVEkiosk or those using a personal VR device to experience this important heritage site in new and exciting ways that only a physical visit to the site can improve on.

## Results and Considerations for Improvement

The primary site for presenting the Kidron Valley Cultural Heritage Asset District is the Greek Orthodox Monastery experienced in the CAVEkiosk multi-panel 3D visualization platform (https://daahl.ucsd.edu/DAAHL/; https://ucsdnews.ucsd.edu/pressrelease/new_3_d_cavekiosk_at_uc_san_diego_brings_cyber_archaeology_to_geisel). Once all the components for each type of media were processed from the field, the main effort centered on connecting all these media assets with a point of interest (POI). POIs were added to one of NASA's high-resolution Blue Marble satellite images, file paths with pictures and audio were added to the POIs, and the system was mostly ready to go. Some initial narration was added as audio, but most of the demo focused on the Mar Saba site, which loaded a point cloud and contained even more POIs within the site.

This project was demonstrated to stakeholders as a rough outline of the system capabilities, and it was generally concluded that more improvements had to be added before a display was ready. Some parts are a matter of fine-tuning, such as preventing motion sickness, adding more archaeological sites, changing input mechanisms, and including more accurate and inspirational narrative to spread awareness about at-risk locations. Other aspects that will require additional work, such as deciding how to better present text, showing content in a more appealing way, making point clouds more visibly appealing, and in general ensuring that the project is enjoyable for a public audience. Currently, this digital cultural heritage project is primarily a large and explorable repository of data in virtual reality, whereas the ultimate goal is to present a motivating and captivating experience for any visitor to appreciate.

The VR portion of the project still needs considerably more work, which means there are many more ideas to be explored. The POI system integrated with the project is fairly intuitive for nontechnical individuals to use, but still requires a user to go through the Unity game engine and manually entire file paths and information. Therefore, although it wouldn't be too difficult for an inexperienced user to be able to create new POIs, the goal is to integrate archaeological site setup into the experience itself, such that archaeologists or nontechnical administrators could easily link data with the system and create new points of interest. This would involve entering some form of password to signify an administrator, and then the ability to create, set up, prepare, and customize media for new locations. An administrator could also create new tour paths, add text to locations, and narrate different locations to convey the importance of each cultural location. If this can be achieved, then anyone will be able to expand and grow the system, giving it a much longer life and significant amount of flexibility.

This project is destined to grow, as there is still a wealth of data and information available for integration with the POI system. Each site must be set up manually, so with the ability for archaeologists to expand it themselves, any sites both new and old could be visualized and presented.

Another requirement is the transition from 3D input mechanisms to simple controllers. Similar CAVE systems have used 3D controllers for an interface, which is why we worked with the Razer Hydras (http://www.razerzone.com/). However, it was later decided that a standard Xbox gamepad would be a better option. This form of controller is more familiar to the average audience and would thus be easier to learn. Additionally, as gamepads do not need to be moved to function, and they can be more strongly secured against theft. Another major obstacle was time frame.

## CONCLUSION

In the Middle East, at-risk world heritage were in the headlines during the height of the horrific activities of the Islamic State (ISIS), which included destruction of famous heritage sites in Syria and Iraq. While political pundits may claim that ISIS has been defeated, the destruction of cultural heritage sites continues by looters, developers, conflicts, natural processes, and other factors. For these reasons, twenty-first-century archaeologists are obliged to consider the protection and conservation of archaeological and heritage sites before they begin to be destroyed. In this chapter we combined innovative approaches to digital archaeology, information technology, virtual reality, and economics to explore new ways of engaging stakeholders in those regions, as well as the public and governmental agencies to help in these efforts. This is only a beginning.

## REFERENCES


Bawaya, Michael. 2010. "Virtual Archaeologists Recreate Parts of Ancient Worlds." *Science* 327: 140–141.

Baycan, Tuzin, and Luigi Fusco. 2011. "Heritage in Socio-Economic Development: Direct and Indirect Impacts." International Council on Monuments and Sites. PDF at openarchive.icomos.org.

Bittel, Jason. 2013. "Get a Whiff of Olden Days with High-Tech Archaeology." *Slate Magazine*, (https://slate.com/technology/future-tense/293

Central Bureau of Statistics. 1968. Census of Population and Housing. Jerusalem Municipality, Table 3.

Chodoronek, Michael. 2015 "The Use and Application of Photogrammetry for the In-field Documentation of Archaeological Features: Three Case Studies from

the Great Plains and Southeastern Alaska." Master's thesis, Department of Anthropology, University of Nebraska, Lincoln.
Dimitriyadis, Irini, Sureyya O. Akyuz, and Feride H. Basgturk. 2013. "A Proposition for a Mechanisms to Provide Supplemental Funding for Cultural Heritage." *Procedia-Social and Behavioral Sciences* 62: 1332–1336.
Dombrowsky, Ines, Ram Almog, Nir Becker, Eran Feitelson, Simone Klawitter, Stefan Lindemann, and Natalie Mutlak. 2010. "How Widely Applicable Is River Basin Management? An Analysis of Wastewater Management in an Arid Transboundary Case." *Environmental Management* 45(5): 1112–1126.
Dumper, Michael. 2002. *The Politics of Sacred Space: The Old City of Jerusalem in the Middle East Conflict*. Boulder, CO: Rienner.
European Investment Bank Institute, Getting Cultural Heritage to Work for Europe, Report of the 2020 Expert Group in Cutlural Heritage, 2015.
Fernea, Elizabeth Warnock, and Mary Evelyn Hocking. 1992. *The Struggle for Peace: Israelis and Palestinians*. Austin: University of Texas Press.
Fox, T. 2015. "UC San Diego's Thomas E. Levy Among Recipients of President's Research Catalyst Award." UC San Diego News Center, December 9. http://ucsdnews.ucsd.edu/pressrelease/uc_san_diegos_thomas_e._levy_among_recipients_of_presidents_research_cataly.
Gannon, Megan. 2014. "'Extraordinarily Rare' Crusade-Era Seal Discovered in Jerusalem." *Live Science*, May 27. www.livescience.com/45893-rare-lead-seal-discovered-ancient-jerusalem.html.
Girardeau-Montaut, Daniel. 2011. *CloudCompare—Open Source Project*. http://www.danielgm.net/cc/.
Goodman, Martin. 2007. *Rome and Jerusalem: The Clash of Ancient Civilizations*. New York: Alfred A. Knopf.
Hirschfeld, Yizhar. 1995. "Spiritualty in the Desert: Judean Wilderness Monasteries." *Biblical Archaeology Review* 21: 5.
Howland, M. D., F. Kuester, and T. E. Levy. 2014. "Structure from Motion: Twenty-First Century Field Recording with 3D Technology." *Near Eastern Archaeology* 77(3): 187–191.
Jewish Virtual Library. 2016. *History of Jerusalem: Jordan's Desecration of Jerusalem (1948–1967)*. http://www.jewishvirtuallibrary.org/jsource/Peace/destoc.html.
Kersel, M., C. Luke, and C. Roosevelt. 2008. "Valuing the Past." *Journal of Social Archaeology* 8(3): 298–319.
Knabb, K. A., J. P. Schulze, F. Kuester, T. A. DeFanti, and T. E. Levy. 2014. "Scientific Visualization, 3D Immersive Virtual Reality Environments, and Archaeology in Jordan and the Near East." *Near Eastern Archaeology* 77(3): 228–232.
Levy, T. E. 2013. "Cyber-Archaeology and World Cultural Heritage: Insights from the Holy Land." *Bulletin of the American Academy of Arts & Sciences* 66: 26–33.
Levy, Thomas E., Thomas Higham, Christopher Bronk Ramsey, Neil G. Smith, Erez Ben-Yosef, Mark Robinson, Stefan Münger, Kyle Knabb, Jurgen Schulze, Mohammad Najjar, and Lisa Tauxe. 2008. "High-Precision Radiocarbon Dating and Historical Biblical Archaeology in Southern Jordan." *Proceedings of the National Academy of Sciences* 105(43): 16460–16465.
Macdonald, Susan. 2011. "Leveraging Heritage, Public-Private, and Third Sector Partnerships for the Conservation of the Historic Urban Environment." International Council on Monuments and Sites. http://openarchive.icomos.org/1303/

Magness, Jodi. 2012. *The Archaeology of the Holy Land: From the Destruction of Solomon's Temple to the Muslim Conquest*. Cambridge and New York: Cambridge University Press.

Microsoft. 2015. C# Programming Guide: Visual Studio. https://msdn.microsoft.com/en-us/library/67ef8sbd.aspx.

Mikropoulos, Tassos A., Nikiforos M. Papachristos, and George Zacharis. 2008. "The Educational Approaches to Virtual Reality Technologies Laboratory (EARTHLAB)." In *Proceeding of International Symposium on Information and Communication Technologies in Cultural Heritage*. Ioannina, Greece: The University of Ioannina, Greecde.

Milken Financial Innovations Lab. 2008. *Financial Innovations for Developing Archaeological Discovery and Conservation*. Jerusalem: Milken Institute.

Milken Financial Innovations Lab. 2011. *Cultural Heritage as an Economic Development Resource in Israel*. Jerusalem: Milken Institute.

Milken Innovation Center-Jerusalem Institute. 2013. *Kidron River Restoration: Project Financing Plan*. Jerusalem: Milken Institute.

Pujol, Laia. 2004. "Archaeology, Museums and Virtual Reality." *Digit·HVM: Revista Digital D'Humanitats* 6. http://www.uoc.edu/humfil/articles/eng/pujol0304/pujol0304.pdf.

Rahmani, L. Y. 1981. "Ancient Jerusalem's Funerary Customs and Tombs: Part Two." *Biblical Archaeologist* 44(4): 229–235.

Richards, Greg. 2018. "Cultural Tourism: A Review of Recent Research and Trends." *Journal of Hospitality and Tourism Management* 36 (September): 12–21.

Skarin, Ingela Palsson. 2011. "A Finance Mode for Built Cultural Heritage—Proposals for Improvements for Future Heritage Economics." Working paper, Lund University.

Rypkema, Donovan D. 2008. "Heritage Conservation and the Local Economy." *Global Urban Development Magazine* (Source: https://www.globalurban.org/GUDMag08Vol4Iss1/Rypkema.htm)

Smith, E. A., and J. Z. Deaton. 2016. "Satellite Pictures Suggest Oldest Christian Monastery in Iraq Razed to Ground." CNN, January 21. www.cnn.com/2016/01/21/world/iraq-monastery-reportedly-destroyed/index.html.

Smith, N. G., L. Passone, S. al-Said, M. al-Farhan, and T. E. Levy. 2014. "Drones in Archaeology: Integrated Data Capture, Processing, and Dissemination in the al-Ula Valley, Saudi Arabia." *Near Eastern Archaeology* 77(3): 176–181.

Unity Technologies. 2001. Unity 3D. https://unity3d.com/.

CHAPTER 8

# Oporto's Historic Center

## *From Historical Research to (Real) Virtual Heritage Visualization*

MARIA LEONOR BOTELHO

### OPORTO'S WORLD HERITAGE

The city of Oporto, built along the hillsides overlooking the mouth of the Douro River in Portugal (figure 8.1), is a city with an outstanding urban landscape. The city developed through its links to the sea, founded by the Romans as Portus (port) and continuously occupied and developed since then. This rich history is reflected in its many and varied monuments, from the cathedral with its Romanesque choir, to the neoclassical Stock Exchange and the typically Portuguese Manueline-style Church of Santa Clara (on World Heritage Convention List). Since 1996, the city has been registered with the World Heritage Committee owing to the exceptional and universal value of its urban fabric and its many historic buildings, which bear remarkable testimony to its development. In Oporto, the area assigned World Heritage Status comprises about 90 hectares, corresponding approximately to the area bounded by the medieval walls of the fourteenth century, the Luíz I Bridge, and the Monastery of Serra do Pilar, in Vila Nova de Gaia.

The Convention for the Protection of the World Cultural and Natural Heritage (UNESCO 1972) has been, on a global scale, one of the most important developments for the dissemination and maintenance of the value and importance of our cultural and natural world heritage. Importantly, the convention states that one of the responsibilities of each country with an area designated as important through the convention is to "ensure the identification,

Maria Leonor Botelho, *Oporto's Historic Center* In: *Digital Cities*. Edited by: Maurizio Forte and Helena Murteira, Oxford University Press (2020). © Oxford University Press.
DOI: 10.1093/oso/9780190498900.003.0009

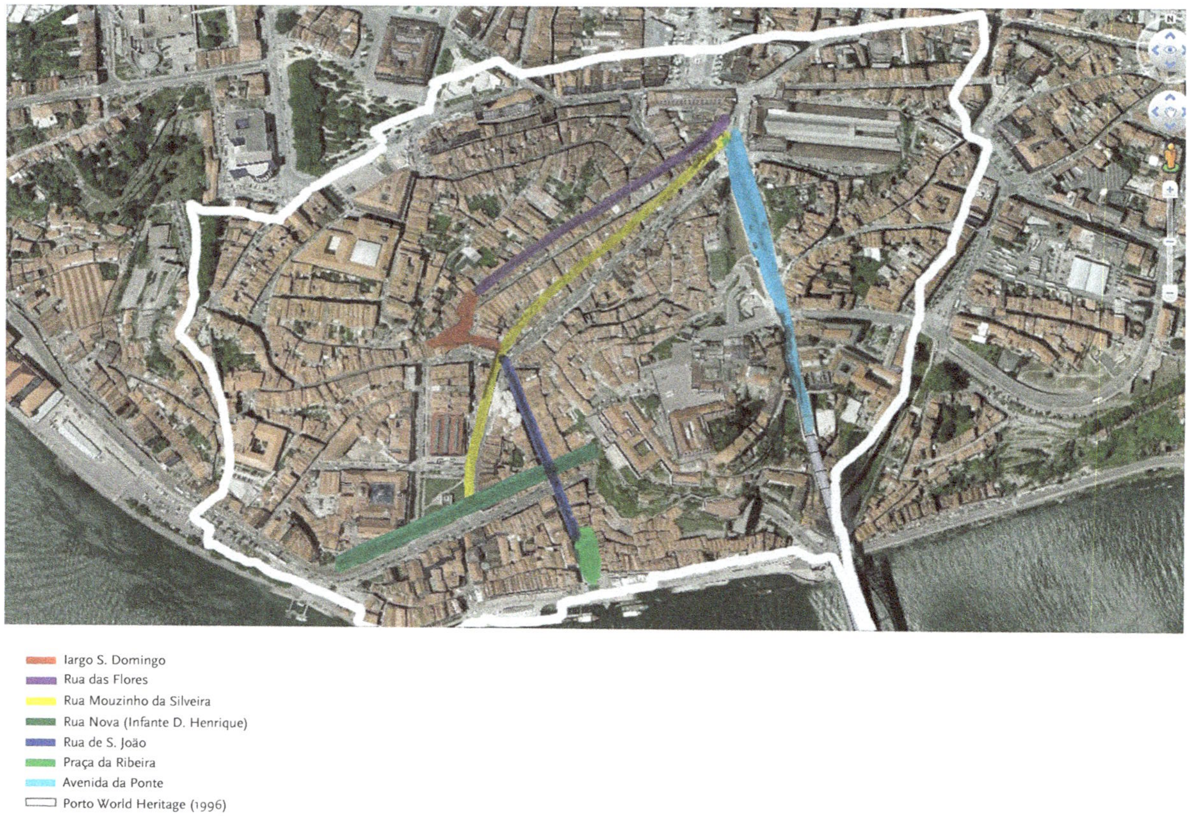


**Figure 8.1** Scheme of urban layering and street opening in Oporto's Historic Center.
*Source:* Botelho and Ferreira 2013.

protection, conservation, appreciation and transmission to future generations of the cultural and natural heritage" (article 4). Moreover, the same document promotes the development of scientific and technical studies or research projects that would benefit World Heritage properties (article 5, paragraph c). The inclusion of the historic center of Oporto, Luiz I Bridge, and Monastery of Serra do Pilar on the list of World Cultural Heritage sites has strengthened the need for integrated and comparative study (CRUARB/CH 1996, 1998) within the academic world, but also with the intention of promoting and disseminating the city's rich heritage to a wider audience.[1]

To facilitate the dissemination and safeguarding of heritage assets, it is essential to have a full and comprehensive understanding of the monuments, or in this case, of the city's development and its history. This understanding operates at many different levels, allowing exploration of the processes and different types of adaptation from centuries of activity that shaped the urban fabric and the resulting monuments—and which we have an obligation to safeguard for future generations (which is beyond merely preserving them).

Today there is a real ability to safeguard our heritage in many different ways, and this chapter, starting from a historiographical and contextualized reading in space and time, focuses on noninvasive techniques for visual reconstruction of Oporto's heritage using physical and digital/virtual media. This is accomplished by analyzing two of the first known examples of attempts to interpret and present Oporto's urban history in one of its multiples layers.

The requirement to communicate the significance and need for heritage conservation, whether natural or cultural, to the host community and tourists is a fundamental objective of the International Cultural Tourism Charter. The charter states that reasonable and well-managed physical, intellectual, and/or emotive access to heritage and cultural development is both a right and a privilege (ICOMOS 1999). To address this need for access, physical or virtual/digital media can (and should) be used, primarily as an instrument by the host community (in this case, as a knowledge tool to draw together our understanding of heritage), but also by the community of visitors to the site (in this context, as a means of promotion/dissemination). This chapter argues, therefore, that it is increasingly important that the creation of content optimize understanding by these two groups of the significant features of heritage assets and the consequent need for protection, recognition, and improvement of the experience.

More recently, the ICOMOS Charter for the Interpretation and Presentation of Cultural Heritage Sites (ICOMOS 2007) assumes that interpretation and presentation are essential elements in the effort to preserve heritage, as well as basic tools for public appreciation and understanding. The guidelines and requirements for management of heritage assets provide a framework for discussion of the historic core of Oporto and the means by which understanding of the city can be investigated. In presenting the conservation of heritage as a communicative act, it is necessary that we value the task of communicating that meaning (or meanings)—that is to say, that the historical layers of the Historic Center of Oporto developed over time to finally reach its unique character that we know today.

In this context, it's important to recognize the reality of the complex and multilayered city of Oporto to understand its evolution.[2] We can protect only what we know, and we can communicate only what we understand. This can be demonstrated with consideration of the medieval core that forms the UNESCO Historical Center, remembering that the protected area roughly coincides with the original layout of the Gothic Wall. This serves as the range of evidence—physical, historical, and pictorial—that has to be included in visualizations of the city.

## OPERTO'S HISTORY OF DEVELOPMENT

The earliest area of medieval Oporto developed on the hill of Pena Ventosa, centered on the cathedral, around which the urban area expanded toward the waterfront. Later development spread toward the Olivais (now known as Vitória). This area is characterized by narrow and winding streets that both served human needs and respected the natural topography. Although retaining much of its medieval character, this area has been altered by later urban developments. In 1406, the Rua Nova (now Rua Infante D. Henrique) was constructed, opened in 1406, during the reign of D. João I (1357–1433), as a broad, straight street that served as a "street-square," and rapidly became one of the most important centers in the city.

Further changes to the urban layout occurred between 1521 and 1525, with the construction of the Rua das Flores by order of King Manuel I (1495–1521) on land almost entirely belonging to the bishop. Retaining much of its original form, it was created to respond to the growing need for access to the sale of products that entered through the Alfândega (customhouse) and the gates of the north wall. Important changes to the urban topography took place in the city after 1769, under de Almadas family city government, within the walled area of the city: the Ribeira Square was developed, and the Rua de São João and the Largo de São Domingos were created, for example.

In the late nineteenth century, the Rua Mouzinho da Silveira was built as an important arterial route to facilitate movement between the waterfront and the São Bento railway station that opened to the public in 1896. Closely related to this is the so-called Avenida da Ponte (Avenida Afonso Henriques), which opened in the mid-twentieth century and was the connecting road between Luiz Bridge and the center of Oporto. The area that comprises the historic center, therefore, reflects a complex sequence of development, with streets added and altered over many centuries. This gives each area a unique character, layer by layer, if read in a horizontal or perspective way, but equally important to any understanding of the area are the structures and street features that mark the changing visual appearance of the city.

## STATE OF THE ART

For Oporto, there are many and varied sources covering a broad range of subjects that provide information and multiple perspectives on the development of the historic center.[3] Importantly, these sources exist for each historical period protected by UNESCO. The following are a few examples of such resources which served as reference studies, with notes as to the rigor of the information they provided for visualizing the city of the past.

First, detailed descriptions by clergymen, such as Manuel Pereira Novaes (Novaes 1912) or Agostinho Rebelo da Costa (Costa 2001), provide detailed portraits of the medieval city as they would have known it in the seventeenth and eighteenth centuries. These detailed descriptions can be complemented with analysis of information gleaned from urban topographic views of the city[4] and cartography (Câmara Municipal do Porto, 1963 and Meireles, 1982)[5]

Second, tourist interest in Oporto began from the end of the first quarter of the twentieth century[6] and yielded early accounts of travels to the city that focus on the city's heritage, even describing the monuments in a technical manner (Passos, 1926). Photography over time provides additional information on the changing appearance of the city. For example, the photographer, engraver, and editor José Antunes Marques Abreu provided a distinct perspective of Oporto's urban landscape which helps highlight how different the city is today from the one he recorded in the 1930's. Other important photographers, who also recorded the history and changing character of Oporto, include Emílio Bïel, Guilherme Bomfim Barreiros,[7] Domingos Alvão, and Foto Beleza (Espólio Fotográfico Português. 2015).[8] Through their works we can become familiar with the early appearance and character of Oporto—sometimes just as background, but often also offering a view of the central character they recorded through the medium of light.

Between 1962 and 1965, what is considered the first detailed work of reference drawing on academic and historical research on Oporto was begun by Artur de Magalhães Basto, Damião Peres, and António Cruz (Peres 1962–65). Their important work had a particularly descriptive (and simultaneously apologetic) tone, and included illustrations, many by Gouveia Portuense, that were 2D reconstructions of monuments and streets, many of which disappeared. Fundamental in drawing together knowledge of the city, their work provided important information on Porto in about 1500 and was highly useful in the creation of the Oporto's Medieval Model.[9]

An important urban study by Jaime Ferreira-Alves, working under the Almadas' family government (Ferreira-Alves 1988), pioneered a new look at the urban area that now comprises Oporto's Historic Center. He studied the changes to the urban landscape of Oporto in the period 1763 to 1804, seeking to define a new area of urban expansion, rebuilding the medieval urban fabric and creating new transport routes linking the riverside area and the gates of the Ferdinand wall, which had become a monument. In order to understand the scope of urban development, Ferreira-Alves thoroughly studied Oporto's medieval development, and he published a number of documentary and iconographic views. A later study by Bernardo Ferrão (Ferrão 1997) took an admittedly exploratory and historicist position, utilizing a projectual and analytical perspective on urban design transformation during the second half of the eighteenth century.

A subsequent study, the *História do Porto* by Luís A. de Oliveira Ramos (Ramos 1995), had the goal of updating the existing historical work by drawing together subsequent investigations that had since been published in countless books and magazines (13–14), among which were the important *Revista da Faculdade de Letras—Série História* and *O Tripeiro*. Using a methodology that involved studying local and municipal history through document research, *História do Porto* draws together the work of several notable authors and stands out for its scientific nature. It is an invaluable historiographical reference that gave a particular emphasis to describing the character of the urban spaces over the centuries. This allows the reader to witness the historic multilayering of the city and confirms its potential for study.[10]

More recently, José Ferrão Afonso undertook a detailed study of the creation and design of Rua de Santa Catarina das Flores in the sixteenth century (Afonso 2000). Through a systematic analysis of sources, the author identified significant data for understanding the urban transformations (which in themselves are reflections of the city's economy and society) that defined the development of Oporto's Historic Center. A subsequent publication by Afonso was a more comprehensive one that sought to understand the role the Church played in Oporto's history during that same century (Afonso 2013). This latter study methodically examined Oporto as a space of representation, assuming that the Church could have had an impact on urban form as well, while simultaneously seeing the space as the result of a long process of overlap (10).

Assuming the street was a key feature of the urban fabric in Oporto between 1386 and 1521, Daniel Borges Braz Afonso (2012) shows how this formed the basis for the center's medieval development. His work argues that there was a contrast between the spontaneous nature of the design for the majority of medieval streets and the regular nature of the streets of royal foundation.[11] To explore the importance of mutual and reciprocal relationships created between buildings/monuments and their immediate surroundings, the author also attempted to understand the role some buildings played in the expansion of its urban surroundings (3).

Some studies of Oporto's urban fabric have examined the changes to the Historic Center beginning in the nineteenth century. The work of António Cardoso (1997) on the architect José Marques da Silva (1869–1947) highlights the addition of São Bento's Railway Station and the projects undertaken regarding the current design of the Aliados avenue. In conjunction with the Management Plan for Oporto's Historic Center (SRU—Porto Vivo; http://www.portovivosru.pt/) and interventions as part of urban rehabilitation of the area, detailed works have been published that focus on specific areas such as the Cathedral (Delgado et al. 2011), the axis Mouzinho-Flores (Delgado et al. 2012), and Aliados (Vale 2013).

Therefore, for Oporto, understanding the urban fabric, and the people who designed it and saw themselves as part of the city, has made a significant contribution to the city's heritage requalification. Through a detailed study of the city, one of the ICOMOS Charter's agendas for the interpretation and presentation of cultural heritage sites states that "continuing research, training, and evaluation are essential components of the interpretation of a cultural heritage site" (principle 7) and so it can be addressed. Studies with an academic basis, and using the know-how of accredited experts, means that we have today a significant body of knowledge that contributes to understanding Oporto's Historic Center, and that knowledge base continues to grow. This wealth of information becomes the basis for future proposals for examinations of the city's past. Moreover, the formation of a scientific school, first within the Academy and then mirrored in accredited institutions and proven at the level of interpretation and asset safeguarding, has resulted in a multidisciplinary body capable of meeting the needs in investigations that address the varied character of the historic information, as well as meeting the challenges posed by the use of new technologies.

## THE FIRST VISUALIZATION MODELS

To demonstrate the potential for using existing historiographical sources and scientific production, we present two case studies that have made a significant contribution to our understanding of Oporto's Historic Center. They serve as teaching tools, as a means of visualizing and consolidating existing data, and for potential tourism use. The case studies, both pioneering in character, represent two different approaches to visualizing Oporto. The first is a physical scale model of the city while the second is a virtual animation. Despite the different approaches, both have careful documentation of the methodology and process involved, which accords with the values enshrined in documents relating to visualizations of the past, such as the London Charter (2009).

## Oporto's Medieval Scale Model

Oporto's medieval interactive scale model,[12] located in the Casa do Infante–Museum (Câmara Municipal do Porto), is a significant example of the role that historical research plays in the protection of heritage sites. The model was built as part of the exhibition organized in 1999 by the Municipal Historical Archive of Oporto, called Um olhar sobre o Porto Medieval (A Look at Medieval Oporto). The scale model is the "first experience of representing a Portuguese medieval city, to the scale 1:5000 and in three dimensions" (Real 1999). The model allows us to examine the urban construction of medieval Oporto, its main centers of development (Penaventosa, Miragaia, and Olivais), its structure, and possible areas for future urban expansion (figure 8.2).

The primary objective for the creation of this pedagogical scale model was that it be based on rigorous research and developed by medievalists who made a careful survey of the documentary sources and developed a program of research which itself is an important aspect of the methodology used. The exhibition at which it was shown featured background information on the accuracy of the research and the implementation requirements that guided its design, especially since there had never been a similar project, except for the Lisboa Pombalina (Real 1999, 3).[13] The exhibition's brochure, never published and accessible only through the project's archive, was a detailed record of the choices made, from the preliminary survey (1992–94), to design (1994), and to refinement/adjustment (1998–99) (Real 1999, 3; Afonso 2000). It was

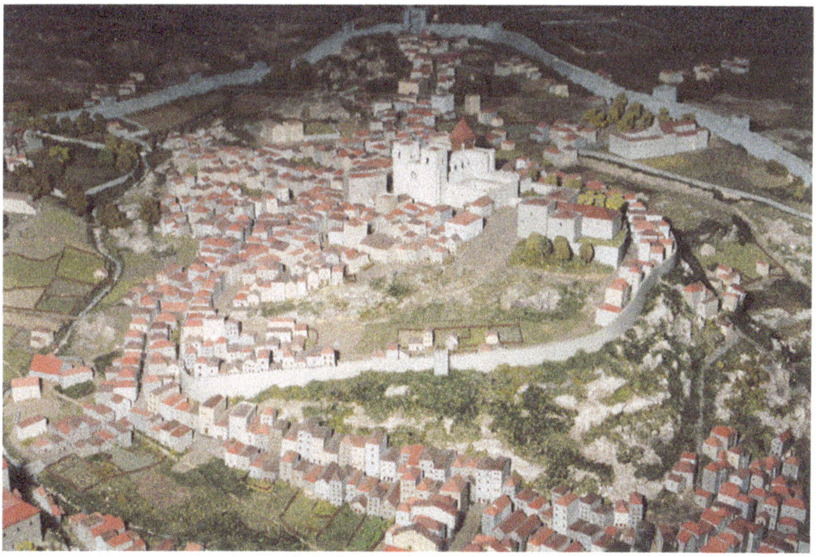

**Figure 8.2** Overview of Oporto's Medieval Scale Model (Casa do Infante).
*Credit:* Photographic reproduction courtesy of defunct file CRUARB.

organized into the following chapters: "How Was the Scale Model Born?"[14], "The Burgo's Reconstitution,"[15] and "The Scale Model's Execution."[16]

To create the model, we consulted a wide range of documentary and iconographic sources (Real 1999, 4), with the goal of developing a rigorous program of data collection, organized by areas of the city (Real 1999, 5). We had access to the Municipal Historical Archives, from which we drew the information that informed the design of the scale model,[17] and we were able, thorough the documentation process, to record how the model was prepared and designed.

The model was designed to be a representation of the entire city in medieval times, rather than focusing on one particular area (Real 1999, 6). Creating a detailed topographic model required understanding the territory's physical form, for which we consulted topographic studies (7–8). It also required grasp of the functional organization of space, the highways, the constructive scale (9),[18] the form and location of the main buildings, the distribution of infrastructure, and some street signage (3). To allow scenic views and to render details accurately, the model produced was of an imposing size, measuring 4 meters x 2 meters.

The model was constructed with dynamic modeling techniques (Real 1999, 25), involving interdisciplinary dialogues between the research team and the model maker, and with production of schematic notes and new computerized cutting machines to aid in accuracy (24). This methodology was particularly critical in defining spatial clusters within the urban landscape, and lent graphical accuracy to the representation of most important buildings (24–26). Though several years from its conception, Oporto's Medieval Model is still a valuable resource for study and an important tourist attraction. The model is still in its original location and its main audience consists of school groups and tourists.

### Virtual Oporto in the Sixteenth Century

The project Porto Virtual no Século XVI (Virtual Oporto in the Sixteenth Century)[19] was built on the new technologies offered by 3D visualizations that allowed us to show the elevations of the city's cathedral and its immediate urban surroundings as they were at the beginning of the sixteenth century. The modeling and animation depict a moment immediately *preceding* replacement of the original Romanesque chevet of Oporto's cathedral with the current one (1606–10) commissioned by D. Fr. Gonçalo de Moraes, bishop of Oporto from 1603 to 1617. The result is good testimony to how one can reconcile rigorous academic knowledge with new technologies and place a specific event within a broader context.

For the virtual project, there was detailed historical research (Afonso and Botelho 2005) that provided that information permitted greater accuracy in modeling the entire built environment, and it was also important for the creation of the animations. The creation of the virtual design, and the complementary historical discussions, storyboards, and videos about the animation process, have been made available online (https://goo.gl/BKMJKk).

The Porto Virtual Século XVI project provides a trip to Oporto's Cathedral, showing it with a totally different appearance from what we know it today (figure 8.3). This is particularly the case with the main facade, but also its north side (where we see the São João Porch, which was subsequently replaced by the current Baroque porch), with the Romanesque chevet, endowed with ambulatory and radiating chapels and the surrounding urban area. The latter has a clearly medieval character, without ample public spaces. This virtual model allows users to explore in detail the cathedral in virtual visits (in high and low resolution) to the main facade, interior, apse, cruise, porch, and cloister.

The approach adopted for the virtual model broadly follows the operating principles of the "crypto-history of art" proposed by Vítor Serrão (2001). This is one approach within the history of art that is attentive to the role of art works that have not survived to the present day, but that may have played a remarkable role in certain circumstances in the past (12). Thus, Serrão notes, "the Crypto-History of Art defines itself as an absolutely new concept for extending a longstanding historiographical practice, but that was never truly well made aware, to integrate the study of the missing heritage in the current methodology of this discipline" (12). In fact, in the current history of art, it is increasingly relevant to study what once existed and has now disappeared or that has been totally transformed by taste through time.

In archeology, we emphasize particularly two of the operating principles proposed by Serrão (2001, 3): (1) the slope of the *deduction*— in other words, the approach to those already missing works, which we try to convey by visual analysis, documentary, stylistic and iconographic meaning, and analysis of the known works' setting; and (2) the slope of the *reconstitution*—in other words, the analysis of a fragment of an artistic ensemble now partially nonexistent in order to unravel its possible initial structure.

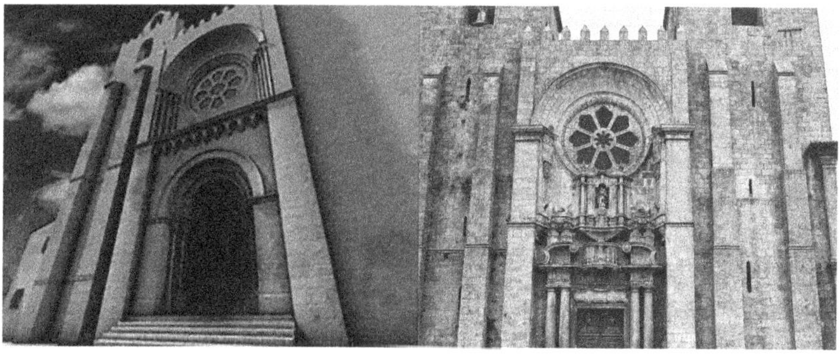

**Figure 8.3** Main façade of Oporto's Cathedral in the sixteenth century and in 2005.
*Source:* www.artes.ucp.pt/citar/portoxvi/imagens_a_d_1.php.

The Virtual Oporto project has therefore opened up a new and innovative approach to the visual representation of one of the city's most important monuments, the cathedral, and its surrounding area. Importantly, this project, with its detailed documentation, has obvious pedagogical potential, either for historical study or for the visual content generated. The methodological rigor of the project in itself is an example of such an approach, demonstrating the gains associated with projects that involve multidisciplinary teams.

## FINAL CONSIDERATIONS: EXISTING MODELS AND THE LONDON CHARTER

The scale model and the virtual visualization are timely examples that demonstrate the importance of academic rigor, basic and fundamental in the construction of materials that may be suitable for tourism (Botelho, and Ferreira, 2014), as well as instruments of study. The development of media, virtual or physical, as tools or devices for representing information, knowledge, and interpretation, should be understood as models, and not as a means of replacing direct experience; we still need the physical and material context to understand the complex and multilayered nature of our urban environment.

Although the principles of the London Charter (2009) are intended for digital reconstruction, we can apply them to both case studies presented here. Even though these projects predate the London Charter, they encapsulate the principles enshrined there—notably the requirement of transparency and the importance of documentation. It should be emphasized that these models were a first approach to the subject, cutting a path ahead for what is now being done in this field. The two Oporto projects are still available, and both have demonstrated their ability to draw together highly qualified individuals in a combined effort. In both cases, the operating principles of the crypto-history of art (deduction and reconstitution) have demonstrated the potential that 3D reconstructions have, whether physical or digital.

As discussed earlier, the designation of World Heritage status for the Historic Center of Oporto includes a requirement to promote the city's heritage, both to the local community and to tourists. There is therefore the question of the capabilities that these projects and any future projects have in terms of meeting those two goals, and to what extent it is possible to update the information through historical research. It can be argued that museum display of the scale model and online access to the Virtual Oporto make them accessible and intelligible to a variety of audiences. The availability of documentation on their development further enhances their usefulness. Since these models were designed and produced, there has been significant additional historical research. The scale model has been updated based on this information. Its digital partner deserved its updating as well, by expanding it to other areas of the city.

In conclusion, this chapter has stressed the importance of varied approaches to visualize and investigate the urban environment of historiographical knowledge. While there are merits to the production of physical models, the use of digital technologies offers greater flexibility for the representation of the complex and evolving character of the urban environment. It is arguably only through such virtual representations that we are able to understand today what made the protected area defined by the Gothic Wall significant. Equally, this discussion has shown that a virtual reconstruction has to be undertaken by an interdisciplinary team, which allows a wider platform for debate and for the constant updating of knowledge on both a spatial and temporal perspective. In turn, the production of virtual models provides a means for communicating knowledge to local and tourist communities, which accords with the covenants for heritage sites set out by UNESCO and ICOMOS.

There are now notable examples of the use of digital technologies to explore the urban history of cities, such as Paris 3D: A Tour of the City Through the Ages (https://www.3ds.com/stories/paris-3d/#Heritage), Discover Medieval Chester (http://discover.medievalchester.ac.uk/), or even City Witness (www.medievalswansea.ac.uk/en/). Thus, this is the time for creation of a more ambitious VR project that can reveal the urban transformations of Oporto's Historic Center in accordance with its historical stratigraphy.

## REFERENCES


Afonso, Daniel. 2012. "A rua na construção da forma urbana medieval: Porto, 1386–1521." MA dissertation, University of Porto.

Afonso, José Ferrão. 2000. *A Rua das Flores no século XVI—Elementos para a História Urbana do Porto Quinhentista*. Porto: FAUP Publicações.

Afonso, José Ferrão. 2013. *"A imagem tem que saltar": A igreja e o Porto no século XVI (1499–1606) Um estudo de história urbana*. Lisbon: FCG & FCT.

Afonso, José Ferrão, and Maria Leonor Botelho. 2005. *A Sé e a sua envolvente no Século XVI. Memória Justificativa*. Projecto Porto Século XVI. Porto: CITAR. http://www.artes.ucp.pt/citar/portoxvi/documentos.php.

Arquivo Histórico Municipal do Porto. 1983–. *Porto. Projectar a Cidade*. 5 vols. Porto: AHCMP.

Botelho, and Teresa Ferreira. 2014. "Tourism, Heritage and Authenticity: The Case of Porto." In *1st International Conference Porto as a Tourism Destination: City Tourism*, 261–271. Porto: CEPESE and Media XXI.

Braga, Maria Helena G., ed. 2001. *Bomfim Barreiros. Fotógrafo de Arte*. Porto: CMP.

Cardoso, António. 1997. *O Arquitecto José Marques da Silva e a arquitectura do Norte do País na primeira metade do séc. XX*. Porto: FAUP.

Câmara Municipal do Porto. 1963. "View of the Oporto's Eastern Part (1844), nº3." In *Edição Comemorativa da Inauguração da Ponte da Arrábida*. Porto: CMP.

Câmara Municipal do Porto. 2015. "Casa do Infante—Museu." In *Balcão de atendimento Virtual—Câmara Municipal do Porto*. https://goo.gl/47YwKk.

Costa, Agostinho Rebelo da. 2001. *Descrição Topográfica e Histórica da Cidade do Porto.* Lisbon: Frenesi.

Costa, Jorge, ed. 2002. *Bibliografia Portuense. Títulos disponíveis no mercado.* Porto: Biblioteca Pública Municipal do Porto.

CRUARB/CH. 1996. *Porto a Património Mundial.* Porto: CMP.

CRUARB/CH. 1998. *Porto a Património Mundial. Processo de Candidatura do Centro Histórico do Porto à UNESCO—Livro II.* Porto: CMP.

Delgado, Ana Paula, Paulo Valença, and Margarida Guimarães, eds. 2011. *Morro da Sé. De Porta a Porta.* Porto: Porto Vivo, SRU.

Delgado, Ana Paula, Paulo Valença, and Margarida Guimarães, eds. 2012. *Eixo Mouzinho/ Flores—Território do Recolhimento e do Mercadejar.* Porto: Porto Vivo, SRU.

Espólio Fotográfico Português. 2015. https://goo.gl/Q5swSt

Ferrão, Bernado. 1997. *Projecto e Transformação Urbana do Porto na Época dos Almadas, 1758/1813. Uma contribuição para o estudo da cidade pombalina.* Porto: FAUP.

Ferreira, Carmo. 2007. *Índice Geral de "O Tripeiro" (junho 1908 a dezembro de 2016).* Porto: Campo das Letras.

Ferreira, José Augusto, and José Marques Abreu. 1928. *Porto: Origens históricas e seus principais monumentos. Cathedral, Santa Clara, S. Francisco e Cedofeita.* Porto: Imprensa Marques Abreu.

Ferreira-Alves, Joaquim Jaime. 1988. *O Porto na Época dos Almadas. Arquitectura. Obras Públicas.* Porto: Câmara Municipal do Porto.

ICOMOS. 1999. International Charter on Cultural Tourism. México: ICOMOS.

ICOMOS. 2007. The ICOMOS Charter for the Interpretation and Presentation of Cultural Heritage Sites.

London Charter. 2009. http://www.londoncharter.org/

Meireles. 1982. *Catálogo dos Livros de Plantas. Instrumentos de Pesquisa para a História do Porto.* Porto: AHCMP.

Museu de Lisboa. 2015. http://www.museudelisboa.pt/.

Novaes, Manuel Pereira. 1912. *Anacrises Historial.* 8 vols. Porto: Bibliotheca Publica Municipal do Porto.

Passos, Carlos de. 1926. *Porto: Noticia histórico-archeologica e artistica da Cathedral e da egrejas de Santa Clara, S. Francisco e Cedofeita.* Arte Portuguesa. Porto: Marques de Abreu.

Passos, Carlos de. 1935. *Guia Histórica e Artística do Porto.* Porto: Livraria Figueirinhas.

Peres, Damião. 1962–65. *História da Cidade do Porto.* 3 vols. Porto: Portucalense Editora.

Porto Virtual Século XVI. 2005. https://goo.gl/BKMJKk

Ramos, Luís A. De Oliveira. 1995. *História do Porto.* Porto: Porto Editora.

Real, Manuel Luís 1999. *Um olhar sobre o Porto Medieval. Roteiro da Exposição.* Porto: Câmara Municipal do Porto.

Real, Manuel Luís, and Rui Tavares. 1993. "Bases para a compreensão do desenvolvimento urbanístico do *Porto.*" In *Porto a Património Mundial, Porto*, 59–82. Processo de candidatura do Centro histórico do Porto à UNESCO – Livro II. 1ª edição, Porto, CMP

Serrão, Vítor. 2001. *A Cripto-História da Arte. Análise de Obras de Arte Inexistentes.* Lisbon: Livros Horizonte.

UNESCO. 1972. World Heritage Convention. Paris: UNESCO.

UNESCO. "Historic Centre of Oporto." World Heritage Convention List. http://whc.unesco.org/en/list/755.

Vale, Clara Pimenta do. 2013. *Avenida dos Aliados e Baixa do Porto. Memória, Realidade e Permanência.* Porto: Porto Vivo, SRU.

CHAPTER 9

# Omnidirectional Strategies for Exploring Ancient Cities and Territories

SARAH KENDERDINE

## INTRODUCTION

Cultural heritage worldwide faces increasing threat from numerous destructive forces, from iconoclasm to climate change. The recent escalation in the annihilation of heritage and its guardians—such as that witnessed in Palmyra, Syria, in 2015—is emblematic of acts of "cultural cleansing" that have been with us for thousands of years. However, such obliteration remains one of the most powerful political and sociocultural weapons of our time (UNESCO 2015). Digital documentation of cultural heritage sites can play a vital role in sustaining our most precious legacies; for example, advanced technologies typically used in archaeology today now include photogrammetry, laser scanning, remote sensing and geographic information systems (GIS), aerial modeling, and gigapixel imaging, to name just a few (Sanders 2015). However, as writer Alexander Stille has observed, the currently abundance of digital data exists alongside the ongoing destruction of cultural heritage in its various forms: "One of the great ironies of the information age is that, while the late twentieth century will undoubtedly have recorded more data than any other period in history, it will also almost certainly have lost more information than any previous era" (Stille 2006).

As UNESCO reminds us, awareness of heritage is a cultural emergency. This awareness is described as both a political and a security imperative. What is crucial to the mission of museums in the project of cultural awareness is responding to the need for successive generations to rediscover and reinvent the past

Sarah Kenderdine, *Omnidirectional Strategies for Exploring Ancient Cities and Territories* In: *Digital Cities*. Edited by: Maurizio Forte and Helena Murteira, Oxford University Press (2020). © Oxford University Press.
DOI: 10.1093/oso/9780190498900.003.0010

themselves—in other words, keeping cultural heritage living (Bokova 2015). Indeed, the very sustainability of culture depends on what Homi K. Bhabha calls "insurgent acts of cultural translation," where art (or exhibition) does not merely recall the past but also renews it—refiguring, innovating, and interrupting the performance of the present. He says that this "'past–present' becomes part of the necessity, not the nostalgia, of living" (Bhabha 1994, 12).

What is demonstrated by the interactive installations discussed in this chapter are strategies by which museum visitors are enabled to virtually re-embody and "perform" digital heritage archives. This "embodied museography" is defined by attributes of immersion, interaction, and participation, and it asks us to re-examine our notions of aura, authenticity, and authorship while enabling audiences to translate the digital record into "living" narratives of engagement (for extended analyses, see Kenderdine 2013a, 2013b). These embodied experiences are achieved through the use of advanced virtual reality (VR) technologies purpose-built for the establishment of presence and the sense of inhabiting "place."

The history of the re-presentation of "place" in virtual reality has been extensively explored in the work of early pioneers in new media art and "future cinema" or expanded cinema (e.g., Shaw and Weibel 2003, Del Favero and Shaw 2001). Projects by virtual heritage researchers—and the language of presence, place, and narrative that has emerged from these experiments—have been subject to numerous analyses and much theorization (e.g., Barceló et al. 2000, Cameron and Kenderdine 2007, Kalay et al. 2007, to name just a few). Through these optics, this chapter explores strategies for representing individual spaces, cultural heritage sites, and whole cultural landscapes connected through particular theses or themes.

Charting the development of these strategies for presence and narrative in immersive VR, this chapter briefly introduces seminal experiments from new media artists that have advanced capacities for panoramic immersion to then focus on interactive installations of world heritage: *PLACE-Hampi* (2006) and *Pure Land: Inside the Mogao Grottoes at Dunhuang* (2012–16). This discussion expands on the exploration of place and presence in these installations to examine postcartographic and "deep mapping" representations of cultural landscapes, and fresh approaches to telling historic narratives via embodied interaction through the forthcoming interactive exhibition of the *Atlas of Maritime Buddhism* (2018).

## MEDIA ARCHAEOLOGY: PANORAMIC IMMERSION

In virtual reality, the panoramic view is joined by sensorimotor exploration of an image space that gives the impression of a "living" environment.

—O. Grau, *Virtual Art*

Panoramic immersion is at the core of many of today's virtual reality paradigms. Extrapolating from the 360-degree panorama has been a basis for developing new methods of representation, embodiment, navigation, and narrative—properties that lie at the heart of many digital cultural heritage works, as well as the current VR explosion.

The mass public screen entertainment of the panorama is the subject of a number of extensive analytical histories[1] and has led Stephen Oettermann to claim the panorama as "the first true mass medium" (Oettermann 1997, 7). The twenty-first century re-emergence of the panoramic format has been described as "the new image vogue" (Parente and Velho 2008, 9), and is based on the desire to extend the immersive quality of expanded fields of vision in order to design virtual space that is truly "inhabited" by the viewer—maximizing immersion in order to elicit "presence."[2] The panorama strategy—immersion via the extension of the lateral field of vision—underwrites many VR and approaches to immersive interaction today, including, for example, head-mounted displays (HMDs). In digital heritage, large-scale panoramic display is typically employed for the display of past environments generated from archaeological and historical data as digital interpretations, or real-world locations—for example, for the display of archaeological sites and documentation, and research visualization.

The panoramic format as a navigable space and an immersive experience is persistent throughout media history to the present (see Kenderdine 2007). In considerations of the current re-emergence of interest in this scheme from contemporary VR producers, this history reinforces one of the primary notions under discussion here: that omnidirectional, navigable, and interactive virtual spaces provide powerful interfaces for digital cultural heritage in the current climate through their capacity to re-create "place."

Since the demise of the nineteenth-century panorama for mass entertainment (Altick 1978), the panoramic schema was reinvented on the cinema screen and the internet. IMAX, for example, exhibits a monumentalism and "declares itself to a global audience" in a "spectacle of gigantism" (Wasson 2007, 83). In such an environment, the spectator is subordinate, and the overwhelming image instinctively invites both awe and trepidation. IMAX may be about the power of the camera to survey everything, but it is simultaneously about our own lack of power to see as is seen, notes Wasson. IMAX augments but also confronts the power of human vision (Wasson 2007, 88). By contrast, HMDs and interactive virtual reality interfaces restore agency to the participants through embodiment and a real-world, natural scale for panoramic vision. Re-scaling immersion in this way takes place not only through what we typically consider "virtual reality experiences" but also through the proliferation of internet technologies for the panoramic form that have been defined by miniaturization and remote consumption of the visual world. These are perfect forms for the digital *flâneur*, the geographer of distributed places, and are rapidly expanding out of the realm of map extensions for simple physical navigation into heavily curated presentations of place. The promise of the infinite panorama in Google Cultural Institute's "World Wonders" project,[3] for example, leads Schiller and Yeo (2014, 45) to observe that "Google is quietly reorganizing cultural spaces on a global scale, to incorporate them into its more encompassing business of information".

Media artists who have experimented with ways of inhabiting the panoramic imaginary frame this discussion of the remediated panoramic form. Large-scale, situated installation artists—such as Michael Naimark (United States), Luc Courchesne (Canada), Bernd Lintermann (Germany), Masaki Fujihata (Japan), and Jeffrey Shaw (Australia)—have all been working within the oeuvre of extended narratives and augmented devices for immersive panoramic images since the mid-1980s, and their works are useful examples in the context of this chapter. Seminal works that invoke the panorama include: *Moving Movie* (1977), *Displacements* (1980–4), and *Be Now Here* (1995–7) by Michael Naimark;[4] *Morel's Panorama* (2003) by Masaki Fujihata;[5] *Place—A User's Manual* (1995), *Place D'Uhr* (2000), and the immersive platforms *EVE* (1993–2004) and *Panoramic Navigator* (1997) by Jeffrey Shaw;[6] *Landscape One* (1997) and *Panoscope 360°* (2000–2001) by Luc Courchesne;[7] *Wall of Death* (1999) by Graham Ellard and Stephen Johnstone;[8] and *Globorama* (2007) by Bernd Lintermann.[9]

These works each feature video sequences and the use of cinematographic devices—but they also extend the capacity of their media by combining immersive architecture with panoramas, thereby conjoining the interactive language of new digital interfaces with a familiar form of engagement (the cinematic image). These artworks have laid foundations for many future possibilities of immersive and interactive cinema and the presentation of place that occupies industry today. As media researcher Timothy Murray (2008) asserts, extrapolating from Deleuze's "time-image" and "movement-image,"[10] new media art requires a fundamental conceptual shift from linear visual projection to nonlinear temporal (and spatial) folds that are intrinsic to the digital form and the expanded cinema.

A prime example of the potential of such infrastructures in a heritage context is Michael Naimark's groundbreaking *Be Now Here* (1994).[11] To create this work, Naimark traveled to heritage sites around the world to record panoramic views and capture 360-degree motion scenes at locations such as Angkor in Cambodia, Dubrovnik in Croatia, Timbuktu in Mali, and Jerusalem in Israel. His immersive display consisted of 3D video projected onto a 360-degree screen, combined with a spatial soundscape and an anthropological approach to both virtual travel and site documentation (Grau 2003, 240–42).

*Be Here Now* predates the virtual heritage work *Sacred Angkor* by nearly ten years (see Kenderdine 2004), but a comparison of these two works is useful for describing issues of presence. *Sacred Angkor* (2004) layers together spherical, stereoscopic panoramas with visualizations of archaeological elements that are not immediately visible at the site. Computer-generated augmentations utilize radar data of the site from NASA; archaeological and architectural plans of respective buildings shown; sparse and occasional recoloring of a once highly-painted bas relief sculpture; and algorithms for displaying water and surface reflections of surrounding scenery in real time.

The differences between *Be Now Here* and *Sacred Angkor* are immediately obvious in terms of the physical architecture of the respective displays: *Sacred Angkor* was designed for The Virtual Room, an eight-sided rear-projected stereo display system around which up to 80 mobile viewers circumambulate a series of slowly rotating panoramas at real-world scale (see Kenderdine 2004, Kenderdine and Hart 2003). This act of circumambulation in nine slowly rotating panoramas has a profoundly somatic affect, and references the typical way in which devotees access sacred sites like Angkor. In an act of dynamic contemplation, visitors to *Sacred Angkor* walk through the landscapes at the correct perspective enhanced by the stereoscopic display and sound fields (Doornbusch and Kenderdine 2004). The slow image sequences can be focused on and followed by the viewer at will. In contrast, *Be Now Here* is designed for a smaller number of viewers who stand on a platform that rotates 360 degrees every two minutes to give the illusion of panning across the images. The viewers are required to walk to keep their position respective to the fixed projection, another effective form of kinaesthetic connection.

*Be Here Now* and *Sacred Angkor* are early examples of immersive works testing strategies for evoking presence, highlighting the contribution of media artists to both the VR project and digital heritage. Another key contributor to the development of panoramic experience is Jeffrey Shaw, whose seminal immersive works are formative to current theories in embodied and interactive panoramas. As the theorist and designer Lev Manovich describes, he "evokes the navigation methods of panorama, cinema, video and virtual reality and 'layers' them side by side" (Manovich 2001, 282). Installations such as *Place* (1995) and *Place - Ruhr* (2000)[12] surround the visitor (who stands on a rotating platform) within a 360-degree panoramic screen, and reframe the traditional panorama within the contemporary modalities of virtual reality. The interface allows the visitor to navigate between the various locations—each of which is depicted in panoramic cylinders that have been distributed throughout the landscape map. Once inside the individual panoramic cylinders, the user confronts a scene augmented by incidental animated effects. These paradigms were further extended to the domain of cultural heritage visualisation in subsequent "Place" collaborations (e.g., *PLACE-Hampi*; see prior mention).

The use of the panorama in virtual, immersive environments provides a lexicon for navigable space that not only is "a topology, geometry and logic of static space" but also is transformed by "new ways in which space can function in computer culture" (Manovich 2001, 280). Virtual derivatives of the traditional panorama demonstrate how closely digitally augmented, hybrid-reality strategies engage with the historical scheme that precedes them. If the purpose of digital cultural heritage research is to develop ways in which cultural heritage assets can be accessed and engaged with, then panoramic virtual worlds are a persuasive interface paradigm for these purposes—indeed, digital heritage specialists and artists are in many senses leading developments in immersive panoramic display. Extension of the panoramic into the establishment of the kinaesthetic and embodied nature of immersive display systems for cultural heritage is demonstrated most clearly in the description of the projects that follows.

## VIRTUAL HERITAGE AS THE EMBODIMENT OF PLACE

*PLACE-Hampi* (2006—see Kenderdine 2013a)[13] conjoins earlier artist experiments to cultural heritage to create an embodied theatre of participation in the drama of Hindu mythology focused at significant archaeological, historical, and sacred locations of the World Heritage Site, Vijayanagara (Hampi), South India. It is an interactive installation that projects 3D panoramas onto a 9-metre diameter, 360-degree panoramic screen from a motorized platform that is controlled by one of the visitors (figure 9.1). The user stands on the platform and is able to steer it 360 degrees, simultaneously rotating a

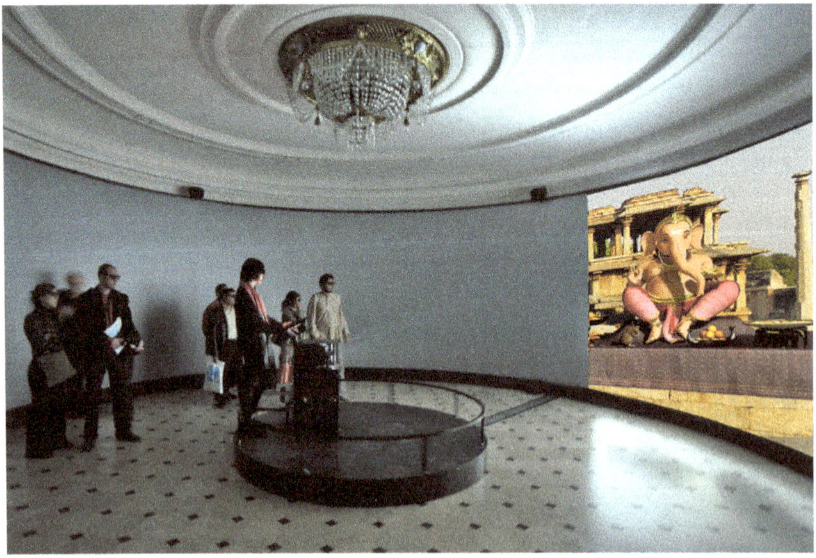

**Figure 9.1** The rotation field of view of the *PLACE 360°* 3D screen with motorized navigational platform showing stereographic panoramas of *PLACE-Hampi* (2006).
*Credit: PLACE-Hampi. Lille 3000*, Opera House, Lille, France 2006. © Sarah Kenderdine & Jeffrey Shaw.

120-degree field of view across the panoramic screen. Enfolded in this interactive installation, visitors explore a virtual landscape of panoramic locations enlivened by animated mythological events that reveal the folkloric imagination of contemporary pilgrims active at the temple complex. Comprising high-resolution augmented stereoscopic panoramas and surrounded by a rich sound field composed from ambisonic recordings, *PLACE-Hampi* articulates an unprecedented level of viewer co-presence in a narrative rediscovery of a cultural landscape (described in further detail in Kenderdine 2013a).

Unlike in *PLACE-Hampi*, the full omnidirectional potential of the tradition of panoramic enclosure is fully realized in *Pure Land: Inside the Mogao Grottoes at Dunhuang* (2012; see Kenderdine 2013b),[14] where the user is completely surrounded by the stereographic image space. *Pure Land: Inside the Mogao Grottoes at Dunhuang* (*Pure Land*) exploits an immersive 3D, 360-degree visualization system (figure 9.2). Up to thirty visitors are able to freely perambulate and interact with a true-to-life-scale virtual Cave 220 (one of the 750 Mogao Grottoe caves, see below) built from laser scans and camera array photography of this World Heritage Site. A single-user interface, operated by one of the visitors or a docent, provides interaction with the digitally rendered cave and allows the user to reveal key elements in the mural paintings on its walls. As well as offering a powerful space for embodied representation, *Pure Land* exploits various digital image-processing techniques, virtual torchlight, magnifying glasses (figure 9.3), 2D and 3D animation, and 3D cinematography to further develop its experiential and interpretative capabilities. The installation was first shown in Hong Kong in 2012, to critical acclaim and subsequent worldwide attention (e.g., Kennicott 2012, Hunt 2013).

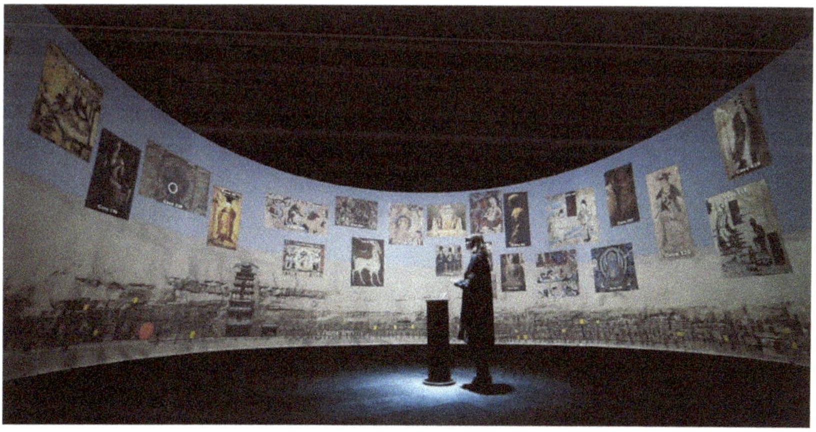

**Figure 9.2** Installed at the beginning of *Pure Land: Inside the Mogao Grottoes at Dunhuang* (2012), a browser displays significant caves distributed along the escarpment.
*Credit: Pure Land: Inside the Mogao Grottoes at Dunhuang. Gallery 360*, Hong Kong, China SAR, 2012. © Applied Laboratory for Interactive Visualization and Embodiment (ALiVE), CityU Hong Kong.

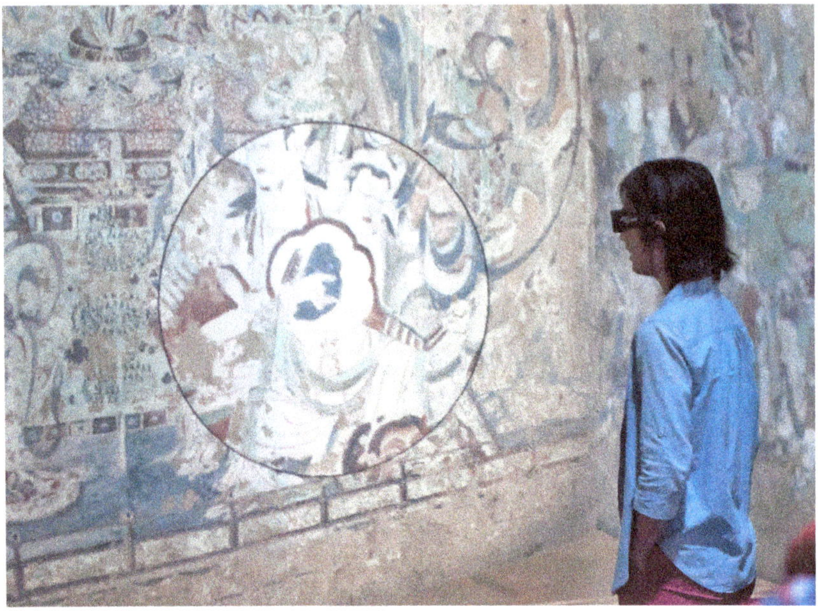

**Figure 9.3** A visitor uses the magnifying glass to inspect a musician in the mural at *Pure Land: Inside the Mogao Grottoes at Dunhuang* (2012).
*Credit: Pure Land: Inside the Mogao Grottoes at Dunhuang. Gallery 360*, Hong Kong, China SAR, 2012. © Applied Laboratory for Interactive Visualization and Embodiment (ALiVE), CityU Hong Kong.

The Mogao Grottoes consist of 750 caves, 492 of them with mural paintings, on five levels hewn into an escarpment in the desert. In total, there are 45,000 square meters of murals and more than 2,000 painted clay figures. Buddha statues, paintings of paradise, and angels adorn the walls, as do images of the patrons or donors who commissioned the paintings. The largest cave is 40 feet high and houses a 30-meter tall Buddha statue installed during the Tang Dynasty (618–906 CE; Larmer 2010). Maintaining careful monitoring, the Dunhuang Academy only opens approximately thirty caves to the public at a time to ensure their long-term preservation. In most caves, the murals and statues are protected (but often optically hindered) by glass panels, and the only lighting is via low-intensity LED torches—one of them held by the guide who is explaining the narrative iconography of the paintings and sculptures. Thus, a real-life visit suffers from restrictive—albeit necessary—limitations. Many believe that Dunhuang's future lies in its digitization program (e.g., Larmer 2010). In the case of Cave 220 in particular, the digital version is the only point of public access owing to permanent closure to the public, on account of the importance of its early Tang mural paintings.

The data from Cave 220 has now been transformed into a range of virtual experiences, including *Pure Land: Augmented Reality Edition* (2012/2016),[15] which uses tracked, tablet-based navigation inside the virtual cave, displayed at 1:1 scale with the real site to simulate revealing the cave in a way that is analogous to having a "window on the world" (Feiner et al. 1992). Depending on the iteration and specificities of the hosting institution, three to four walls are arranged to share the same scale as the real cave and are covered with 1:1 scale prints of Cave 220's "wireframe" polygonal mesh. This provides users with visual cues as to what they can explore with their tablet screens. Infrared cameras track the position and orientation of the visitor tablets as computers render in the appropriate views of the digital cave onto the screen of the tablet in real time—they can even hold the tablets aloft to reveal the painted ceiling of the cave. Enabled by an intricate spatial tracking system, the tablet screen then shifts from being considered as an object in and of itself to functioning as a mobile framing device for the staging of a "virtual" rendering (figure 9.4).

This is not a passive televisual environment but, rather, an interactive performance. Cave 220 is being exactly mapped between real space and the digital model. The conjunction between the exhibition walls and the lifelike cave rendering seen "through" the tablet "window" operates in the border between the indexically real and the phantasmally virtual—between re-embodiment and disembodiment. *Pure Land: Augmented Reality Edition* thus weaves a set of subtle paradoxes into its web of virtualization and actualization, and these paradoxes feed the kinesthetic excitement that is clearly evident in visitors' astonished enjoyment of this installation. It thus aligns

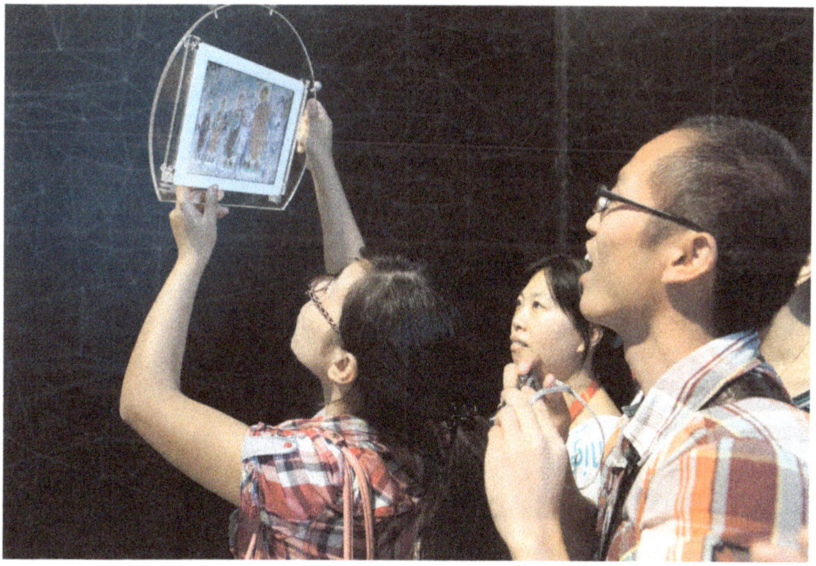

**Figure 9.4** Visitors crowd around a digital tablet giving augmented views of Cave 220.
*Credit: Pure Land: Augmented Reality Edition*. 9th Shanghai Biennale, 2012-2013. © Applied Laboratory for Interactive Visualization and Embodiment (ALiVE), CityU Hong Kong.

with the technologies of telepresence that virtually transport the viewer between the present location and another place—in this case, from the exhibition space to Dunhuang.

Other *Pure Land* iterations, each presenting Cave 220 in uniquely immersive ways, include, *Pure Land Unwired* (2015),[16] again at real-world scale, allowing users to physically walk inside the digital model wearing an untethered HMD; and *Cave Dome*,[17] presented for the first time at the 2015 World Economic Forum, Dalian, as a fully interactive version of the application for a 9-meter diameter digital dome (figure 9.5).

The *Pure Land* projects and *PLACE-Hampi* have both been theoretically situated and analyzed through the optics of the archaeology of performance (e.g., Pearson and Shanks 2001); alongside interpretive archaeology, embodiment, and phenomenology (e.g., Tilley 2008); as well as post-processual archaeology and its extensions such as symmetrical archaeology (e.g., Witmore 2006, Webmoor 2005); and the archaeology of presence (Giannachi et al. 2012), among others. One of the possible extensions to this research lies in the domain of post-cartographic frameworks, which for the purposes of this chapter are called the methods of "deep mapping." The remainder of this chapter looks at the frameworks for place-making in virtual heritage and extrapolates to post-cartographic understandings of mapmaking to arrive at deep mapping as a framework for inclusive interfaces for heterogeneous assemblages of geospatial and historic data.

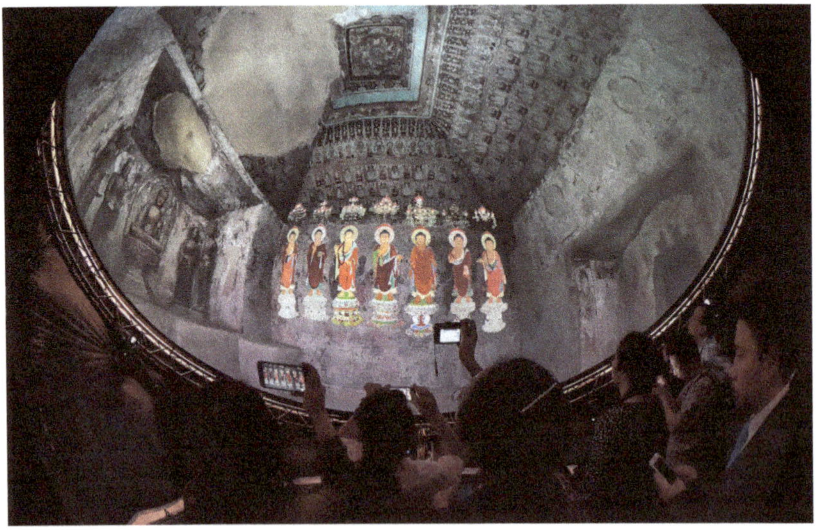

**Figure 9.5** Visitors engage with the real-time application of Cave 220 in the interactive digital dome.
*Credit: Cave Dome.* World Economic Forum, Dalian, China, 2015. © Laboratory for Innovation in Galleries, Libraries, Archives and Museums (iGLAM), University of New South Wales, Sydney, Australia.

## PLACE-MAKING

Today, "place" defined simply as a geographic location is meaningless. As cultural cartographer Edward Casey points out, over the past 500 years, place had been reduced to simply a "location" with "modifications"—that is, the space of human enterprise (Casey 1998, 200–201). Following this reasoning, cultural geographer Jeff Malpas maintains that conceptual problems of new media for cultural heritage lie in misguided definitions of "place." Malpas (2008) notes that:

> [P]laces that have cultural heritage significance, and so that have a relevant "sense of place," are precisely those that have a special character or identity of their own and that also play a special role in shaping the identity of certain societies and cultures. The role of new media in relation to such a "sense of place" can then be understood as one assisting in remembering, reconstructing, recording, and representing of these significant locales. (Malpas 2008, 201)

Using new technologies, many practitioners believe it is possible to transport or translate meanings of place to different times and locations as maps or virtual renderings. The understanding of place currently at the heart of most digital and virtual heritage work has been reinforced by contemporary theory that distinguishes (to an extent) the intangible aspects of place from their material bases:

> [O]ne can view the many projects dedicated to creating digital versions of cultural heritage sites—whether it be Cairo, Stonehenge or Gettysburg—as predicated on the idea that what matters to that place is not the actual location but rather the meaning that attaches to the location, a meaning that can, as it were, be "read from" the location, and then recreated digitally, thereby recreating the place, and the sense of place with it. (Malpas 2008, 204)

However, for Malpas, simply visual virtual *re-creation* is a mistaken aim. Instead, he proposes that new media tools should providing new ways to *enter into* existing places by "drawing upon neglected sensory modalities" (2008, 207), or bringing forth those aspects that go unseen within the heritage landscape (developments represented by both *PLACE-Hampi* and the *Pure Land* projects). In a sense, Malpas is talking about reinserting the material into conceptions of place, thereby acknowledging that one of the crucial functions of digital cultural heritage is facilitating access and preservation of real-world heritage sites.

Heritage displays, such as the installations discussed here, also reassert the role of the material/physical into the meaning and examination of place through metaphors of mapping that are predicated on embodiment: *PLACE-Hampi*, for example, demonstrates new methodologies of corporeal mapmaking. Such works are not only topographic readings of landscape presented in 3D visual panoramic enclosures but also acoustic maps and multimodal tactile terrains. As cultural cartography, these works introduce augmentation of the photographic scenography using computer graphics to provide an interpretive framework for a particular culture/body space as a topological map of place. These works reach beyond other representational forms and cartographies through the somatic engagements they provoke. The works render the experience of cultural heritage as an emergent, personalized map in a landscape in motion. As archaeologist Christopher Tilley states:

> Being in th[e] landscape, being part of th[e] landscape, inevitably result[s] in the kind of intimate knowledge of it that is largely lost today. Walking the past in the present is an attempt to regain at least some of that intimacy and lost experience. (Tilley 2008, 268)

Inhabiting the map through new media technologies and modalities of embodiment takes us beyond other forms of abstracted representation provided by texts, photographs, paintings, sketches and topographic maps. Taking one's body into (real) landscapes allows it to mediate the encounter, facilitating new forms of knowledge that are not accessible through remote-sensing abstractions of that landscape presented in traditional forms of graphical user interface (GUI), texts and plans (Tilley 2008). Walking has three main characteristics: it is material; it can be used to "gather" views, impressions and knowledge of place; and it has a temporal component (Tilley 2008, 268). One can extrapolate this "walking" of the real landscape to my immersive works that provide similar opportunities through the strategies of immersion employed, wherein physical modalities such as walking and driving are embedded in the user experience. Space itself is a paradox in the ontology of interactive narrative. While agency is real, the immersive environment of interactivity is virtual (Mitra and Schwartz 2001). Virtual space is not a passive menu of scenarios from which to choose. It is a place that calls for active responses from the user. Thus, the experience of space by users in virtual environments is largely conditioned by the navigational tools at their disposal and how sympathetic these tools are to the agency that users need to perform.

Edward Casey's work provides inspiration for the representation of landscape in his analysis of both maps and painting (2002), and art in general (2004, 2005). He describes a fourth form of landscape mapping (following cartography, chorography, and topology) as body-mapping—that is, in practical work, the bodily gestures of the artist. In my interactive works described in this chapter, where the engagement with "place" is emergent, it may be possible to describe the acts of the participants in a similar way to that of a painter:

> The gesticulations already incorporate into the body a sense of the circumambient landscape. Just as this landscape is retraced in bodily motions, so these same motions leave traces on the canvas that, rather than *representing* the landscape's precise contours, *reimplace* them on the pictorial surface. (Casey 2004, 261)

Following a phenomenological position on space and place, Casey further observes:

> Just as the body is indispensable in the full experience of human dwelling—that is, in built place—so it now demonstrates itself to be sine qua non in the full experience of mapped places. (Casey 2002, 212)

*PLACE-Hampi* and *Pure Land* enable spectators to physically map the terrains with their bodies through the act of walking, enabling users to "travel" through the virtual world using an interactive interface and testing physical navigation as an interpretive mechanism. With consideration of bodily positions and movements in relation to images and their positioning in space (the virtual space depicted, as well as the real space inhabited by the participants), an "embodied state of awareness go[es] far beyond the sense of sight" and becomes a "fundamental part of the experience involving an awareness of both self and locomotion" (Tilley 2008, 41).

## POST-CARTOGRAPHIC FRAMEWORKS FOR SPACE AND PLACE

This consideration of bodily engagements with virtually rendered places as a form of corporeal cartography references not only the changing nature of the concept of place but also the rise of contemporary post-cartographic frameworks for considering how the act of mapping actively engages with place. While conceptions of place require a balance between the material and immaterial (as discussed in the previous section), the "spatial turn" occurring within the humanities also demands that we extend our conceptions of mapping and cartography beyond the positive epistemologies of geographic information science (GIS) (Kitchin et al. 2011).

The practice of carto-criticism over the last quarter century has focused on debunking cartography's ideology of objectivity (starting with Harley 1988, 1989). Postmodern interpretation of cartography has exposed the discipline's hegemony defined by political and often colonial mandates. No map can be considered "neutral"; they are all portrayals of the commissioning agents desire to "create" knowledge (of places) rather than simply to portray them. A discourse on creative mapping modalities considers representations of landscapes as phenomenological encounters that mobilize viewers inside the image/memory/imagination space during a form of mapmaking,[18] with contemporary psycho-geographies of place being re-energized through new media and exemplified by work in locative media and emotion mapping (for example, the Bio Mapping project by artist Christian Nold and the work of new media artist collective BlastTheory).[19]

These interpretations of the map as moving, embodied, social, technical, and emotive are an opportunity to present alternative visual solutions to specific problems of depicting and describing places, and signal further major shifts underway in the conception of mapping technologies and conceptual frameworks that are being demonstrated in critical cartography (Presner et al. 2014). Enter deep mapping:

> Reflecting eighteenth century antiquarian approaches to place which included history, folklore, natural history and hearsay, the deep map attempts to record and represent the grain and patina of place through juxtapositions and interpenetrations of the historical and the contemporary, the political and the poetic, the factual and the fictional, the discursive and the sensual; the conflation of oral testimony, anthology, memoir, biography, natural history and everything you might ever want to say about a place. (Pearson and Shanks 2001, 64–65)

Digital cultural heritage atlases can be described as "deep maps"—contingent, semantically rich, and potentially performative and discursive (Pearson and Shanks 2001; Biggs 2010, following Clifford McLucas's 2001 artistic project *Deep Maps*[20]). As critical cartographers Kitchin and Dodge (2007) have argued, maps are always in the state of becoming, brought into being through embodied social and technical practices to solve relational problems such as plotting, planning, navigating, and so on. Maps then emerge through a mix of creative, reflexive, playful, tactile, and habitual practices as a co-constitutive production—a production that is constantly in motion. Conceiving of maps through a deep-mapping schema posits that they are never fully formed (never static) but, rather, emerge in navigational processes where emergent narrative formations are possible (dynamic). In quite fundamental ways, such a reimagining of maps changes the focus of cartography, moving it away from notions of accuracy, design, aesthetics, and power and toward emphasizing the complex, contingent interactions between the researchers collecting the data, the designers of the information visualization system, its users, geotemporal data, and the nonphysical aspects of place.

Developments in geographic information systems (GIS) and the transdisciplinary expansion or penetration of spatial science over the past few decades have had a profound impact on many facets of government administration, industrial infrastructure, commerce, and academia. The expansion of spatial sciences and technologies can be compared to the discoveries brought about by the microscope, the telescope, and the printing press in terms of the ways in which they offer unprecedented avenues for perceiving spatial phenomena and communication across fields. However, the dialogue between geographic information science and cultural heritage narratives has been limited to "off-the-shelf" GIS in historical mapping projects (Bodenhamer et al. 2013, 171). It is increasingly recognized that the spatial turn in cultural heritage research must extend beyond the mimetic views of geo-browsers such as Google Maps and Google Earth (Presner et al. 2014, 87; November et al. 2010). In their introduction to the above 2013 report, titled *Spatial Narratives and Deep Maps,* Bodenhamer and colleagues state that "an emphasis on absolute space based on Euclidean coordinate systems frustrates the humanist's effort to understand how spaces change over time and how spatial relativities emerge and develop" (2013, 4). New developments, they argue, are required of spatiotemporal tools in order to enable scholars and communicators such as museums to "incorporate dynamic time elements into analyses that are spatially contextualised" (2013, 173). It is now possible to create narrative forms that join spatial story-telling to more traditional humanities and curatorial research. Here the rewards include, according to Bodenhamer et al., "significant extensions of work already underway in domains such as literary and cultural studies (e.g., narrative topographies, the spatial imaginaries, and novel mappings" 2013, 173). These new approaches and major advancements in spatial technologies driving current work in the fields of digital humanities and critical cartography open up pathways for new interpretive possibilities in museums. This is demonstrated by the establishment of spatial research centers and projects, such as Stanford's Spatial History Project, Columbia's Spatial Information Design Lab, and Hypercities (Presner et al. 2014, 51).

## NARRATIVE AND DEEP MAPPING

Digital heritage and immersive interaction specialists from the University of New South Wales, University of California Berkeley, City University of Hong Kong, the Hong Kong Maritime Museum, Fudan University, the Australian National Maritime Museum, and Chronus Art Center are currently collaborating[21] on the representation of the Electronic Cultural Atlas Initiative's (ECAI) *Atlas of Maritime Buddhism*.[22] Through the exhibition of this vast dataset as an immersive, interactive "deep map", the *Atlas* seeks to resolve the fundamental challenges of narrative coherence for museum audiences exploring digital cultural atlases. The project is based on a unique heritage dataset of spatially and temporally enabled sources, diverse in type and format, which provides historic evidence for the spread of Buddhism from India to Korea through the seaports of Southeast Asia.

The ECAI *Atlas of Maritime Buddhism* is a database of great academic significance (Tu et al. 2014, Blundell and Zerneke 2014), and the importance of the Maritime Silk Road has never been more apparent. The recent creation of the Maritime Silk Road Institute in China is aligned with a multi-billion-renminbi endeavor by the government of China under Xi Jinping to "re-create" the Maritime Silk Road as a major trading network through South and Southeast Asia (Bromby 2013). The "One Belt One Road" initiative, as it is known, is the combination of both maritime and overland routes, intended for economic expansion. Reflecting this renewed focus on the Maritime Silk Road, the *Atlas* counterbalances prevailing narratives that foreground the overland Silk Road and neglect the importance of maritime and Buddhist entrepreneurship in an expansive network of trade.

The *Atlas* comprises multiple overlapping chronological events, supporting diverse types of historic evidence from disparate spatial locations represented by approximately 170 generalized information layers. With contributions from researchers around the world, it includes geospatial coordinates; gazetteers for hundreds of sites; images of archaeological sites and artifacts; religious and geopolitical empires and zones of influence; inscriptions and transcriptions of Sanskrit texts; historic maps; accounts by Buddhist monks and ambassadors; records of trade; hydrographic data; monsoon records; and shipwreck data sets. Given this vast heterogeneity, the *Atlas* requires a new form of visual, cartographic and time-space narrative strategy that is currently outside traditional forms of interpretation (Presner et al. 2014; Ridge et al. 2013).

The present deep mapping project engages the *Atlas* in telling the thus far untold story of maritime history in an accessible and compelling form for museum visitors, while connecting to sociopolitical and economic dynamics of the Asian region today. Although spatiotemporal data sets are typically visualized as hot spots on a map conveying static narratives of place (Straumann et al. 2014), the proposed deep-mapping schema converts the map of information from mimetic object (based on a perceived territory) into a navigational one where "everything is on the move" (November et al. 2010). The schema will support narrative construction by creating appropriate visualization tools for museum users through the experimental application of the world's first deep-mapping data browser—an interface for exploring narrative patterns, processes, and phenomena in the *Atlas* in a 360-degree, 3D (omnidirectional) virtual environment.

Within an interactive visualization framework, it is necessary to refine the tensions between the user's freedom to explore an array of diverse navigational pathways and narrative elements and the need for narrative coherence to give meaning to the discoveries that users make (Drucker 2014, Kosara and Mackinlay 2013). Conveying narrative with visualization often propels designers and curators to choose an order in which to present visualizations, resulting in linear storylines. Furthermore, while evidence exists that narrative sequencing in traditional stories can affect comprehension and memory, little is known about how sequencing choices affect narrative visualization where interaction makes sequencing open-ended and emergent (Hullman and Diakopoulos 2011, Segel and Heer 2010). To resolve these issues, a deep-mapping schema is proposed which will provide users with a coherent experience through identifiable visual and navigational modes of exploring the data. By extracting relationships embedded in the texts, archaeological, and other historic materials of the *Atlas*, it is possible to create spatial narrative forms that do not rely on linear frameworks (Bodenhamer et al. 2013).

Deep mapping, as outlined, is a response to a critical need in museums for appropriate visual and interactive paradigms that give visitors access to burgeoning digital cultural heritage (Sula 2012). Scientific visualizations have been used extensively in museums in recent years to show things that cannot otherwise be seen. These visualizations often depict complex, abstract phenomena at extreme scales using colors that have no inherent meaning. For example, one research study at the Exploratorium—a science museum in San Francisco—showed that many "visitors grossly misinterpreted the scale and use of colour in a nanoscale image" (Ma et al. 2012). Similar studies have documented learners' difficulty in interpreting visualizations from fields as disparate as genetics and astrophysics, indicating that new modalities of data-driven storytelling need to be found. The *Atlas of Maritime Buddhism* project employs a deep-mapping narrative approach to respond to this need, and also aims to produce a new navigation paradigm for many other cultural and historical data sets currently in development.

## CONCLUSION

The projects described in this chapter offer a counterpoint to the exaggerated rhetoric surrounding immersive virtual reality and extensive industry engagement with VR technologies, particularly the head-mounted display. As Facebook's CEO Mark Zuckerberg stated in 2014: VR "is really a new communication platform." "By feeling truly present, you can share unbounded spaces and experiences with the people in your life." This enthusiasm is reflected and reinforced by the exponential growth of investment in VR tools and experiences in recent times. For example, in 2014, Zuckerberg's company spent US$2 billion to purchase the VR technology company Oculus—well known for the VR HMD that had attracted Zuckerberg, the Oculus

Rift. Alongside such developments as the Samsung's Gear VR and Google Cardboard, we see a proliferation of services for creating, aggregating, and streaming content. Startups such as Jaunt are occupying new market spaces by developing the latest in immersive 360-degree, 3D video capture for HMD (e.g., STEAM)[23] and providing platforms for distribution of aligned content (e.g., WITHIN).[24]

The majority of these technologies are focused on the presentation of place, and developers are acknowledging that they struggle with the issues of narrative engagement, socialization, interaction, and, significantly, presence. That is, the industry is expressing a need to define a new language that accompanies the shift from cinema to VR. As Job Stauffer, Head of Creative Communications for Telltale Games, says, "If you look at VR and the language of presence, it's another totally different language. So for us to translate everything we do from the cinematic language to the language of presence, it's a whole new ball game" (Orland 2016). CNET critic Richard Trenholm (2016) refers to presence as "[v]irtual reality's fragile magic ingredient," while VR expert Caecilia Charbonnier lists four pillars of presence as the illusions of: "being in a stable space," "self-embodiment," "physical interaction" and "social collaboration" (Orland 2016).

Virtual heritage practitioners have extensively explored these territories in numerous ways over the past fifteen years, and the projects presented in this chapter demonstrate how a sophisticated language around issues of narrative, presence, and place has already emerged in the field of digital cultural heritage visualization.

## REFERENCES


Altick, R. 1978. *The Shows of London*. London: Cambridge University Press.

Avery, K. J. 1995. "The Panorama and Its Manifestation in American Landscape Painting, 1795–1870." PhD dissertation, Department of Art History and Archeology, Columbia University.

Barceló, J., M. Forte, and D. Sanders, eds. 2000. *Virtual Reality in Archaeology*. British Archaeological Reports, International Series, No. 843. Oxford: Archaeopress.

Bhabha, H. K. 1994. *The Location of Culture*. London: Routledge.

Biggs, I. 2010. "Deep Mapping: An Introduction." In *Mapping Spectral Traces*, edited by K. E. Till, 5–8. Exhibition catalog. Blacksburg: Virginia Tech College of Architecture and Urban Studies.

Blundell, D., and J. Zerneke. 2014. "Early Austronesian Historical Voyaging in Monsoon Asia: Heritage and Knowledge for Museum Displays Utilizing Texts, Archaeology, Digital Interactive Components, and GIS Approaches." *International Journal of Humanities and Arts Computing* 8(Suppl.): 237–252.

Bodenhamer, D. J., T. M. Harris, and J. Corrigan. 2013. "Deep Mapping and the Spatial Humanities." *International Journal of Humanities and Arts Computing* 7(1–2): 170–175.

Bokova, I. 2015. "Fighting Cultural Cleansing: Harnessing the Law to Preserve Cultural Heritage." *Harvard International Review* 36(4): 40–45.

Bromby, R. 2013. "Down the Maritime Silk Road." *Australian Business Review*, December 6. http://www.theaustralian.com.au/business/in-depth/down-the-maritime-silk-road/story-fnjy4qn5-1226776242929.

Bruno, G. 2002. *Atlas of Emotion: Journeys in Art, Architecture, and Film*. New York and London: Verso.

Cameron, F. R., and S. Kenderdine, eds. 2007. *Theorizing Digital Cultural Heritage: A Critical Discourse*. Cambridge, MA: MIT Press.

Casey, E. S. 1998. *The Fate of Place: A Philosophical History*. Berkeley: University of California Press.

Casey, E. S. 2002. *Representing Place: Landscape Painting and Maps*. Minneapolis: University of Minnesota Press.

Casey, E. S. 2004. "Mapping the Earth in Works of Art." In *Rethinking Nature: Essays in Environmental Philosophy*, edited by B.V. Foltz and R. Frodeman, 260–272. Bloomington: Indiana University Press.

Casey, E. S. 2005. *Earth-Mapping: Artists Reshaping Landscape*. Minneapolis: University of Minnesota Press.

Comment, B. 2000. *The Painted Panorama*. New York: Harry N. Abrams.

Debord, G. 1955. "Introduction to a Critique of Urban Geography." Translated by K. Knabb. First published 1955. In *Critical Geographies: A Collection of Readings*, edited by H. Bauder and S. Engel-Di Mauro, 23–32. Kelowna, BC: Praxis.

Del Favero, D., and J. Shaw, eds. 2001. *(dis)LOCATIONS*. ZKM Digital Arts Edition DVD-ROM. Karlsruhe: ZKM and Centre for Art & Media.

Deleuze, G. 1986. *Cinema 1: The Movement-Image*. Translated by H. Tomlinson and B. Habberjam. Minneapolis: University of Minnesota Press.

Deleuze, G. 1989. *Cinema 2: The Time Image*. Translated by H. Tomlinson and R. Galeta. Minneapolis: University of Minneapolis Press.

Doornbusch, P., and S. Kenderdine. 2004. "Presence and Sound: Identifying Sonic Means to 'Be There.'" In *Qi Connectivity and Complexity*, edited by R. Ascott, 67–70. Plymouth: Planetary Collegium.

Drucker, J. 2014. *Graphesis*. MetaLABprojects. Cambridge, MA: Harvard University Press.

Feiner S., B. MacIntyre, M. Haupt, and E. Solomon. 1993. "Windows on the World: 2D Windows for 3D Augmented Reality." In *UIST '93 Proceedings of the 6th annual ACM Symposium on User Interface Software and Technology*, 145–155. New York: Association for Computing Machinery, Inc. (ACM).

Giannachi, G., N. Kaye, and M. Shanks, eds. 2012. *Archaeologies of Presence: Art, Performance and the Persistence of Being*. Abingdon: Routledge.

Grau, O. 2003. *Virtual Art: From Illusion to Immersion*. Cambridge, MA: MIT Press.

Harley, J. B. 1988. "Silences and Secrecy: The Hidden Agenda of Cartography in Early Modern Europe." *Imago Mundi: The International Journal for the History of Cartography* 40(1): 57–76.

Harley, J. B. 1989. "Historical Geography and the Cartographic Illusion." *Journal of Historical Geography* 15(1): 80–91.

Hullman, J., and N. Diakopoulos. 2011. "Visualization Rhetoric: Framing Effects in Narrative Visualization." *IEEE Transactions on Visualization and Computer Graphics* 17(12): 2231–2240.

Hunt, K. 2013. "Buddhas in 3-D: Technology and the Battle to Preserve Asia's Heritage." *CNN.com*, January 15. http://edition.cnn.com/2013/01/15/world/asia/china-digital-caves/.

Huhtamo, E. 2004. "Peristrephic Pleasures, or The Origins of the Moving Panorama." In: *Allegories of Communication: Intermedial Concerns from Cinema to the Digital*, eds. J. Olsson and J. Fullerton. Rome: John Libbey, 215–248.

Huhtamo, E. 2013. *Illusions in Motion: Media Archaeology of the Moving Panorama and Related Spectacles*. Cambridge, Mass.: MIT Press.

Hyde, R. 1988. *Panoramania!: The Art and Entertainment of the "All-Embracing View."* Exhibition catalog. London: Trefoil.

Kalay, Y., T. Kvan, and J. Affleck, eds. 2007. *New Heritage: New Media and Cultural Heritage*. Abingdon: Routledge.

Kenderdine, S. 2004. "Stereographic Panoramas of Angkor, Cambodia." In *VSMM2004: Proceedings of the Tenth International Conference on Virtual Systems and Multimedia*, ed. H. Thwaites, 612–621. Amsterdam: ISO Press.

Kenderdine, S. 2007. "Speaking in Rama: Panoramic Vision in Cultural Heritage Visualisation." In *Theorizing Digital Cultural Heritage: A Critical Discourse*, edited by F. R. Cameron and S. Kenderdine, 301–332. Cambridge, MA: MIT Press.

Kenderdine, S. 2013a. *PLACE-Hampi: Inhabiting the Panoramic Imaginary of Vijayanagara*. Heidelberg: Kehrer Verlag.

Kenderdine, S. 2013b. "Pure Land: Inhabiting the Mogao Caves at Dunhuang." *Curator: The Museum Journal* 56(2): 199–218.

Kenderdine, S. 2016. "Embodiment, Entanglement, and Immersion in Digital Cultural Heritage." In *A New Companion to Digital Humanities*, edited by S. Schreibman, R. Siemes, and J. Unsworth, 22–24. New York: Wiley Blackwell.

Kenderdine, S., and T. Hart. 2003. "This Is Not a Peep Show! The Virtual Room at the Melbourne Museum." In *International Cultural Heritage Informatics Meeting: Proceedings from ichim03*, edited by X. Perrot, CD-ROM. Pittsburgh, PA: Museum Archives and Informatics. http://www.archimuse.com/publishing/ichim03/003C.pdf.

Kennicott, P. 2012. "'Pure Land' Tour: For Visitors Virtually Exploring Buddhist Cave, It's Pure Fun." *Washington Post*, November 29.

Kitchin, R., and M. Dodge. 2007. "Rethinking Maps." *Progress in Human Geography* 31(3): 331–344.

Kitchin, R., M. Dodge, and C. Perkins. 2011. "Introductory Essay: Conceptualising mapping." In *The Map Reader: Theories of Mapping Practice and Cartographic Representation*, edited by M. Dodge, R. Kitchin, and C. Perkins, 2–7. Oxford: Wiley Blackwell.

Kosara, R., and J. Mackinlay. 2013. "Storytelling: The Next Step for Visualization." *Computer* 46(5): 44–50.

Larmer, B. 2010. "Caves of Faith." *National Geographic*, June. https://www.nationalgeographic.com/magazine/2010/06/dunhuang-caves/.

Ma, K. L., I. Liao, J. Frazier, H. Hauser, and H. N. Kostis. 2012. "Scientific Storytelling Using Visualization." *IEEE Computer Graphics and Applications* 32(1): 12–19.

Malpas, J. 2008. "New Media, Cultural Heritage, and the Sense of Place: Mapping the Conceptual Ground." *International Journal of Heritage Studies* 14(3): 197–209.

Manovich, L. 2001. *The Language of New Media*. Cambridge, MA: MIT Press.

Mitra, A., and R. L. Schwartz. 2001. "From Cyber Space to Cybernetic Space: Rethinking the Relationship between Real and Virtual Spaces." *Journal of Computer-Mediated Communication* 7(1): n.p.

Murray, T. 2008. *Digital Baroque: New Media Art and Cinematic Folds*. Minneapolis: University of Minnesota Press.

Nold, C. 2004. "Bio Mapping / Emotion Mapping." http://www.biomapping.net/.

November, V., E. Camacho-Hübner, and B. Latour. 2010. "Entering a Risky Territory: Space in the Age of Digital Navigation." *Society and Space* 28(4): 581–599.
Oettermann, S. 1997. *The Panorama: History of a Mass Medium*. New York: Zone Books.
Orland, K. 2016. "'People Will Fall for It Like a Drug'—Game Devs on the Future of VR." *Ars Technica*, March 2. http://arstechnica.com/gaming/2016/03/people-will-fall-for-it-like-a-drug-game-devs-on-the-future-of-vr/.
Parente, A., and L. Velho. 2008. A Cybernetic Observatory Based on Panoramic Vision." *Technoetic Arts: A Journal of Speculative Research* 6(1): 79–98.
Pearson, M., and M. Shanks. 2001. *Theatre/Archaeology*. London: Routledge.
Presner, T., D. Shepard, and Y. Kawano. 2014. *HyperCities: Thick Mapping in the Digital Humanities*. Cambridge, MA: Harvard University Press.
Ridge, M., D. Lafreniere, and S. Nesbit. 2013. "Creating Deep Maps and Spatial Narratives Through Design." *International Journal of Humanities and Arts Computing* 7(1–2): 176–189.
Sanders, D. H. 2015. "Advances in Virtual Heritage: Conditions and Caveats." In *2015 Digital Heritage International Congress*, vol. 2, edited by G. Guidi, R. Scopigno, J. Carlos Torres, and H. Graf, 643–646. Piscataway: IEEE.
Schiller, D., and S. Yeo. 2014. "Powered by Google: Widening Access and Tightening Corporate Control." In *RED ART: New Utopias in Data Capitalism*, Leonardo Eletronic Almanac 20 (1), edited by L. Aceti, S. Jaschko, and J. Stallabrass, 44–57. San Francisco: Leonardo/ISAST and MIT Press.
Segel, E., and J. Heer. 2010. "Narrative Visualization: Telling Stories with Data." *IEEE Transactions on Visualization and Computer Graphics* 16(6): 1139–1148.
Shaw, J., and P. Weibel, eds. 2003. *Future Cinema: The Cinematic Imaginary After Film*. Electronic Culture: History, Theory, and Practice. Cambridge, MA: MIT Press.
Stille, A. 2006. "Are We Losing Our Memory? or The Museum of Obsolete Technology." *Lost Magazine* 3. http://www.lostmag.com/issue3/memory.php.
Straumann, R. K., A. Çöltekin, and G. Andrienko. 2014. "Towards (Re)Constructing Narratives from Georeferenced Photographs through Visual Analytics." *Cartographic Journal: The World of Mapping* 51(2): 152–165.
Sula, C. A. 2012. "Visualizing Social Connections in the Humanities: Beyond Bibliometrics." *Bulletin of the American Society for Information Science and Technology* 38(4): 31–35.
Tilley, C. 2008. *Body and Image: Explorations in Landscape Phenomenology*. Walnut Creek, CA: Left Coast Press.
Trenholm, R. 2016. "Virtual Reality's Fragile Magic Ingredient." *CNet*, February 20. http://www.cnet.com/au/news/vr-sundance-secrets-of-virtual-realitys-fragile-magic-ingredient
Tu, T. N., T. D. Thien, and L. R. Lancaster. 2014. *Buddhist Culture and Technology: New Strategies for Study*. Vietnam Buddhist University Series 26. Hanoi: Vietnam Buddhist University Publications.
UNESCO. 2015. "Director-General Irina Bokova Firmly Condemns the Destruction of Palmyra's Ancient Temple of Baalshamin, Syria." *UNESCO News*, August 24. https://en.unesco.org/news/director-general-irina-bokova-firmly-condemns-destruction-palmyra-s-ancient-temple-baalshamin/.
Wasson, H. 2007. "The Networked Screen: Moving Images, Materiality and the Aesthetics of Size." In *Fluid Screens, Expanded Cinema*, edited by J. Marchessault and S. Lord, 74–95. Toronto: University of Toronto Press.

Webmoor, T. 2005. "Mediational Techniques and Conceptual Frameworks in Archaeology: A Model in 'Mapwork' at Teotihuacán, Mexico." *Journal of Social Archaeology* 5(1): 54–86.

Witmore, C. L. 2006. "Vision, Media, Noise and the Percolation of Time: Symmetrical Approaches to the Mediation of the Material World." *Journal of Material Culture* 11(3): 267–292.

Zuckerberg, M. 2014. Facebook page post, March 25. https://www.facebook.com/zuck/posts/10101319050523971.

# PART THREE

# *Hermeneutics and Epistemological Boundaries*

CHAPTER 10

# Çatalhöyük as an Open Site?

## *On the Openness of Virtual Reconstructions of Archaeological Sites to a Multiplicity of Interpretations*

ZEYNEP AKTÜRE

### THE PRINCIPLE OF TRANSPARENCY IN VIRTUAL ARCHAEOLOGY AND THE CASE OF ÇATALHÖYÜK

In the past two decades, employment of digital technologies in research and presentation of archaeological heritage sites has widened. The international community of experts has responded to this widening by formulating common rules of conduct, building on the tradition of scholarly debate since the end of nineteenth century that is based on common principles to be employed before, during, and after physical interventions at cultural heritage sites (e.g., The Venice Charter, ICOMOS 1964; *Management Guidelines*, Feilden and Jokilehto 1993; Operational Guidelines, WHC 2015). The series of principles agreed upon for virtual interventions has progressed from more general ones, aiming to develop coordination mechanisms for digitization programs (Lund Principles, EU CORDIS 2001), toward addressing more narrowly the use of 3D visualization in research and communications of cultural heritage (London Charter, Denard 2009, first draft in 2006) and their implementation in digital archaeology (Principles of Seville, IFVA 2012, first draft in 2010). These principles were further developed through expert meetings that produced revised versions in several world languages, responding to novel concerns parallel to the expansion promised by the employment of developing digital technologies in the management of cultural heritage resources.

Zeynep Aktüre, *Çatalhöyük as an Open Site?* In: *Digital Cities*. Edited by: Maurizio Forte and Helena Murteira, Oxford University Press (2020). © Oxford University Press.
DOI: 10.1093/oso/9780190498900.003.0011

## Transparency in Computer-Based Visualizations of Cultural Heritage

These revisions point to the need for openness in all virtual processes, as expressed in the principle of transparency, which was the topic of an expert meeting in 2006 before the drafting of the London Charter. The glossary of the London Charter describes intellectual transparency as "[t]he provision of information, presented in any medium or format, to allow users to understand the nature and scope of 'knowledge claim' made by a computer-based visualization outcome" to lay down the principle of clarifying to users what a computer-based visualization seeks to represent (e.g., the existing state, an evidence-based restoration or an hypothetical reconstruction of a cultural heritage object or site) and the extent and nature of any factual uncertainty (principle 4.4). This should be done through dissemination of documentation on the evaluative, analytical, deductive, interpretative, and creative decisions made in the course of visualization to reveal explicitly the relationships between research sources, implicit knowledge, explicit reasoning, and visualization-based outcomes (principle 4.6). Users should thus be able to identify clearly the nature and importance of significant, hypothetical dependency relationships between elements and to understand the reasoning underlying such hypotheses (principle 4.10).

The naming of the corresponding article in the Principles of Seville as "scientific" transparency may be taken to acknowledge that testing the validity, and therefore the scope, of the conclusions produced by digital visualizations of archaeological sites and remains would "depend largely on the ability of others to confirm or refute the results obtained" (principle 7). Therefore, even for scientific transparency to other researchers and professionals in the field, "it is necessary to create a large globally-accessible database with projects that offer optimum levels of quality" (principle 7.5).

## Transparency as a Strategy in the Çatalhöyük Research Project

The Çatahöyük archaeological research project, as defined and directed by Dr. Ian Hodder of Stanford University, California, between 1993 and 2018, is a unique case in Turkey that aims at creating such a database among the primary goals of the project, which was achieved through digital technologies by creating a project website (Çatalhöyük—Excavations of a Neolithic Anatolian Höyük, ÇRP 2016). This publication strategy was verbalized by Hodder (1996) as the exploration of the possibility of a post-processual field method arranged around three themes: problematization of categories and codes imposed from outside and developed within disciplinary discourses, construction of local and relational understandings at all points in the field and analytical process, and feedback from local participation in wider and multivocal debates. This method was devised by acknowledging the relationality of meaning, which required creation of a circuitry between project participants and different types of data to achieve conclusions that would, nevertheless, always be momentary, fluid, and flexible owing to possibilities of new relations. Those possibilities would be created by producing information that could be questioned and approached from different angles interactively, by a diversity of people who could participate in the discourse about the archaeological process through access to a multivocal, plural, open, or transparent project website.

Yet, already in 1998, Hodder (1998, 137–138) acknowledged the existence of many groups that did not have access to the required technologies and admitted that there was no such thing as "open multivocality," since those users who did have access also needed a certain level of knowledge to participate in hypertext presentations. Hodder further underlined that links and nodes in a hypertext were created by its producer, who had to make choices about the target audiences and given messages. His conclusion was that "[a]s much as the Web and hypertext allow a greater diversity and openness of communication, the onus remains on the producer and writer to be reflexive about the impact of 'the text' in the world" (Hodder 1998, 137).

### Scope and Objective: Transparency as a Limit to Openness?

Hodder's remarks on the simultaneous ideal and impracticability of achieving all-embracing multivocality, plurality, openness, and transparency in research and the presentation of Çatalhöyük through digital and other media are in line with the main argument in this chapter: that all works (including the ÇRP website) encourage *and* limit their possible and probable interpretations by their very physicality (as exemplified in hypertext presentations).

The relevance of this idea of "openness" for digital visualizations of Çatalhöyük is argued in the next section of this chapter, on the basis of a diversity in expert opinion on the settlement's rank in the transition from settled village to urban agglomeration, which makes the site an ideal case for discussing scientific transparency in the presence of a multiplicity of expert interpretations. A theoretical framework for assessing "open *and* limited multivocality" is then presented based on Umberto Eco's idea of the "open work" as applied in literary and visual works, including motion pictures, and on Siegfried Kracauer's idea of "cinematic materiality," in view of the fact that many archaeological visualizations are in the format of moving images. In the concluding section of this chapter, three of the abundant digital visualizations of Çatalhöyük, conceived by the ÇRP team members and others for a variety of purposes and target audiences, are briefly presented to illustrate the applicability of Eco's and Kracauer's ideas to this type of production, with the goal of assessing whether visualizations may encourage any or all of the alternative expert interpretations of the site.

Although two of the three projects, and the images selected to highlight the relevance of Eco's and Kracauer's ideas for recent cognitive neuroscientific research on the connection of visual perception with bodily practices, are authored by experts involved in the ÇRP, an overall assessment of the intended "openness" of the project website to all-embracing multivocality is beyond the scope of this chapter, especially in light of Hodder's 1998 acknowledgment of the impossibility of actively involving all nonexperts in the archaeological interpretation process. Instead, this chapter attempts to assess the openness of selected digital visualizations of the site to accommodate the existing diversity of expert interpretations on Çatalhöyük's rank in the transition from settled village to urban agglomeration, which would correspond to the most probable ones among all possible interpretations of the Çatalhöyük material if taken in its entirety, as will be further elaborated.

None of the images selected for discussion in this chapter has pronounced transparency assertions about its knowledge claim or for creating a space for multivocality in the presented project outcomes or for embracing all the expert interpretations of the site. Yet this chapter follows Eco's idea that, regardless of the producer's intentions, all works (including digital visualizations) are by nature open to a multiplicity of interpretations to a greater or lesser degree, and so the chapter tries to identify, in the selected visual works, the physical correspondence of the idea of openness, as well as the recently adopted principle of scientific transparency in computer-based visualizations of cultural heritage as potential limits for over-interpretation.

## COMPETING INTERPRETATIONS OF THE NEOLITHIC SITE OF ÇATALHÖYÜK

The openness of research data on Çatalhöyük for a multiplicity of interpretations is important, owing to the centrality of the site in debates on the emergence of an urban way of living. Çatalhöyük is included in the UNESCO World Heritage List mainly based on evidence it provides of a transition from settled village to urban agglomeration (WHC 2012). Yet there still exists a diversity of opinion on the ranking of the site along this developmental track (figure 10.1).

### Mellaart's Çatalhöyük: A Neolithic Town in Anatolia

Çatalhöyük was discovered in 1958 by the British archaeologist James Mellaart (1925–2012), who was then a visiting scholar at the British Institute

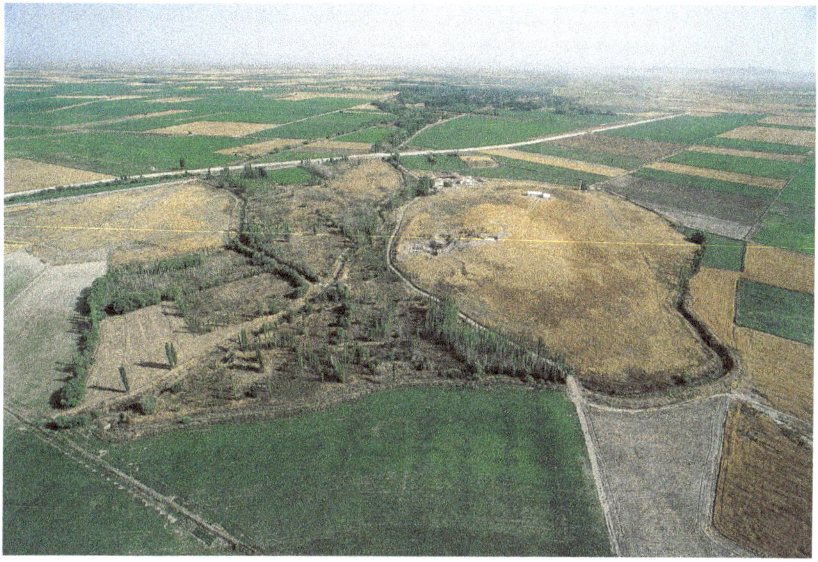

**Figure 10.1** An aerial view of the Çatalhöyük archaeological site.
*Credit:* © Hakan Öge/ATLAS Magazine, 2002.

of Archaeology in Ankara, with the mission 'to prove that Anatolia had played a pivotal role in prehistory" (Balter 2006, 8). At the time, the image that the archaeological community generally held of the Anatolian Plateau in prehistory was as a backwater, since the real activity was believed to have taken place in Near Eastern Neolithic sites like Jericho, where some of the earliest-known farming villages had been unearthed in the 1950s. At Jericho, more than 10,000 years of archaeological deposits had been interpreted as revealing a settlement history that started as "a seasonal camp for hunter-gatherers and then continued to build up during the Neolithic period, the Bronze Age, and the Iron Age" (10).

This sequence formed the conventional opinion on the origins of cities in Southwest Asia: hunter-gatherers, intensifying the exploitation of wild plants and animals, settled in more permanent camps to domesticate plants and animals. Development of cultivation and herding eventually led to a more stratified social order. The multiplication and spread of farming villages initiated the Agricultural or Neolithic Revolution, as the first great breakthrough in the development of human societies. However, enlargement of the villages thanks to increasing innovations and adaptations to the environment, was not enough for the second great breakthrough, the Urban Revolution, which was symbolized and stimulated by the local invention of writing and larger-scale irrigation technology (Soja 2003, 27).

According to the Marxist prehistorian Vere Gordon Childe (1892–1957), these inventions were made possible by farmers' production of food surpluses for a new urban population of specialized craftsmen, merchants, priests, officials, and clerks (Childe 1971, 30–31). The next economic revolution, therefore, was only possible thanks to the Agricultural Revolution, which puts the two in a linear causality relationship. Childe mapped the growth of urban literate societies in a gradual spread from what was termed the Fertile Crescent, extending from the Taurus Mountains in Anatolia to the Zagros in Iran and south through the Levant and the lower Nile Valley (Goody 2006, 504–505). The earliest Neolithic (i.e., farming) villages are evidenced in Greece a few thousand years later than in this area, and much later in the rest of Europe.

Mellaart had come to Anatolia to find archaeological support for the hypothesis that Anatolia served as the land bridge over which the earlier development of farming in Southwest Asia would have spread westward to these later sites. He excavated at Çatalhöyük between 1961 and 1965, unearthing some two hundred multilayered Neolithic houses that corresponded to about 3% of the larger East one of the two mounds at the site (figure 10.1; see also figure 10.2). Under the influence of the sheer size of the mound, and the sophistication of the symbolic language of the small finds and wall paintings, Mellaart soon started to call Çatalhöyük a city (figures 10.3 and 10.4). His publication of this hypothesis in the *Scientific American* (1964) brought Çatalhöyük to worldwide attention, before his later work, *Çatal Hüyük: A Neolithic Town in Anatolia* (1967).

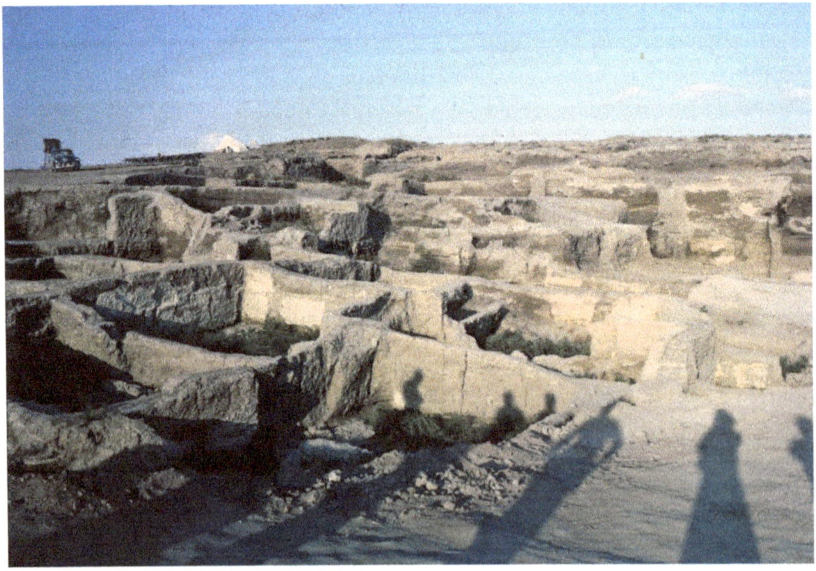

**Figure 10.2** Çatalhöyük archaeological site after James Mellaart's excavations (1961–965).
*Credit:* © Çatalhöyük Research Project (CC BY-NC).

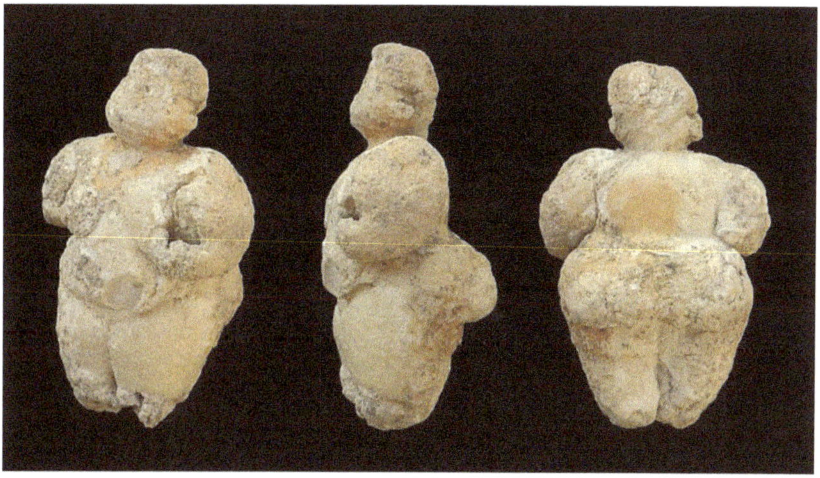

**Figure 10.3** Figurine from Çatalhöyük.
*Credit:* © Çatalhöyük Research Project (CC BY-NC).

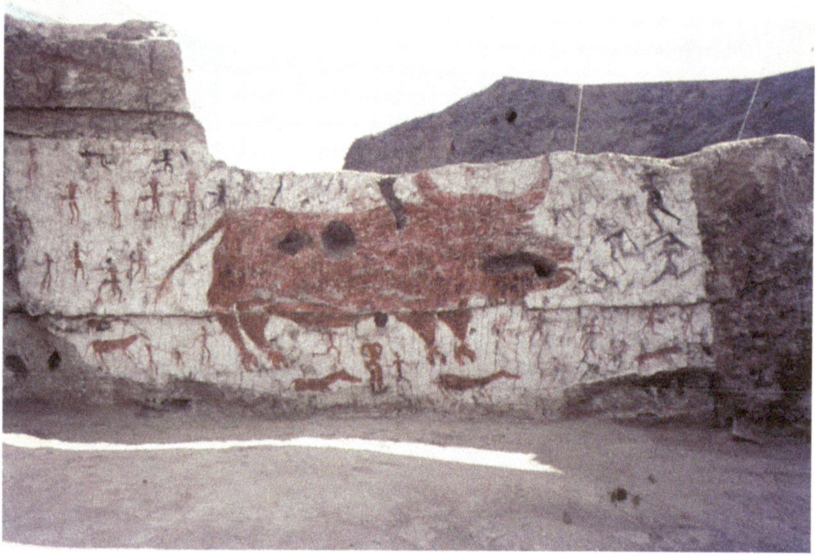

**Figure 10.4** In situ wall-painting from Çatalhöyük, now on display in the Museum of Anatolian Civilisations in Ankara, Turkey.
*Credit:* Photograph by Ian Todd, © Çatalhöyük Research Project (CC BY-NC).

## Jacobs's Çatalhöyük: A Pre-Agricultural City of Hunters

Mellaart's conclusions were based on an image of the central Anatolian Plateau as the most culturally advanced region of the Neolithic world, with a developing network of settlements and a local urban society of some 10,000 inhabitants at Çatalhöyük where trade (of obsidian for tool-making, as well as production of grains, legumes, and animals) was well developed. In *The Economy of Cities* (1969), Jane Jacobs (1916–2006) picked up from this point and argued that every city had a direct economic ancestry or parentage in an older city or cities, hence linking "contemporary urbanism and the 'spark of city economic life' to a virtually unbroken chain of urban societies stretching back over 10,000 years to her choice for the first known city, Çatal Hüyük in south-central Anatolia" (Soja 2003, 26). Her ancestral site for Çatalhöyük was New Obsidian, a "pre-agricultural city of hunters" centered on the crucial obsidian trade that was capable of generating economic growth from its own resources and an expanding division of labor, also for the subordinate networks of agricultural, pastoral, trading, and service settlements. In this way, Jacobs's hypothesis reversed the causal link between Childe's two economic revolutions by suggesting an urban origin for the Agricultural Revolution—an idea supported also by the political geographer and urban theorist Edward W. Soja (2003).

The archaeological evidence Soja lists as the material basis for this alternative hypothesis is exactly the same as that for evolution from hunting-gathering into a city-based life through two revolutions—that is, the absence of massive stone fortifications, a dense agglomeration of attached adobe units without openings on their walls or in-between them in the form of streets or paths, and the human burials beneath their floors. Some of the units identified by Mellaart as shrines were decorated with wall paintings, plaster reliefs, cult statues, animal heads, and bull horns (figure 10.5). Soja notes the gradual

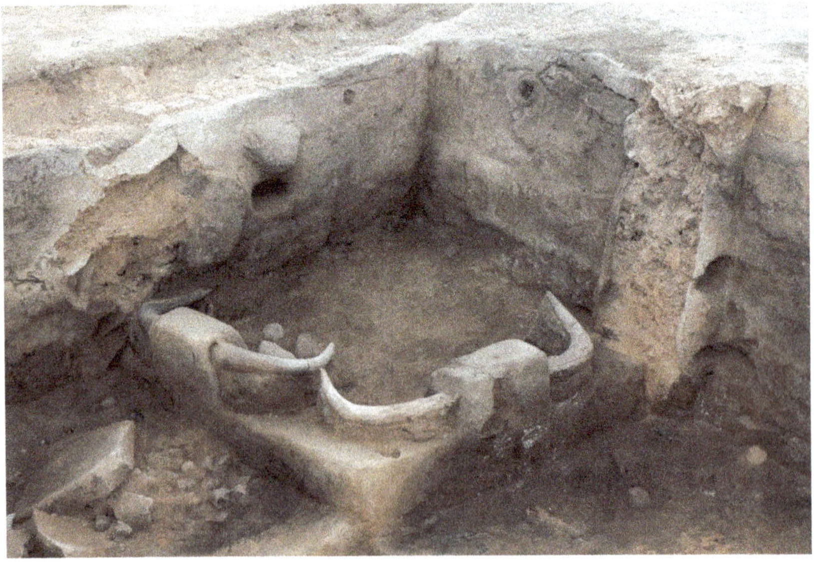

**Figure 10.5** Wild bull horn installation in Building 89 of Area 4040 at Çatalhöyük.
*Credit line:* Photographer: Jason Quinlan © Çatalhöyük Research Project (CC BY-NC).

prominence, over declining hunting scenes, of female statues, as well as realistic and abstract symbols interpreted as pertaining to fertility, agricultural production and urban life, in the upper/later plaster layers. In addition to a large opening in the densely built area, which is interpreted by Soja as a public square that may have served as a marketplace, critical presences that breath an urban life into this ruined physical setting were:

> an extraordinary collection of highly skilled artists, craftworkers, manufacturers, and merchants, among them the first known weavers of cloth and tapestry, the first crude pottery-makers and carvers of elaborated wooden bowls, the first metalworkers along with the beginnings of a substantial weapon industry, and carvers of some of the best religious statuary of the Neolithic, including the now famous "Mother Goddess." (Soja 2003, 30)

The competition between these two hypotheses regarding the emergence of an urban way of living has been going on ever since (e.g., Taylor 2012, Smith et al. 2014), well after the restart of archaeological research at Çatalhöyük in 1993 under Hodder's direction.

### Hodder's Çatalhöyük: An Egalitarian Community

One of Ian Hodder's major objectives in resuming work at Çatalhöyük was to conserve and present to the public the material unearthed during Mellaart's excavations. In due course, experts on the research team have contended, on the basis mainly of micro-artifact distributions, that houses at Çatalhöyük were not shrines but, rather, ordinary houses that were divided into cleaner and less clean areas—which reduced Çatalhöyük to a simpler community than suggested by Mellaart—without any stratum showing different characteristics. Soja's public square serving as a marketplace has been, likewise, reinterpreted as a large opening with dump residues, while the absence of evidence of administrative buildings and elite homes or quarters has led to an image of Çatalhöyük's society of some 8,000 people as an egalitarian one, with no centralized government and hierarchy (figure 10.6). A wide range of functions clearly drawn into the house—from burial, ritual, and art to storage, manufacture, and production—has been taken to indicate the autonomy of the Çatalhöyük household in a sedentary community. Decorations in Mellaart's "shrines" are now thought of as intended to protect and communicate with the dead ancestors buried in the houses (Hodder 2006, 2014). This new interpretation is now widely circulated through various channels, including Jacques Malaterre's highly popular TV documentary *The Rise of Man* (2007), targeting a secondary education audience.

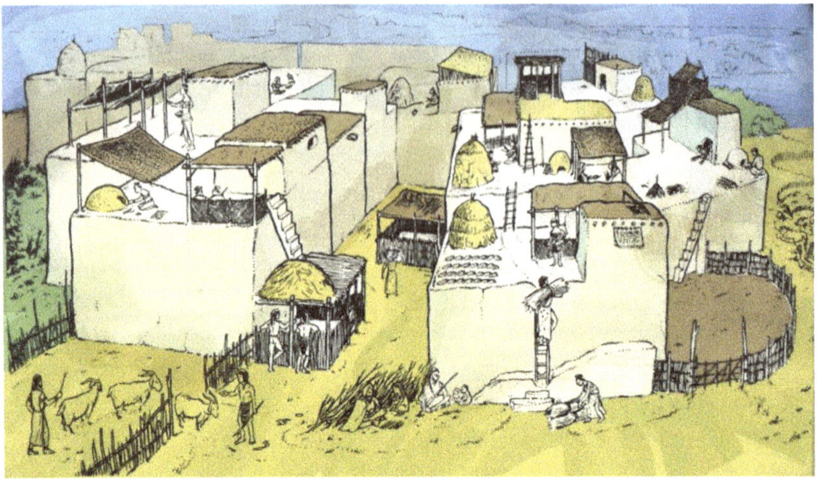

**Figure 10.6** Çatalhöyük as illustration by John Swogger in the framework of the Çatalhöyük Research Project.
*Credit:* © Çatalhöyük Research Project (CC BY-NC).

This plurality of hypotheses, past and present, about the settlement ranking of Çatalhöyük reveals the importance of openness of research data and conclusions on the site for participatory multivocal debates, in approval of the publication policy reported earlier, after Hodder, for the Çatalhöyük Research Project website. Research data and conclusions shared on the website have been produced using advanced computer-based technologies for various tasks, in scales ranging between the magnetometric study of the mound and the lowland around for reconstruction drawings and computer-mapping of the micro-artifact distribution of the organic residue with the largest water floatation system so far used in archaeological excavations for collecting very small particles. Computer-aided research on the remains for their conservation and management has also focused on a variety ranging from construction materials such as plaster with wall paintings to stratigraphy, geomorphologic and hermeneutic analysis, and experimentation with construction technologies.

## ECO AND KRACAUER ON THE OPENNESS OF CREATIVE WORKS TO A MULTIPLICITY OF INTERPRETATIONS

The outcomes of these different research types in digital archaeology have a different semantic relation with the represented material reality, as demonstrated in the three digital works selected for discussion in this chapter. The discussion focuses on whether or not the selected works are open for alternative interpretations of the Çatalhöyük archaeological site regarding its evidencing a rural and/or an urban way of living— this openness not necessarily being the intention of the creators of the works. The focus for this discussion has been formulated within a theoretical framework outlined on the basis of Umberto Eco's idea of "open work" and Siegfried Kracauer's idea of "cinematic materiality."

## Umberto Eco's Notion of the Open Work and Its Relevance for 3D Archaeology

Umberto Eco (1932–2016) is among the philosophers who conceptualize the process of dissemination and interpretation as the concrete, material, and perceptibly enjoyable result of the author's underlying project, which is achieved by the reader's reaction to the physical stimuli of the work as an object, not only intellectually but also physically (Eco 1989, 176–177). Interpretation, therefore, depends on the intentions of the author, the reader, and the work; and it is, for this reason, an open process wherein the work, as a physical entity, has autonomy both from the intention of the author and from the participatory freedom enjoyed by the reader. As much as its cultural and linguistic background, the internal coherence of a work should serve as a parameter for its interpretation, to avoid making it speak of what we want it to tell (Eco 1994, 60). In this way, the work as a physical entity serves as a dam blocking the flood of interpretations (Casetti and Grespi 1998, 267)—as a system (like that of equations) or structure (like that of natural laws) that imposes some constrains upon its interpreters (Eco 1994, 6).

This viewpoint is built on the premise that "a theory of interpretation—even when it assumes that texts are open to multiple readings—must also assume that it is possible to reach an agreement, if not about the meanings that a text encourages, at least about those a text discourages" (Eco 1994, 45). This reveals the work as "the space where some interpretations can be proven invalid. There are cases when, even if we cannot decide between various possible interpretations, we can still eliminate wrongful ones" (Violi 1998, 37) as over-interpretations (Colini 1992, Eco 1994). This locates the probable interpretations encouraged by a work between a single established meaning, conveyed through the translation of every signal to a single message, and an infinity of all possible meanings that can be derived from the work, regardless of its internal integrity.

Eco's information theory for intentionally open works such as *Ulysses* (1922) or *Finnegans Wake* (1939) by James Joyce (1882–1941) explains the relation between a work's physical body and its interpretation by "a rupture with (or departure from) the linguistic system of probability, which serves to convey established meanings, in order to increase the signifying potential of the message" (Eco 1989, 58). According to Eco, authors like Joyce introduce forms of organized (and controlled) disorder into a traditional system of communication so as to increase its capacity for conveying information that would allow for a multiplicity of interpretations (Eco 1989, 60). In this oscillation between the rejection and preservation of the (easily communicable) traditional linguistic system, Eco sees a dialectic tension between form and the possibility of multiple meanings, which constitutes the very essence of openness in a work. As to Eco's distinction between interpretation processes of textual and visual works:

> The linguistic sign "provokes" the reader's rational apprehension by opening up a semantic field in need of contextual delimitation, while the cinematic image acts directly upon the spectator's senses and emotional state, thereby postponing the comprehension of the image itself at a later stage. In the two arts, the modalities of reception are therefore inverted. In cinema, the emotional impact precedes rational-cognitive operations, while in literature emotional stimuli are produced only after the cognitive process has taken place. (Casetti and Grespi 1998, 258)

This may also be taken as a warning against the "power of images," also in digital visualizations that may act directly upon the viewer's senses and emotional state before he or she has had a chance to grasp the nature and scope of this "knowledge claim"—in the case corresponding information is indeed provided to attain scientific transparency, as agreed in the London Charter and the Principles of Seville. This early view of Eco's on the spectator as perceiving the film through experience with her or his own body was founded on Gilbert Cohen-Séat's encephalographic experiments, which had demonstrated total involvement of the spectator with the represented image, with a slight reactive movement and acceleration of the heartbeat in the physiological realm as well (Casetti and Grespi 1998, 276). In this way, the filmic discourse provokes in the spectator *feelings only felt as real*, which is the reason to decode it for producing meaning (Casetti and Grespi 1998, 260; emphasis in the original).

While assessing the contribution of the neurosciences to archaeological research, archaeologist Maurizio Forte and neuroscientist Vittorio Gallese point to the origin of the idea that the body might play an important role in visual experience in the notion of empathy (*Einfühlung*) as coined in aesthetics by the German philosopher Robert Vischer (1847–1933) in 1873 (Forte and Gallese 2015, 49). Recent explorations of empathy by cognitive neuroscience has enabled connection of the perception of architectural spaces, object, and artifacts to bodily practices, through the activation of motor, somatosensory, and emotion-related brain networks by vision, that engages bodily responses of the viewer. The embodied simulation of the movements that could be performed in a perceived space, or of the practices connected to the making of perceived artefacts and the experiences they afford, among the outcomes of neuroscientific work, have informed the 3D-Digging project at Çatalhöyük under Forte's direction (Forte et al. 2012) (compare figures 10.7 and 10.8).

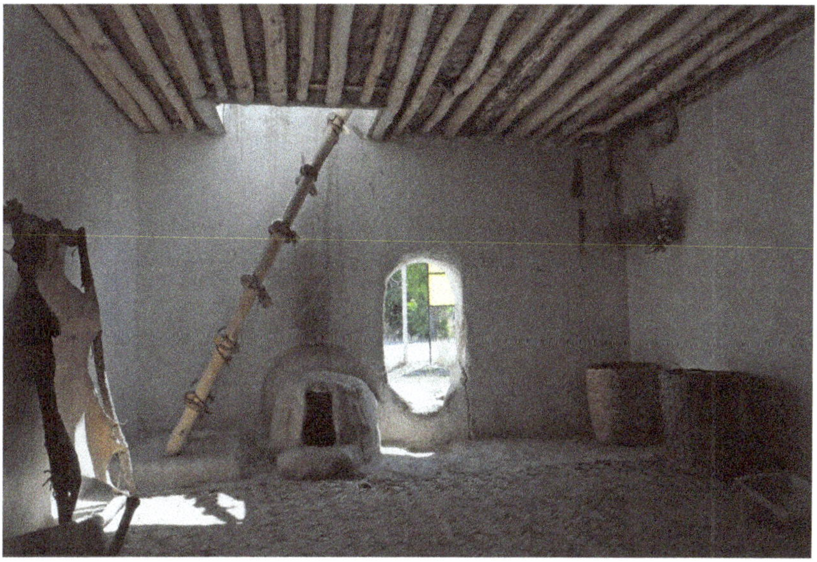

**Figure 10.7** An interior view from the experimental house at Çatalhöyük.
*Credit:* © Grant Cox.

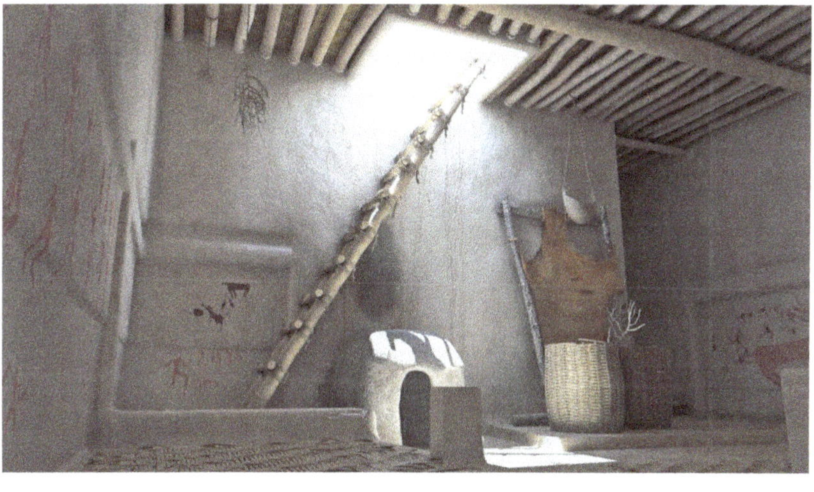

**Figure 10.8** A digital reconstruction of the Shrine of the Hunters at Çatalhöyük by Grant Cox.
*Credit:* © Artas Media—Grant Cox.

The project started in 2009 with detailed digital recording and documentation of the overall East mound and of all stratigraphic layers and architectural elements of a single housing unit that has an estimated sequence of over fifty floors in a life span of some fifty-five to sixty years. Employment of stereo projectors has enabled generation of 3D models of layers and micro-stratigraphies during the excavation process, which were made accessible through a tele-immersive system after post-processing. In this way, instead of promoting a conclusive single interpretation, possibilities have been multiplied through the visualization of complex "potential relationships generated by ornaments, sculptures, architectural features, burials, wall paintings, textures and colours and all the objects designed to create an imaginative world inside the building" (Forte and Gallese 2015, 36). The key question for the neuroscientist working on such an archaeological record has been how to assess the similarities between ancient and modern minds in the spatial embodiment (i.e., visual patterns, symbols, colors, shapes, textures), building on the idea of genetic universals (compare figures 10.9 and 10.10).

## Siegfried Kracauer's Notion of Cinematic Materiality and Its Relevance for Propaganda Films

Siegfried Kracauer (1889–1966) had earlier introduced the idea of similar bodily engagement with the materiality of moving images in cinema studies, with his accent on the physical, tactile dimension of film spectatorship that was built on a perception of the film as a corporeal, material being that stimulates the material layers of its viewer by seizing the human body with skin and hair (Rutherford 2004). This material dimension of cinema assumes a life of

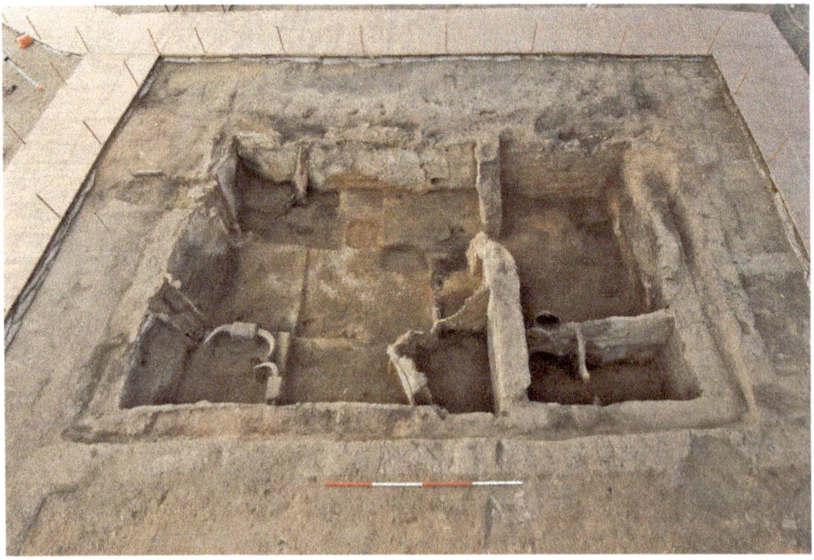

**Figure 10.9** Building 89 in Area 4040, with wild bull horn installation as displayed under a shelter at Çatalhöyük.
*Credit:* Photograph by Jason Quinlan, © Çatalhöyük Research Project (CC BY-NC).

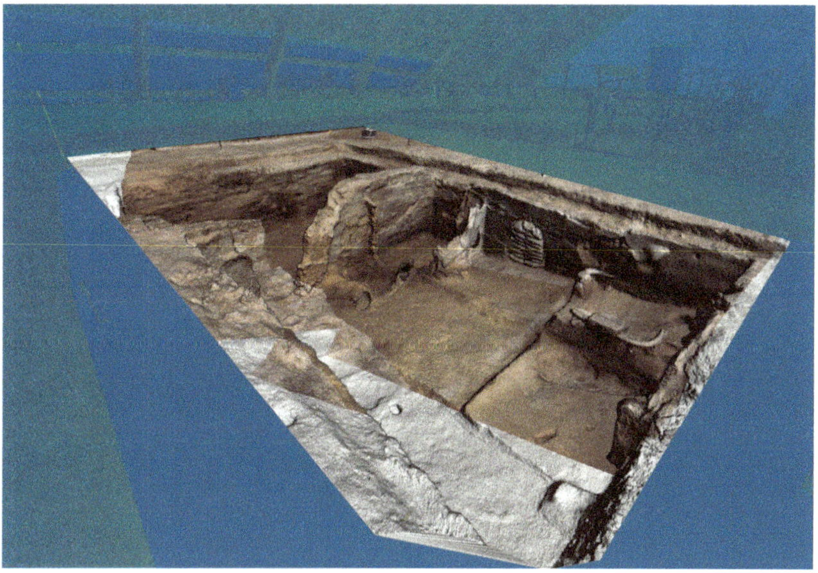

**Figure 10.10** 3D digital documentation and modeling of Building 89 in Area 4040 of Çatalhöyük by a team from the University of California, Merced, under the direction of Maurizio Forte.
*Credit:* © Maurizio Forte, Nicolò Dell'unto, & Nicola Lercari—UC Merced.

its own, and triggers in the viewer associations and memories of the senses, in what Kracauer calls "psychophysical correspondences" through a palpable sensuous connection between the body of the viewer and the viewed. Following from the related concept of "mimetic innervation" as coined by Walter Benjamin (1892–1940), this enables a movement away from the sensorium into the world outside of the film, and an ingestion or incorporation of the world outside of the film, through the interface provided by the mise en scène as an energetic field. An example is *Ulysses' Gaze* (1995) by Theo Angelopoulos (1935–2012), whose loosely composed, porous, permeable, open-ended narratives leave gaps and fissures for the world outside the film to stream into, thanks to the palpable sensuous connection between the film and its spectator (Rutherford 2004). The slippery realm of personal experience, the heterogeneous social space, and the unpredictable dynamics of public life stream through the gaps and fissures in the work, creating a field of possibilities for its interpretation (figure 10.11).

In this way, while giving a form to the complex web of linguistic signs, the cooperative spectator also reflects his or her own image in a film, filling in the gaps in reference to a partial (personal, group, class, ethnic community, sect, etc.) encyclopedia to produce an array of meanings that may occasionally conflict with each other, thus challenging the idea of genetic universals. A striking example of this was the crisis between Italy and China on the occasion of the

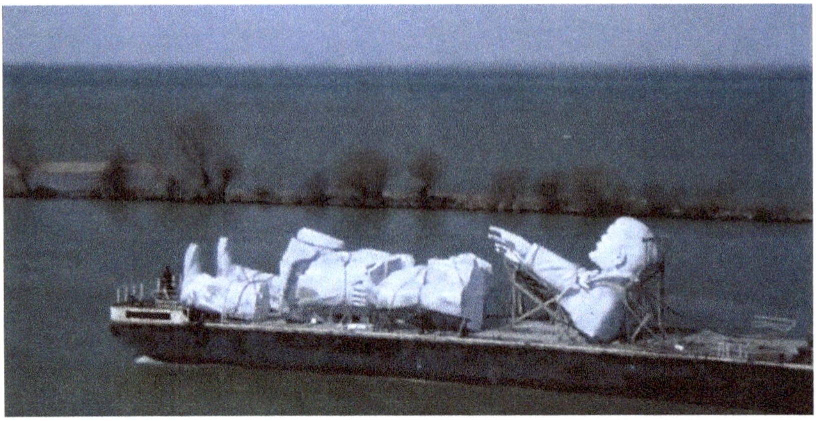

**Figure 10.11** The frame selected by Anne Rutherford from the film *Ulysses' Gaze* (1995) to illustrate how Theo Angelopoulos engages his spectator with the memory, experience, and chance location of the viewer upon the flux of history in the Balkans.
*Credit:* Screen capture from *Ulysses' Gaze* © Theo Angelopoulos, 1995.

Venice Film Festival (1974) screening of *Chung Kuo* (1972) by Michelangelo Antonioni (1912–2007). While the Italian audience that saw the film on TV remembered its attitude of warm and cordial participation in the great event of the peaceful Chinese people, for the Chinese the film was "an inconceivable act of hostility, an insult to the Chinese people" (Eco 1977, 9). On hearing the shot-by-shot commentary of the film by a young Chinese movie critic during the Venice screening, Eco realized how "words and images acquire different meanings according to the cultures which interpret them" (Eco 1977, 9), owing to the difference in the symbolic superstructures through which each culture represents to itself the same political and social problems. Eco's analysis of the film revealed Antonioni's poetic content-form (privileging nuances and gray areas over clear-cut oppositions, pale and cold colors over the bright, dissymmetry and tension over frontal representation and symmetrical distance shots, aspects of everyday life over the theme of revolution, realistic representation of both human subjects and places over symbolism of revolutionary art, etc.) as equally mediated culturally as the film's ideological framework, highlighting it as a main source of the confrontation (figure 10.12). This reveals the possibility that, depending on their content-form, digital visualizations of Çatalhöyük may be open to conflicting interpretations in reference to different partial encyclopedias.

All these analyses reveal the reader/viewer as an active participant in the interpretation process, completing the work of the author/director within the limits set by the work itself. In an exceptional group of works labeled by Eco as "works in motion" (such as *Mobile* for two pianos [1957–58] by Henri Pousseur [1929–2009], mobile sculpture by Alexander Calder [1898–1976],

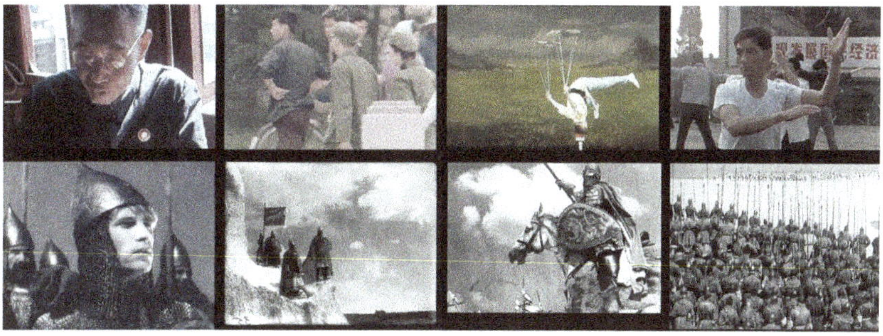

**Figure 10.12** A comparison of the content-forms of Michelangelo Antonioni's *Chung Kuo* (1973) and Sergei Eisenstein's *Alexander Nevsky* (1938).
*Credit:* Michelangelo Antonioni, *Chung Kuo* Cina (1972), http://worldscinema.org/2012/06/michelangelo-antonioni-chung-kuo-cina-1972/; Olivier's *Henry V* and Eisenstein's *Alexander Nevsky*, http://www.second-reel.com/annex/choruslines/olivier-nevsky.shtml.

and the unfinished *Livre* [*Book*] by Stéphane Mallarmé [1842–1898]) the artists allowed the public or chance to arrange some constituents of their works that, therefore, do not have a single, definitive physical order but, rather, a multiplicity of possible orders. In physically static works, information conveyed by the work makes the reader confront all the probabilities at once before establishment of any order (i.e., interpretation of any meaning). In the linguistic system of probability, which serves to convey established meanings, every signal translates into a message that can be referred back to a particular code, according to which every signal signifies one and only one thing. In open works, the message is at once rich in information and yet very difficult to decode, as information becomes the source filtered for a selection (message), and therefore order, and therefore meaning (Eco 1989, 63). At this point, information theory becomes a theory of communication. This difference between information and meaning is crucial for Eco. Certain elements of disorder (such as those introduced by Joyce) may increase the level of information conveyed by a message (Eco 1989, 53): "the larger the amount of information, the more difficult its communication: the clearer the message, the smaller the amount of information" (Eco 1989, 57).

Examples of the latter from the field of cinema may be found in propaganda films that often build on the political nature of our relationship with the image of the ancient past in its symbolic link to modern society (Grunes 2007). A well-known example is the film *Alexander Nevsky* (1938) by Sergei Eisenstein (1898–1948), made in response to a personal request by the Soviet leader Joseph Stalin (1878–1953) during a period of armed conflict with Nazi Germany. In preparing the Soviet public for an alliance with Western powers against Germany, the film pictures Nevsky as the leader of the victorious Russians in Novogorod against the occupying German Teutonic knights, thanks to his historically not verified alliance with their once enemy, the

Mongols. More than this story, however, the mise en scène of the film has been analyzed as essentially fulfilling the propaganda function, with its most conventional images of great bodies of men and horse taking on a posed and static quality with sharp contrasts, as in a sequence of torchbearers at a midday smoke smudge or all the people in the film walking along a horizon (Nugent 1939). This would correspond to Eco's definition of low information content that is transmitted by the visual language of the film that seizes its target audience with skin and hair for propaganda purposes.

## A DISCUSSION ON THE OPENNESS OF THREE VIRTUAL RECONSTRUCTIONS FROM ÇATALHÖYÜK

As such, propaganda films seem to find their parallel in written works, such as Mellaart's *Scientific American* (1964) article that presented Çatalhöyük as the world's oldest-dating city on the basis of conventional images such as the sheer size of the mound or sophistication in the symbolic language of the small finds and wall paintings. As discussed by the Çatalhöyük research team repeatedly (e.g., Leibhammer 2000, Swogger 2000), complementary visual media and language have had a great share in the construction and dissemination of this or other arguments about the past. In this section, these formal characteristics of propaganda films are referenced together with those of open works and works in motion, while comparing the openness of three digital visualizations on Çatalhöyük to a multiplicity of interpretations, including those on the settlement's rank. Necessarily, the following selection and discussion reflect the author's own intention and personal encyclopedia as the reader of the presented works. The selection aims to respond to (though without fully covering) the technological improvements in digital visualization media and dissemination possibilities in the past two decades.

### Çatalhöyük at HABITAT II (1996)

The earliest-dating example is from a commercial CD-ROM, *Housing in Anatolia* (Atlas Media 1996), published on the occasion of the Housing and Settlement in Anatolia exhibition that opened in Istanbul, Turkey, during the Second United Nations Conference on Human Settlements (HABITAT II, 1996). The CD-ROM has a digital image of Çatalhöyük on its cover, which is taken from a two-minute digital movie produced by the Information Technologies and Electronic Research Institute (BILTEN for Bilgi Teknolojileri ve Elektronik Araştırma Enstitüsü) of the Scientific and Technological Research Council of Turkey (TUBITAK for Türkiye Bilimsel ve Teknolojik Araştırma Kurumu). The movie is one of the thirty-six videos in the CD-ROM, only a few of which are digital visualizations, and it does not circulate on the internet (figure 10.13).

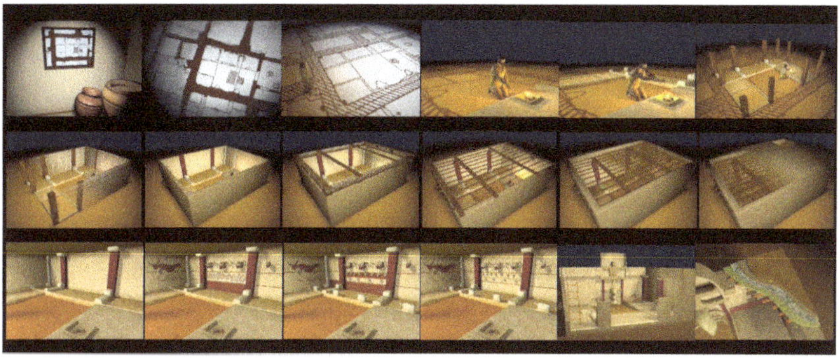

**Figure 10.13** Digital Çatalhöyük visualization from *Housing in Anatolia* CD-ROM (1996).
*Credit:* © TUBITAK BILTEN & Atlas Multimedia.

The video starts with an image of two large terra-cotta vessels in front of a wall with a framed technical drawing of a building unit, and it continues by zooming in and entering into the frame to show a larger plan with more units of similar type, in a very conventional drawing technique for architectural projects. Then appears a bearded male figure in animal skin, on a part of the drawing corresponding to a hatched wall. As the figure walks into the initially framed unit, the first feature to acquire three-dimensionality is a fireplace over a raised platform, toward which the figure proceeds. As the figure warms his hand over the fire, other platforms start to project upward from the ground, however without following from the information given in the plan drawing except in the rough contours of the gradually enclosing interior space. As the projection of walls starts, the viewing angle moves above the modeled unit, the multi-unit base drawing disappears and visualization of the single unit continues, component by component, up to the roof finishing. Layers of material below the roof finishing are presented in the final plan of the video, after visualization of the interior wall decoration layers (including the one on the CD-ROM's cover), some of which feature bull heads, apparently on the basis of archaeological evidence presented by Mellaart as pertaining to shrines.

In terms of intellectual transparency as described in the London Charter, there is no provision of information that would clarify for the viewer whether this computer-based visualization seeks to represent an existing state, an evidence-based restitution, or a hypothetical reconstruction. However, the choice of a conventional architectural drawing and modeling language would seem to claim accuracy for the presented construction details, leaving no room for factual uncertainty—even concerning the archaeologically unevidenced roof structure. Details themselves are also of a familiar and conventional type, as in propaganda films, though the form chosen for their communication lacks, perhaps intentionally, the physicality and tactility that could have seized the viewers with skin and hair. Danish director Lars von Trier's gameboard-like mise en scène in *Dogville* (2003) may be the closest parallel to the opening mise en scène of the work in cinema, which results in a total loss of a sense of materiality. Without prior knowledge from elsewhere about Çatalhöyük, it is impossible to grasp from this work alone that the represented bull heads are "real" ones and that the space has clean and less clean areas, as suggested by Hodder, owing to the low information content that is transmitted by the abstract visual language of the work that excludes information on the possible uses of the decorated spaces.

Preference for an easily communicable traditional technical visual language would seem to offer a linguistic (instead of a visual) system wherein every signal translates into a message that can be referred back to a particular code, according to which every signal signifies one and only one thing (like timber posts, beams, joists, etc.), without requiring a selection through filtering on the part of the viewer. As to the settlement type and rank, while the starting multi-unit base plan may be taken as a message signifying a settlement of similar attached units without streets, the mise en scène with the bearded male figure walking straight on a part of the drawing corresponding to a hatched wall is far from suggesting roof entrances, despite momentary visualization of a ladder, and focus on the construction details and decorations of a single unit eventually fails in giving an idea about the larger settlement.

### Stanford Figurines Project at Çatalhöyük (2016)

Although similarly focusing on a single interior space with a similar shortcoming, the second example selected for this discussion, which is a digital image from the collections of the Stanford Figurines Project (SFP 2016), is stronger than the *Housing in Anatolia* video in terms of the conveyed sense of space, which may open grounds for discussing the type of environment the visualized interior may belong to (figure 10.14). This is thanks to the selected visual language that differs from the conventional architectural drawing and modeling technique of the first example by rendering details that introduce disorder into the visual communication system, increasing the level of communicated information toward cinematic materiality, as in the almost tactile irregularities on the floor, wall, and roof surfaces that suggest the possibility of clean and less clean areas in the represented space.

These views help in making a palpable sensuous connection between the body of the viewer and the viewed, as suggested by Kracauer's idea of cinematic materiality, which is further strengthened by dramatic light and shadow effects. Through the fissure opened by the contrast between the surface irregularities and the sharp and stylized bull horns on the walls and the perfectly weaved baskets on the floor, the outside world would seem to stream into the visualization as the viewer fills in the narrative gaps in the image (concerning

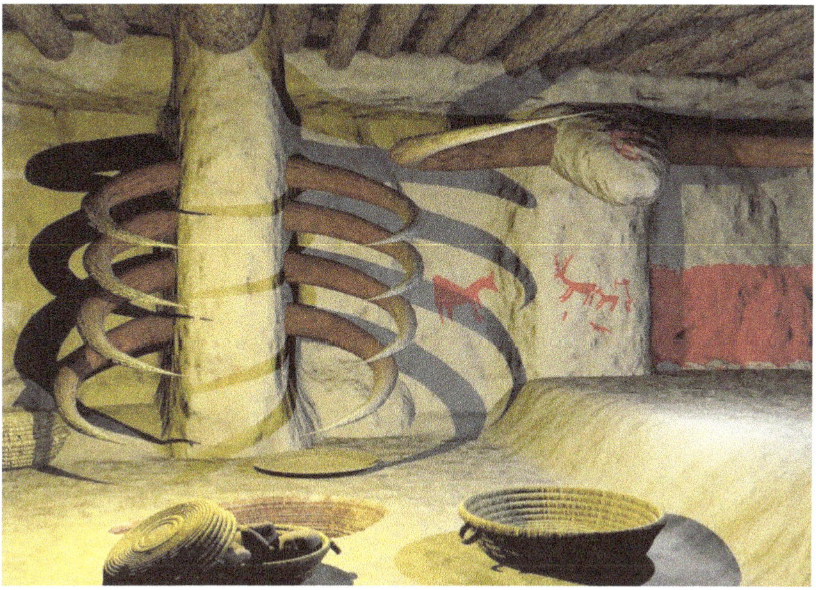

**Figure 10.14** Inside Çatalhöyük houses, from the Stanford Figurines Project.
*Credit:* Bucrania_3.jpg, Stanford Figurines Project., https://web.stanford.edu/group/figurines/cgi-bin/omeka/items/show/6841.

the light source, contents of the baskets, a hole in the ground, the height of space, etc.) in reference to her or his personal encyclopedia and through empathy. This reveals the position of the viewer as an active participant in the process of interpretation and the status of the image as a more open one than the first, which may possibly lead to some viewer's reading artisanal refinement into it. Yet, interpreting this message alone as a signal for urbanity would be an "over-interpretation" in the sense defined by Eco, since it is not supported by the overall image in its internal coherence.

Moreover, unlike the existence of the first visualization within the confines of a CD-ROM that may be considered as the larger work serving as a parameter for its interpretation, this example is accessible through internet search, among shared documents (including also other categories such as publication, fingerprint analysis, assemblages, museum displays, interacting with figurines, technical drawings, etc.) of the Stanford Figurines Project linked to the Çatalhöyük Research Project. This expands the reference work for any interpretation of the visualization, first to the first and then to the second project websites. The Stanford Figurines Project is described as proposing "to expand the ways in which archaeological material is analyzed" by investigating the procedural, dynamic, and contextualized character of some 1,500 figurines (including the "Mother Goddesses") as archaeological objects from a multiplicity of perspectives through constructing multiple narratives that "take into account the different materials and locational contexts of the figurines" (SFP 2016). The selected visualization comes from a collection of images of various types and dates that provide a context for the figurines.

This line of research fits perfectly into the Çatalhöyük Research Project website's reported policy of multivocality, plurality, openness, and transparency. As such, both websites would rank as "works in motion," as coined by Eco, within the confines of which conclusions are described by Hodder as always momentary, fluid, and flexible. This aspect of its interpretation context may compensate for the lack of information that would clarify for the viewers the "knowledge claim" made by the selected visualization for the provision of scientific transparency, as described in the London Charter, except in the case of encountering it out of its publication context—for example, through image search.

### Çatalhöyük in the Second Life Interface (2007)

Among the examples selected for this discussion, the nature and scope of the "knowledge claim" thus made has been clarified verbally only for the third and last one (figure 10.15). This is a virtual reconstruction of Çatalhöyük in the real-time, interactive, three-dimensional virtual settlement Second Life (SL) that was created in 2003 by San Francisco–based Linden Lab:

> SL is an internet-based graphical interface, three-dimensional computer-mediated communication space in which residents can buy and sell land to create virtual landscapes of recreation, communication and commerce. A downloadable client software program allows users to interact with each other over the internet using "avatars" which they design and may modify the appearance of at any time." (Harrison 2009, 82)

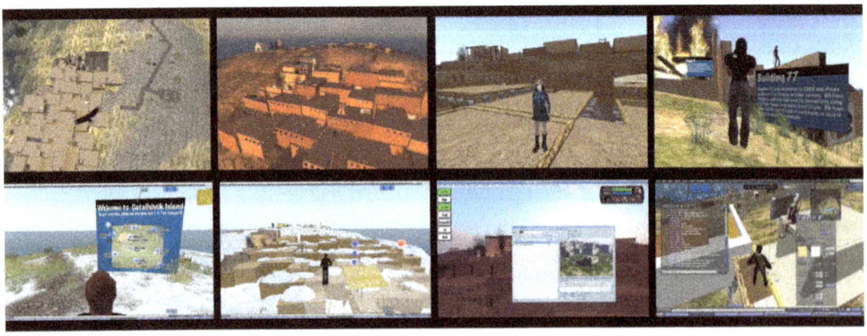

**Figure 10.15** Images from the OKAPI Island in virtual settlement Second Life.
*Credit:* © Colleen Morgan (CC-BY), https://middlesavagery.wordpress.com; © Ruth Tringham and and MACTiA/CoDA, https://mactia.wordpress.com; © Burkhard Schröder, http://www.burks.de/burksblog/2010/01/31/catalhoyuk.

In Second Life, Çatalhöyük was created as an island landscape in 2007, and was named after the project's sponsor, OKAPI (i.e., the Open Knowledge and the Public Interest program of the University of California, Berkeley) in extension of another project called Remixing Çatalhöyük, which aimed to share research data on Çatalhöyük through software with a Creative Commons (CC) license (NCPTT 2009). The project was the brainchild of anthropologist Ruth Tringham, a founder of the People in Multimedia Authoring Centre for Teaching Anthropology (MACTiA) at UC Berkeley. Tringham collaborated with a team of graduate and undergraduate students to explore Second Life's potentials for embedding archaeological research, with its database and interpretations, in a game-like environment that brings three-dimensional imagery into thinking about places in the past (NCPTT 2009). For this purpose, an existing model of Çatalhöyük was duplicated and opened up to alternate interpretations of space and architecture, by "avatars" that were enabled by the software to come and construct their own buildings by experimenting with the provided tools, as in a virtual installation for the San Francisco Modern Art Museum done by Michael Shanks.

This reveals the OKAPI Island to be a "work in motion," as described by Eco, which distinguishes it from the majority of digital reconstructions that are end products of processes engaging archaeologists and other participants to a relatively limited extent. In this way, the reader/viewer in the interpretative process described by Eco identifies with the author, which expires all possibilities for a definitive interpretation of Çatalhöyük on the basis of its visualization as OKAPI Island. This is because its opening to active participant experimentation has thus transformed it from a digital archaeological site model into an open knowledge source in the public interest, which may fit better into the semiotic theory of Jacques Derrida (1930–2004) (e.g., Derrida 1978).

## CONCLUSION: MULTIPLE INTERPRETATIONS OF THE ÇATALHÖYÜK MATERIAL

Morgan (2009, 476) explains how her own experience of redecorating a room interior using the Second Life toolbox challenged her perception of Çatalhöyük, resulting in an engagement with some questions she bypassed as an excavator, and changed the way she interacted visually with the physical site and the way she recorded it. This was because, instead of a photorealistic full-sensory immersion, the environment offers telepresence, or virtual reality, originating from online text-based networks, which is more closely related to the literature-based genre of hypertext (Morgan 2009, 470–471). Morgan describes the working medium as:

> stylized in such a way as to remain decipherable to the residents, and perfect, photo-realistic accuracy is not only unachievable, but is undesirable and jarring

to the greater visual fabric of the space. Archaeologists need not employ specialist designers, or become the most perfect 3D modelers in order to experiment with reconstructions. (Morgan 2009, 479)

This leads us back to the importance of the earlier distinction made by Eco between the linguistic sign and the cinematic image that has the potential to seize the viewer's body with skin and hair. Morgan's conclusions come from an expert who has been immersed in the physical material reality of Çatalhöyük archaeological site season after season and, therefore, her personal encyclopedia is well equipped with first-hand knowledge and experience with the signified of each visual signifier in the Second Life toolbox. As reported after Hodder in the introduction, the diversity of experiment in this medium would depend largely on the reflexivity of its creator and designer of the toolboxes. For Morgan's type of author/readers among Çatalhöyük Research Project participants, the stylized medium of Second Life would indeed function as a system of verbal language that enables their contribution in the open knowledge source in motion.

On the other hand, for OKAPI Island visitors from outside the discipline and without any direct encounter with the archaeological site itself, the possibility of a photorealistic full-sensory immersion could be a unique opportunity for making up for the thinness of their personal encyclopedias to be able to participate actively in the interpretation process, by moving away from the sensorium into the material world through associations and memories of the senses. Tele-immersive models constructed by post-processing on site scans within the scope of the 3D-Digging project would seem to address this exact issue. Crucial importance given to lengthy video recordings of individual figurines in the Stanford Figurines Project or high-resolution 3D scans of sample settlement units in the Çatalhöyük Research Project are experiments by ÇRP participants in this potential offered by the cinematic image.

In the final analysis, the centrality of the site in debates on the emergence of an urban way of living would appear not to be a main concern for the producers of the selected three digital visualizations of Çatalhöyük in their content-form decisions, nor would the principle of transparency. While the HABITAT video would seem to convey coded information on the construction system of a decontextualized Çatalhöyük building, in the other two examples viewers are rather immersed in visual media for open-ended experimentation with archaeological knowledge production, instead of being exposed to visual representations of archaeological expert knowledge. This information content of the visualizations would seem to both expand *and* limit the interpretative possibilities of the viewers, perhaps to the degree of their failing to appreciate the attributes of "outstanding universal value" that enabled inclusion of the site on the UNESCO World Heritage List.

## REFERENCES


Atlas Multimedia. 1996. *Housing in Anatolia "Story of 12 Thousand Years."* CD, Atlas Multimedia.

Balter, Michael. 2006. *The Goddess & The Bull. Çatalhöyük: An Archaeological Journey to the Dawn of Civilization*. Walnut Creek, CA: Left Coast Press.

Casetti, Francesco, and Barbara Grespi. 1998. "Chapter 13—Cinema and the Question of Reception," translated by Veronica Pravadelli. In *Umberto Eco's Alternative: The Politics of Culture and the Ambiguities of Interpretation*, edited by Norma Buchard and Veronica Pravadelli, 257–279. New York: Peter Lang.

Çatalhöyük Research Project (ÇRP). 2016. "Çatalhöyük—Excavations of a Neolithic Anatolian Höyük." http://www.catalhoyuk.com/.

Childe, Vere Gordon. 1971. *What Happened in History*. Middlesex: Penguin.

Colini, Stephano, ed. 1992. *Interpretation and Overinterpretation*. Cambridge: Cambridge University Press.

Denard, Hugh, ed. 2009. "London Charter for the Computer-Based Visualisation of Cultural Heritage. Draft 2.1." http://www.londoncharter.org/fileadmin/templates/main/docs/london_charter_2_1_en.pdf.

Derrida, Jacques. 1978. *Writing and Difference*. Translated by Alan Bass. Chicago: University of Chicago Press.

Eco, Umberto. 1977. "De Interpretatione, or the Difficulty of Being Marco Polo [On the Occasion of Antonioni's China Film]." Translated by Christine Leefeldt. *Film Quarterly* (Special Book Issue) 30(4): 8–12.

Eco, Umberto. 1989. *The Open Work*. Cambridge, MA: Harvard University Press.

Eco, Umberto. 1994. *The Limits of Interpretation*. Bloomington: Indiana University Press.

European Commission (EU) Community Research and Development Information Service (CORDIS). 2001. "The Lund Principles: Conclusions of Experts Meeting." ftp://ftp.cordis.europa.eu/pub/ist/docs/digicult/lund_principles-en.pdf.

Feilden, Bernard M., and Jukka Jokilehto. 1993. *Management Guidelines for World Cultural Heritage Sites*. Rome: ICCROM.

Forte, Maurizio, Nicolo Dell'Unto, Justine Issavi, Lionel Onsurez, and and Nicola Lercari. 2012. "3D Archaeology at Çatalhöyük." *International Journal of Heritage in the Digital Era* 1(3): 351–378.

Forte, Maurizio, and Vittorio Gallese. 2015. "Embodiment and 3D Archaeology: A Neolithic House at Çatalhöyük." In *Breaking Barriers: Proceedings of the 47th Annual Chacmool Archaeological Conference; November 7–9, 2014*, edited by Robyn Crook, Kim Edwards, and Colleen Hughes, 35–55. Chacmool: Chacmool Archaeological Association of the University of Calgary.

Goody, Jack. 2006. "Gordon Childe, the Urban Revolution, and the Haute Cuisine: An Anthropo-Archaeological View of Modern History." *Comparative Studies in Society and History* 48: 503–519.

Grunes, Dennis. 2007. "Ulysses' Gaze (Theodoros Angelopoulos, 1995)." *A Short Chronology of World Cinema*, excerpt published by the author. http://grunes.wordpress.com/2007/03/19/ulysses'-gaze-theodoros-angelopoulos-1995/.

Harrison, Rodney. 2009. "Excavating *Second Life*: Cyber-Archaeologies, Heritage and Virtual Communities." *Journal of Material Culture* 14(1): 75–106.

Hodder, Ian. 1996. "TAG 1996—Intro Talk: Glocalising Catal: Towards postprocessual methodology." http://www.catalhoyuk.com/TAG_papers/ian.htm.

Hodder, Ian. 1998. "The Past as Passion and Play: Çatalhöyük as a Site of Conflict in the Construction of Multiple Pasts." In *Archaeology Under Fire: Nationalism,*

*Politics and Heritage in the Eastern Mediterranean and Middle East*, edited by Lynn Meskell, 124–139. London: Routledge.
Hodder, Ian. 2006. "This Old House." *Natural History*, June. http://www.naturalhistorymag.com/htmlsite/0606/0606_feature.html
Hodder, Ian. 2014. "Çatalhöyük'te 9 bin yıl önce hem yerleşik hem de hükümetsiz bir yaşam vardı." *CNN Türk*, September 22. http://www.cnnturk.com/haber/turkiye/catalhoyukte-9-bin-yil-once-hem-yerlesik-hem-de-hukumetsiz-esit-bir-yasam-vardi.
International Council on Monuments and Sites (ICOMOS). 1964. "International Charter for the Conservation and Restoration of Monuments and Sites (The Venice Charter)." http://www.icomos.org/charters/venice_e.pdf.
International Forum of Virtual Archaeology (IFVA). 2012. "Principles of Seville—International Principles of Virtual Archaeology." http://www.arqueologiavirtual.com/carta/wp-content/uploads/2012/03/BORRADOR-FINAL-FINAL-DRAFT.pdf.
Jacobs, Jane. 1969. *The Economy of Cities*. New York. Random House.
Joyce, James. 1992. *Ulysses*, with Introduction by Declan Kiberd. London: Penguin.
Joyce, James. 1999. *Finnegans Wake*, with Introduction by John Bishop. London: Penguin.
Leibhammer, Nessa. 2000. "Chapter 11: Rendering Realities." In *Towards Reflexive Method in Archaeology: The Example at Çatalhöyük*, edited by Ian Hodder, 129–142. Cambridge: McDonald Institute for Archaeological Research.
Malaterre, Jacques, dir. 2007. *Le sacre de l'homme*. France: uFilm, France 2 (FR2), Le Tax Shelter du Gouvernement Fédéral de Belgique, Umedia.
Mellaart, James. 1964. "A Neolithic city in Turkey." *Scientific American* 210(4): 94–104.
Mellaart, James. 1967. *Çatal Hüyük: A Neolithic Town in Anatolia*. New York: McGraw-Hill.
Morgan, Colleen L. 2009. "(Re)Building Çatalhöyük: Changing Virtual Reality in Archaeology." *Archaeologies: Journal of the World Archaeological Congress* 5(3): 468–487.
National Center for Preservation Technology and Training (NCPTT). 2009. "Second Life as an Archaeological Tool (Podcast 5)." https://www.ncptt.nps.gov/blog/second-life-as-an-archaeological-tool/.
Nugent, Frank Stanley. 1939. "Aleksandr Nevski (1938) The Screen: Eisenstein's 'Alexander Nevsky' Opens at the Camea—New Films at Paramount, Criterion and Rialto." *New York Times*, March 23.
Rutherford, Anne. 2004. "Precarious Boundaries: Affect, *mise en scène* and the Senses in Angelopoulos' Balkans Epic." *Senses of Cinema* 31. http://sensesofcinema.com/2004/feature-articles/angelopoulos_balkan_epic/. 10----[Originally published in Richard Cándida Smith, ed., *Art and the Performance of Memory: Sounds and Gestures of Recollection*. New York: Routledge, 2002]
Smith, Michael E., Jason Ur, and Gary M. Feinman. 2014. "Jane Jacobs' 'Cities First' Model and Archaeological Reality." *International Journal of Urban and Regional Research* 38(4): 1525–1535.
Soja, Edward William. 2003. "Chapter 3: Putting Cities First: Remapping the Origins of Urbanism." In *A Companion to the City*, edited by Gary Bridge and Sophie Watson, 26–34. Oxford: Blackwell.
Stanford Figurines Project (SFP). 2016. "About the Project." https://web.stanford.edu/group/figurines/cgi-bin/omeka/about.
Swogger, John-Gordon. 2000. "Chapter 12: Image and interpretation: The Tyranny of Representation?" In *Towards Reflexive Method in Archaeology: The Example at*

*Çatalhöyük*, edited by Ian Hodder, 143–152. Cambridge: McDonald Institute for Archaeological Research.

Taylor, Peter J. 2012. "Extraordinary Cities: Early 'City-ness' and the Origins of Agriculture and States." *International Journal of Urban and Regional Research* 36(3): 415–447. doi: 10.1111/j.1468-2427.2011.01101.x.

University of California, Berkeley. 2008. "Remixing Çatalhöyük." Archaeological Research Facility. http://arf.berkeley.edu/project/remixing-çatalhöyük.

University of California, Berkeley. 2016. MACTiA Multimedia Authoring Center for Teaching in Anthropology. https://mactia.wordpress.com.

Violi, Patrizia. 1998. "Chapter 1: Individual and Communal Encyclopedias." In *Umberto Eco's Alternative. The Politics of Culture and the Ambiguities of Interpretation*, edited by Norma Buchard and Veronica Pravadelli, 25–38. New York: Peter Lang.

World Heritage Committee (WHC). 2012. "Neolithic Site of Çatalhöyük." http://whc.unesco.org/en/list/1405/.

World Heritage Committee (WHC). 2015. "The Operational Guidelines for the Implementation of the World Heritage Convention." http://whc.unesco.org/en/guidelines/.

CHAPTER 11

# Virtual Cities as Memoryscapes

## *The Case of Lisbon*

MARIA ALEXANDRA GAGO DA CÂMARA, HELENA MURTEIRA, AND PAULO SIMÕES RODRIGUES

### INTRODUCTION

In 1857, as a result of the second exhibition of the Societé Française de Photographie, which took place in Paris, the poet and literary critic Théophile Gautier (1811–1872) wrote the following:

> Originally held with enthusiasm, photography arose, as any new invention, considerable criticism. Undoubtedly, there were those well-intentioned spirits who wished to see in this new and remarkable creation a danger to the arts, as they feared that human intervention could now be replaced by a machine. It is a concern which has no solid grounds. . . . Photography is to the arts, as Faust is to Wagner, its creator. It is converging, critic, provides texts, and may fill the walls of an office of documents, but is not conclusive, and therefore awaits for the outlook of the master. What does that all mean? That it doesn't know. It simply is, nothing more. (Gautier 2015, 125, 126, 129)[1]

Similarly with the invention of photography in the nineteen century, in a more recent past and still to this day, digital technology and its employment in the creation of virtual worlds—namely with the purpose of propagating scientific, cultural, and educational development in the fields of history, art history, and archaeology—has provoked negative responses from those who fear that the use of technology may dehumanize one's interaction with the past and, therefore, with oneself. [2] The concern frequently put forward is based on

Maria Alexandra Gago da Câmara, Helena Murteira, and Paulo Simões Rodrigues, *Virtual Cities as Memoryscapes* In: *Digital Cities*. Edited by: Maurizio Forte and Helena Murteira, Oxford University Press (2020). 
DOI: 10.1093/oso/9780190498900.003.0012

the notion that the dimension of the historical information made accessible to us by technology may be so overwhelming that, akin to Jorge Luís Borges Babel's Library, the repository of all possible texts, it would be impossible to comprehend and interpret it.

This fear is grossly exaggerated. In the same way as Théophile Gautier's conclusions about photography through the metaphor of Faust's servant, the digital as technology is a medium or an instrument which still relies on its interaction with human intelligibility. Its ability to gather, process, and connect information, and to respond to our questions and needs, still highly depends on our ability to provide information, ask questions, and enunciate requirements. However, it would be naive of us to assume that digital technology is merely a highly sophisticated means of reproduction, storage, and diffusion of information. In the same way as photography altered our ability to perceive the visible world, by conferring on it new angles and perspectives, scales, dimensions, volumes, frameworks and chromatisms (Benjamin 1992, Malraux 2000), digital technology is shaping our means of acquiring knowledge. It is doing so by establishing a new dialectic between the user and the information, from which new interactions occur between the two, the users themselves, the data composing the body of information, the real and virtual data, and the data already captured by digital tools (Parry 2012).

In regard to virtual worlds, more specifically to the virtual re-creations of the past, digital technology is introducing a larger complexity of interpretations—namely by creating new relationships between the elements, the objects and the recreated actions, and the coeval documentation and the established knowledge—thus building and revealing new meanings and critical interpretations through a digital perspective (Parry 2007). When little or nothing physical remains from the past, the memory of that past latent in the few documents and physical evidence which survived the passage of time may be brought to light, re-interpreted and transformed into knowledge through new interactions, thanks only to digital technologies and digital hermeneutics (Forte 2014). We are therefore confronted with a cybernetic memory, or a cyber-memory. When that cyber-memory corresponds to a city's past— namely a city that has disappeared or been radically altered by the passage of time—that same cyber-memory becomes a landscape of a memory, a *memoryscape*.

*Landscape* is a conceptual category which classifies a complex system in permanent change resulting from the relation between the human and nature, from the interaction between biophysical and cultural factors (Matos and Baptista 2013). In the more restricted disciplinary scope of artistic studies, a landscape is seen as the representation of a point of view which guides the sensorial experience of a specific physical reality, the outcome of the interaction between the place, its perception, and its representation (Collot 2013). When the factor which forms the structuring point of view of the sensorial experience is the memory, be it collective or individual, its representation can be called *memoryscape* (Nutall 1991).

The memoryscapes represent points of view which arrange the past according to events significant to collective and individual memories from the present. When we apply the concept of memoryscape to cities, the structuring points of view are frequently represented by areas, spaces, or buildings which formed its imaginability over time. According to Kevin Lynch, the imaginability of a city is its capacity to evoke a strong image in an observer "in a defined and intense way" (Lynch 1999). This is an image that allows us to understand a city beyond time as a model of great continuity, in which the areas, the spaces, and the buildings that grant the city its imaginability are also mnemonic devices, making that image perceptible, coherent, comprehensible, and legible—a memoryscape.

For this reason, we argue that virtual recreations of cities are memoryscapes: by creating the simulacrum of a specific historic moment of the city, they establish a way of viewing its past, caused by the subsequent event which changed it (whose future occurrence is always implicit) and configured by the areas and buildings recreated (Stoichita 2011). These virtual recreations also form part of the memory of the city, because the memory of its existence was preserved by historic documents and/or because the areas and buildings of the past still exist, even if altered, in the present-day city. As simulations of the past capable of producing a sensorial experience more complex and complete than the literary description (which is exclusively intellectual and thus abstract) or the visualization of drawn or painted images (which are fixed and thus non-interactive and non-immersive), the virtual recreations of cities project that past onto the present in such an impressive way that they tend to form hyper-realities and become normative memories (Halbwachs 1968).

## LISBON AS VIRTUAL REPRESENTATION

Let us now focus on the case of the city of Lisbon, capital of Portugal, the subject of a number of virtual reconstruction/recreation projects. Despite all of them having distinct goals and using different technologies, they all had as a reference point the same determining event in Lisbon's history which radically changed it: the earthquake of November 1, 1755.

### The Historical Context

Lisbon's geographic location, alongside the mouth of the river Tagus and close to the Atlantic Ocean, favored an early maritime-port vocation that brought to the city growth and status. From the small maritime port of Phoenician times to the considerable Roman and Muslim Iberian city, Lisbon's development was relentless and steadfast. Its growth persisted after its conquest by the first king of Portugal in 1147. Lisbon became Portugal's first city by the late thirteenth century and from the mid-fifteenth century to the mid-sixteenth century, during the height of Portugal's overseas expansion, the city's population doubled, reaching more than 100,000 inhabitants by the 1550s (Rodrigues 2008).

Roman Lisbon (Olisipo) was a walled-in area extending from the top of a hill where the castle was later built down to the riverfront. From the Moorish occupation (early eighth century), the city began expanding outside its walls, particularly with the growth of the Alfama neighborhood to the east. From this period also dates a reinforcement of the urbanization process of the valley west of the Castle Hill. This area grew in population and building fabric from the twelfth to the fourteenth centuries, but it was from the fifteenth century and the start of the Portuguese colonial expansion that the city center clearly shifted to this location. In the early sixteenth century, after King Manuel I moved the royal residence from the Castle Hill to the south end of the valley, just opposite the riverfront, this shift was sealed. The open field opposite the new royal palace was surrounded by state and administrative buildings, and the Terreiro do Paço (Palace Courtyard) became the political and economic heart of Lisbon and its most iconic urban ensemble until the 1755 earthquake (figure 11.1).

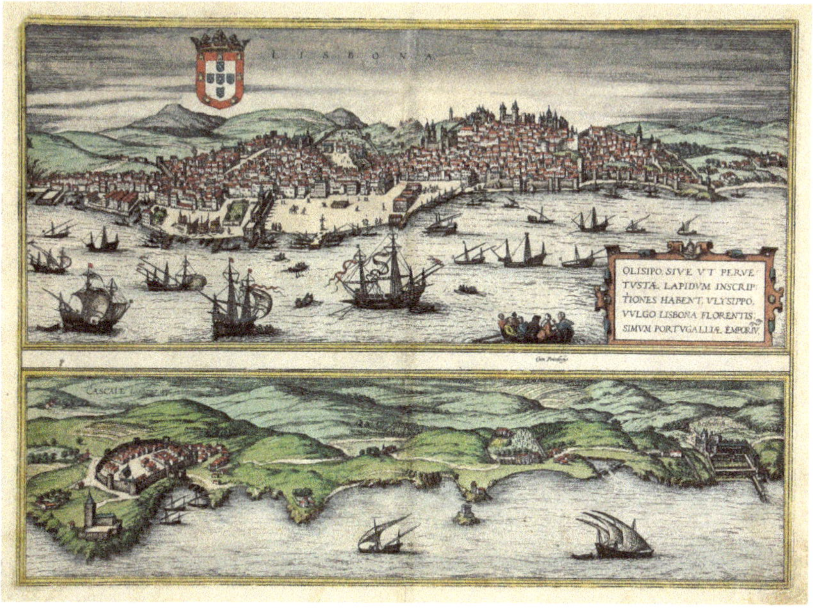


**Figure 11.1** *Civitates Orbis Terrarum* (1572). Engraving by Georgius Braunius and Frans Hogenberg. Lisbon in the early sixteenth century.
*Credit:* Colecção do Museu de Lisboa /Câmara Municipal Lisboa –EGEAC (Collection of the Lisbon Museum/ Lisbon City Council-EGEAC).

The Ribeira (waterfront) Palace experienced two documented major renovations until its destruction in 1755. The first one took place at the start of the Iberian Union (1580–1640), during the sovereignty of King Phillip II of Spain (Phillip I of Portugal). The renewed royal palace performed a distinctive role in the representations of Lisbon that the popularity of engravings helped to disseminate throughout seventeenth- and early eighteenth-century Europe. In the first half of the eighteenth century, the royal palace was once more largely improved and extended, this time by King John V. Depictions of Lisbon portraying its emblematic Ribeira Royal Palace continued to be disseminated throughout Europe until the end of the eighteenth century. As such, this building complex remained in Europe's imagination long after its ruin in 1755 (figure 11.2).

From the sixteenth to the mid-eighteenth century, Lisbon was mainly its central area, the long valley between the Rossio square at the north end of the city and the Palace Courtyard at the south end, overlooking the Tagus River. The San Francisco Hill, to the west of the valley, was also included in the city center, particularly after the linking of both areas by the opening of the Rua Nova do Almada (Almada New Street) in the early seventeenth century. The old Castle Hill remained as a historical anchor of Lisbon's fast-growing city core. In its city center, Lisbon grew as a relevant European urban cosmos

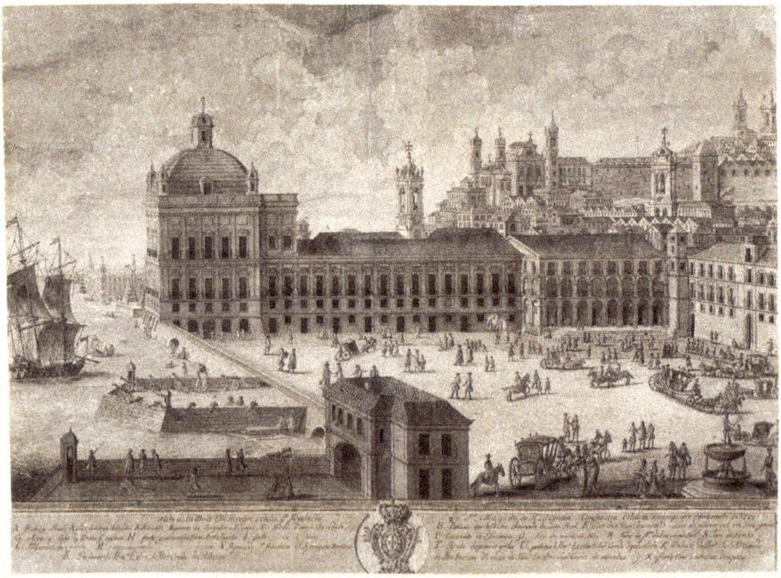

**Figure 11.2** The Palace Courtyard at the eve of the 1755 earthquake (n. d.). China-ink drawing attributed to Francisco Zuzarte.
*Credit:* Colecção do Museu de Lisboa /Câmara Municipal Lisboa –EGEAC (Collection of the Lisbon Museum/ Lisbon City Council-EGEAC).

with a distinct architectural, social, and cultural character (Gschwend and Lowe 2015).

The late seventeenth and early eighteenth centuries were times of change. After regaining its independence in 1640 and the end of the Restoration War with Spain in 1668,[3] the Portuguese Crown tried to bridge the economic gap with its European counterparts. These efforts implied town planning actions envisaging the modernization of the architectural appearance of some city areas and the renovation or construction of key infrastructures—namely the construction of quays, the pavement and enlargement of some of the main roads, and the building of a monumental aqueduct. Albeit essentially medieval in its urban layout, Lisbon was aware of the most recent European town planning ventures and moved forward. Lisbon's city council was active in safeguarding the public interest and of taking a more rational approach to town planning (Murteira 1999). Simultaneously, the Roman Baroque taste favored by King John V was enlivening the architectural fabric where sober solutions of the military engineering had prevailed (Murteira 1999, Pimentel 2012).

The 1755 earthquake halted this town planning dynamic, erased a substantial part of the memory of Lisbon, and opened a new chapter in the city's history.

With an estimated magnitude of 8.75 to 8.90 on the Richer Scale and a probable duration of 8 to 10 minutes, the 1755 Lisbon earthquake remains the most powerful ever recorded in Europe (Gutenberg and Richter 1949, Paice 2008). On November 1, 1755, Lisbon suffered the devastating consequences of several seismic shocks and a tsunami that broke over the riverbank into bordering neighborhoods. Adding to that destruction, a strong fire raged through the city for approximately one week. It was precisely in the valley between the Rossio Square and the Palace Courtyard, as well as in the San Francisco Hill (also known as Chiado Hill) to the west, that the earthquake was most destructive. Lisbon's central area was mainly reduced to ruins and ashes. The Castle Hill also suffered severe damage.

According to contemporaneous accounts, roughly 10% of the architectural fabric was completely ruined and more than half was left in no condition to be inhabited (Mendonça 1758). A staggering 75% of the parish churches and 85% of the city's convents were severely damaged or completely ruined. The catastrophe also destroyed a great number of palaces and all the city's hospitals and prisons (Mendonça 1758, França 1987).

Although contemporaneous estimates of the number of victims were diverse and mostly exaggerated, recent studies suggest that the death toll in Lisbon may have reached 30,000 to 40,000 people, which represented about 12% of the city's population at the time (Pereira 2009).

The cultural, artistic, administrative, financial, and economic life of the city was greatly hampered afterwards, with the loss of an extensive number of records, books, and other written documents, works of art, currency, and

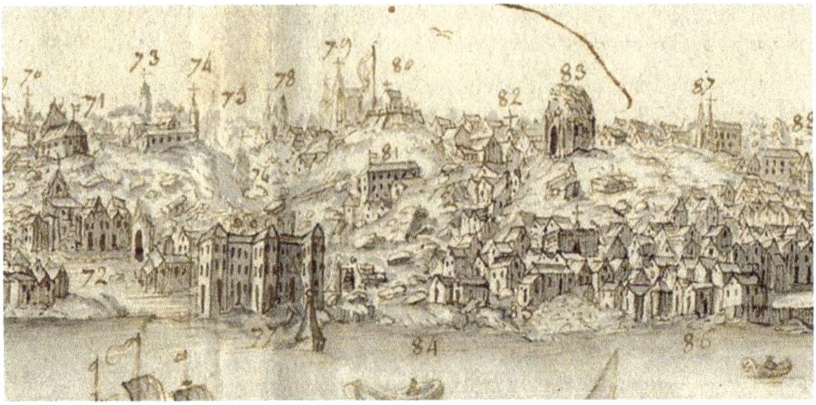

**Figure 11.3** View of Lisbon ruined by the 1755 earthquake, detail of The Palace Courtyard, by Bernardo de Caula (n.d.). Quill pen drawing with sepia and gray ink.
*Credit:* Bilioteca Nacional de Portugal (National Library of Portugal).

goods. The catastrophic impact of the 1755 earthquake in Lisbon is exemplarily portrayed in the view of the city drawn by the military engineer Bernardo de Caula; from the Tagus, a recurrent perspective in the representations of Lisbon since the sixteenth century, Lisbon's central area is pictured as a sequence of ruins and debris (figure 11.3).

This perception of complete annihilation is equally visible in the various testimonies of the time,—namely in the letters sent by the British community of merchants residing in Lisbon to their families in Great Britain. This was the echo of a tragedy that resonated throughout mid-eighteenth-century Europe, further intensified by the human impotence in face of these occurrences.

The earthquake of 1755, whose devastating effects were felt from the south of the Iberian Peninsula to the west of North Africa, equally became an event of considerable scientific and philosophical focus; it reinforced criticism of the Optimist view of the world put forward by Gottfried Leibniz, especially in the case of Voltaire,[4] and contributed to the development of seismology as a discipline within its own right (Almeida 2009, Murteira 2017).

This brief journey into Lisbon's remote history allows us to understand the impact that the earthquake of 1755 had on the history and identity of the country. It equally allows us to comprehend the actual scale and meaning of the memory which, between the myths and facts, has reached us.

The city which succeeded the old, between Rossio and the Palace Courtyard (north/south) and between the bottom of the Castle Hill and the San Francisco Hill (east/west), obeyed a new urban layout in accordance with the Cartesian principles of town planning. The most damaged areas of the Castle Hill were not included in this plan. The oldest area of the city was thus rebuilt following its old layout still imprinted in its ruins (figure 11.4).

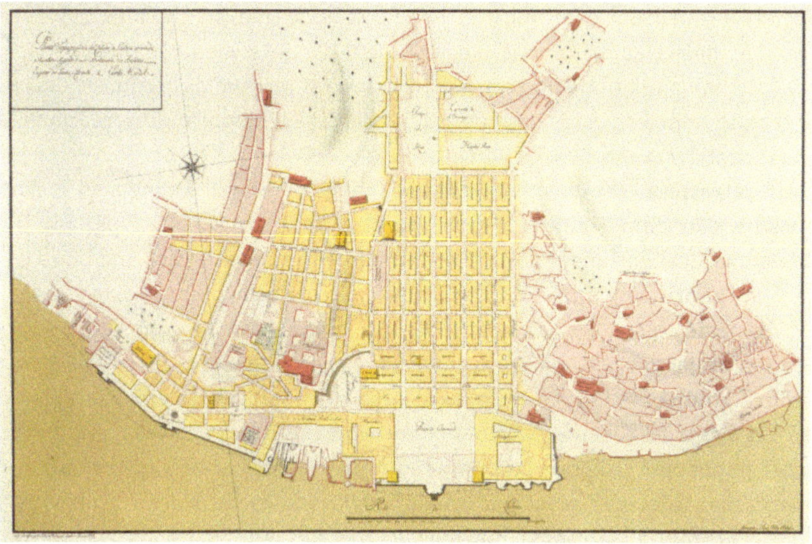

**Figure 11.4** Copy of the adopted plan for Lisbon by Eugénio dos Santos and Carlos Mardel. Lithograph by João Pedro Ribeiro (1947).
*Credit:* Colecção do Museu de Lisboa /Câmara Municipal Lisboa –EGEAC (Collection of the Lisbon Museum/ Lisbon City Council-EGEAC).

## The Rebuilt Lisbon

This new Lisbon, implemented by the Marquis of Pombal, who was the secretary of state for King Dom José I (King Joseph I), and planned out by Portuguese military engineers, profoundly altered Lisbon's physiognomy and the city's relation with its inhabitants. Lisbon's reconstruction plan was only finalized a century later, in the mid-nineteenth century, and the resulting ensemble remained dominant in Lisbon's urban context until the beginning of the twentieth century.

The Pombaline Lisbon is not just an autonomous area of study and one of the richest in historiographic terms; it has also been used as a research field that keeps the study of the modern and the contemporary city (from the nineteenth century to the present day) separate from the city of the *ancien régime* (sixteenth to eighteenth centuries) (França 1987, Tostões and Rossa 2009, Leal 2004). The memory of the city lost in the earthquake established a unique historical defining point of view, where the time interval between the sixteenth century and 1755 stands out. With few exceptions (André 2012), the Lisbon of the *ancien régime* has been approached and understood as a separate temporal process but internally continuous, in which the transformations identified or projected are seen as a consequence of the same urban logic (Murteira 1999). This can be seen in the study of the evolution of the Ribeira Palace, where the architectural changes, even the most radical ones, such as Filippo Terzi's tower, built by order of Portugal's Phillip I (Phillip II of Spain), acquire a consistent continuity in the relationship that the city and, in particular, the palace complex, was establishing with the river since the sixteenth century; it's a relationship seen as a continuous process of becoming deeper and more impressive, for reasons of political and symbolic representations of the monarchy (Senos 2002, Martinho 2010).

In retrospect, the limit to which the memory of pre-earthquake Lisbon goes back in time is defined by the documentation available, especially the iconographical documentation, since an image is more effective and more immediate at bringing into the present the absence of the past (Belting 2014).

The same reason might explain the fact that the visual reconstruction/recreation projects of Lisbon are focused on a specific area of the city and on a limited group of buildings—that is, the riverfront area, with emphasis placed on the Praça da Ribeira (Ribeira Square) and the Ribeira Palace, and on the most important religious buildings, such as the Sé (Lisbon's Cathedral) and the Carmo Convent. The city's iconography, established since the sixteenth century through paintings, drawings, engravings, and plans, focused on the areas and urban elements mentioned here, recording its evolution during the period taken into consideration. In this way, the collective memory which has conditioned the way of viewing Lisbon's past configured by virtual recreations has been established, as we shall now demonstrate.

## THE CASE STUDIES

### 3D Reconstruction of Lisbon

The 3D Reconstruction of Lisbon Before the Earthquake project was developed by the Lisbon Museum from 2005 to 2010. It was based on a large reconstructive model of pre-1755 Lisbon made by Ticiano Volante, under the guidance of the Lisbon historian Gustavo Matos Sequeira, and was to be displayed in the 200th anniversary commemorations of the 1755 earthquake. This model was continuously improved up until 1959, and was on display in the Pimenta Palace, a branch of the Museum of Lisbon. The project's design, the historical research, and the iconographical survey were carried out by a team at the museum, coordinated by its former director, Ana Cristina Leite. The three-dimensional modeling was developed by the company Sdesign_world—Agência de Design Global (figure 11.5).

The project aimed to re-create the city during the first half of the eighteenth century through twenty-three emblematic urban areas or buildings which disappeared or were altered by the 1755 earthquake: the Terreiro do Paço (Palace Courtyard), the Ribeira Palace, the Customs House, the Terreiro do Trigo (Wheat Yard), the Corte Real Palace, the Rossio, the Estáus Palace, the San Domingos Convent, the All Saints Royal Hospital (figure 11.6),

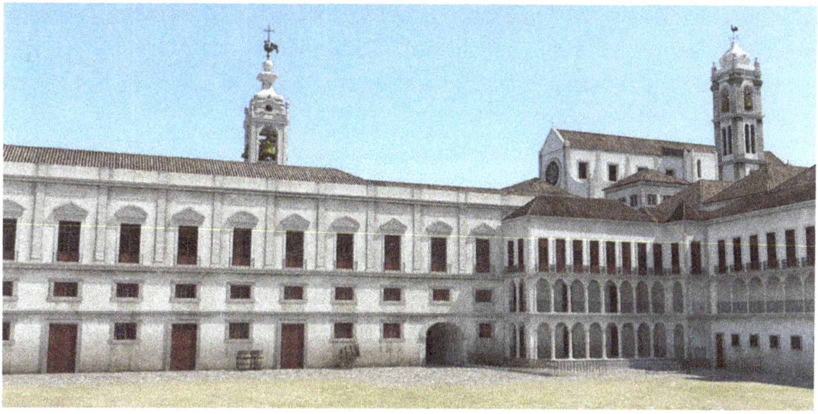

**Figure 11.5** Detail of the recreation of the Ribeira Royal Palace. For the 3D Reconstruction of Lisbon Before the Earthquake project.
*Credit:* Colecção do Museu de Lisboa /Câmara Municipal Lisboa –EGEAC (Collection of the Lisbon Museum/ Lisbon City Council-EGEAC).

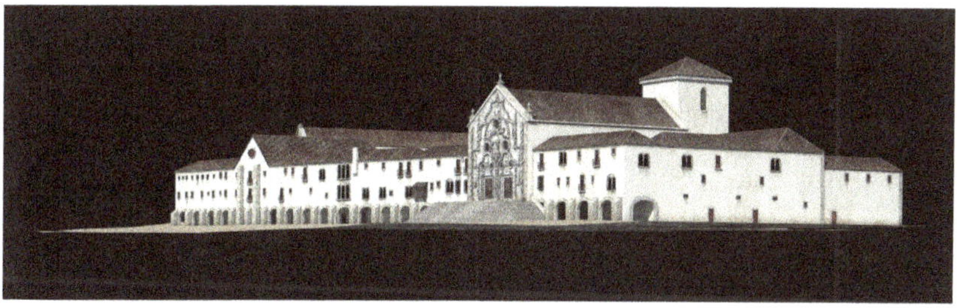

**Figure 11.6** Re-creation of the old All-Saints Hospital in Rossio. For the 3D Reconstruction of Lisbon Before the Earthquake project.
*Credit:* Colecção do Museu de Lisboa /Câmara Municipal Lisboa –EGEAC (Collection of the Lisbon Museum/ Lisbon City Council-EGEAC).

the Santo Antão-o-Novo Convent, the San Francisco Convent, the Carmo Convent, the San Roque Church, the San Bento da Saúde Convent, the Casa dos Bicos, the Chafariz del Rei, the Chafariz de Dentro, the old Cathedral, the Graça Convent, the San Vicente Convent, the Saint Engrácia Church, the Necessidades Palace, and the Rua Nova dos Ferros. These elements were accompanied by a video recreation of some of the early eighteenth-century Lisbon routes, as it is the case of the route going from the Palace Courtyard to the Rossio, through the Rua Nova dos Ferros. The results of the project were displayed in the Pimenta Palace branch of the Lisbon Museum. The 3D interactive content was installed in multimedia touch-screen kiosks, enabling the visitors to take part in an individual exploration of the recreated Lisbon.[5] The technology used allowed only in situ access to the model, and the interaction with the user was limited to the script provided—that is, information about the buildings, displayed separately, and the urban routes recreated. As happened to other projects using traditional 3D rendering techniques, the model encapsulated the research hypothesis at a given point in time: the beginning of the project.

Virtual heritage and virtual archaeology were in their infancy in Portugal in 2005. This project had the merit of being one of the earliest efforts to translate the memory and history of Lisbon through 3D visualization tools. As such, it also represented a significant effort to gather scattered historical evidence and to visually display the research results to a wide audience. The undergoing reformulation of the Palácio Pimenta branch of the Lisbon Museum has halted its display to the public.

## Lisbon Apps

Four years later, in 2014, the Lisbon Museum launched the app "Rewind Cities Lisbon" for Android and IOS systems, sponsored by Samsung and the Lisbon transport company Metro/Carris, and developed by the company IT People. Similar to the "Streetmuseum" Iphone app introduced by The London Museum in 2010, "Rewind Cities Lisbon" allowed visitors to access information about Lisbon and through augmented reality to view on location old representations of some areas of the city superimposed on their present-day aspect. Unfortunately, this app is no longer in use.

The Tourism Board of Lisbon provides one other possibility to visit the city's past through the Lisbon Story Center—Memory of the Cities, at the interpretive center for the history of Lisbon, located in the Praça do Comércio (Commerce Square—riverfront area) and inaugurated in 2012.[6] The center holds interactive and technological equipment that tells the history of the city of Lisbon from its mythical origins in Ulysses's travels on up to the present day. The narrative of the city's past is structured around central characters and events, such as the 1755 earthquake, presented in the form of an immersive theatrical experience. The pre-earthquake Lisbon is shown through scenographic installations of parts of the old city and through reproductions of iconographic elements of Lisbon in the seventeenth century and early eighteenth century. Aimed at a nonexpert audience and therefore with a very broad and straightforward approach, the Lisbon Story Center is part of Lisbon's main tourist attractions. The public feedback has generally been positive, with the occasional criticism from a more demanding audience (*vide* Tripadvisor). An updating of the storytelling and of its interactive character would be advantageous, particularly taking into account that this is the only operating museum multimedia display of the history of Lisbon.

## Lisbon in Tiles

The project Lisbon in Tiles Before the 1755 Earthquake was developed by the Art History Institute of the Universidade Nova de Lisboa (New University of Lisbon), under the scientific coordination of Pedro Flor.[7] It was started in 2010 and was funded by the Fundação para a Ciência e Tecnologia (Foundation for Science and Technology, or FCT). This project was unique in that it focused on the iconographic, historic, and material study of one single object— O Grande Panorama de Lisboa (the Great Lisbon Panorama)—a large tile panel (115 x 2247 cm) from the first quarter of the eighteenth century. The tile panel presents a panoramic view of the city of Lisbon during this period (figure 11.7). It belonged to the old palace of the Earls of Tentugal in Lisbon, and is currently on display at the National Tile Museum.

One of the project's outputs was the re-creation of the tile panel in a web environment. Strictly speaking, it doesn't aim at re-creating the city of Lisbon through the iconographic reference of the tile panel, but the panel itself is used to explore its potential as a source for the history of the city of Lisbon. The project was only recently accessible to the public via the website http://lisboaemazulejo.fcsh.unl.pt/, after an artistic and historical investigation was carried out to document the dozens of churches, convents, palaces, manors, forts, bastions, bridges, streets, and neighborhoods of Lisbon which can be found on this magnificent panel of tiles. Aimed at an expert audience, the

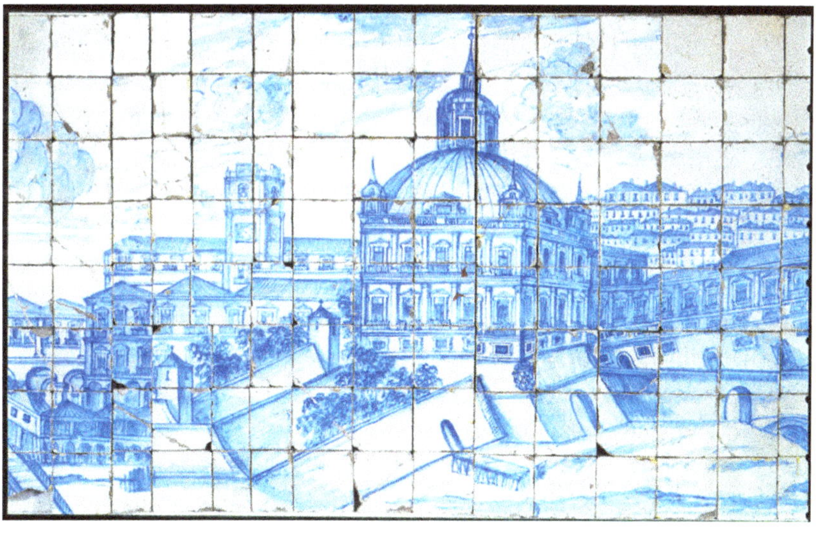

**Figure 11.7** Terreiro do Paço (Palace Courtyard). Detail from the tile panel *Grande panorama de Lisboa* (*Great View of Lisbon*), c. 1699–1700. Blue and white ceramic tiles, 115 x 2247 cm. The tile panel is from the old Palace of the Earls of Tentúgal, Lisbon.
*Credit:* Museu Nacional do Azulejo (National Tile Museum), Portugal.

project does not have a virtual-archaeology component. It is, rather, an academic research project that uses an interactive website to better display its research results.

## The Royal Opera House

In 2005, again for the commemorations of the 250 years since the 1755 earthquake, Alexandra Gago da Câmara, one of the authors of this chapter, responded to an exciting challenge, that of putting forward a 3D proposal for the re-creation of one of the most important buildings lost in the earthquake, the Royal Opera House. This short-lived Baroque building was opened to the public March 31, 1755, and was destroyed on November 1 of the same year. It was the last extension to be added to the renovated early eighteenth-century Ribeira Royal Palace. The evidence that survived for this building is scarce. There are no visual representations of its interiors and exterior, apart from an engraving of its ruins and a controversial cross-section and ground floor plan.[8] The project was accomplished through the use of Second Life technology, which allowed a swift and low-cost process and its continuous updating.[9]

## City and Spectacle Project

The success of the Royal Opera House project led to a more ambitious undertaking, the project City and Spectacle: A Vision of Pre-Earthquake Lisbon, which was developed from 2008 by the Centre for Art History and Artistic Research (CHAIA) of the University of Évora, in partnership with Beta Technologies, the same company that worked on the Opera House project.[10] City and Spectacle has been developed under the coordination of the authors of this chapter. It aims to re-create/simulate the memory of the Lisbon destroyed by the 1755 earthquake, in order to develop a research lab model on the history of the city (Câmara et al. 2009, Murteira et al., 2015, Câmara et al. 2016, Murteira and Rodrigues 2016). Given its wide scope and the need to rethink its digital format and funding options, the project was temporarily suspended in 2014. It will be resumed in 2020.

Its team includes researchers from the field of art history, specializing in urban history, architecture, and landscape, as well as specialists in virtual worlds and in applying IT resources to research and the public presentation of history. In 2008, the choice of Second Life/OpenSimulator technology was based upon the fact that it presented itself as a reliable, fast, and low-cost tool for meeting the goals of the project, which are to test, through a 3D interactive and immersive representation, a long research path in the history of the city. The team needed a technological framework that allowed a real-time and collaborative working process and that was not bound to the time-consuming and redundancy risks of other 3D graphics-development tools popular in virtual reconstructions and recreations at the time. Therefore, virtual-worlds technology seemed the best option (Murteira et al. 2017).

The City and Spectacle project focuses on the Lisbon cityscape immediately before the 1755 earthquake—more specifically, with regard to the area of the city which disappeared in post-earthquake reconstruction. It will include a re-creation of the urban layout, with the architectural fabric and the interior of some important buildings of the ruined city, such as the Ribeira Palace, the Patriarchal, the Tagus Royal Opera, the Corpus Christi Convent, and the All Saints Hospital. It will also include audio and animation components—namely a simulation of the city sounds of early eighteenth-century Lisbon and re-creations of operas, bullfights, processions, and other significant events of the time.

Using manuscripts and printed iconographic and cartographic sources, the project acts as a re-creation of a memory of Lisbon, of the possible memory given the scarce and fragmented documentation available. Since the city suffered some significant alterations from the mid-seventeenth century to the early eighteenth century, extra care is necessary so that all the buildings and urban areas coincide chronologically in terms of appearance and physical condition (Murteira 1999). A less precise chronological limit could lead to the problem identified in the model displayed in the Lisbon Museum, which does not correspond to the city's appearance in 1755 but, rather, to different situations in regard to buildings and urban areas within the period between the last quarter of the seventeenth century and 1755.

Taking into account the extent of this project, it was decided to first re-create the most iconic area of the city: the vast complex of the old Ribeira Palace in the Terreiro do Paço (figure 11.8), including the adjacent Royal Opera House and the Patriarchal Square and Church. The lack of consistent documental evidence for the whole ensemble stressed the need to clearly identify the sources used for the working hypothesis put forward, as asserted in documents such as the London Charter (2009) and the Seville Principles (2011). In this context, it was vital to distinguish between a simulation of a reliable iconographic source and a mere supposition based upon the fragmentary documentary evidence available. All the buildings were re-created in their corresponding urban context; the references to the sources that were used for the re-creations were made available and the time frames of the recreations were precisely defined.

The primary and secondary sources used are available in a virtual museum installed in one of the recreated rooms of the old palace (figure 11.9). This virtual museum combines information on the history of Lisbon and on the project itself through a display of the most significant iconography used and a presentation of the underlying concept of the project and its several stages.

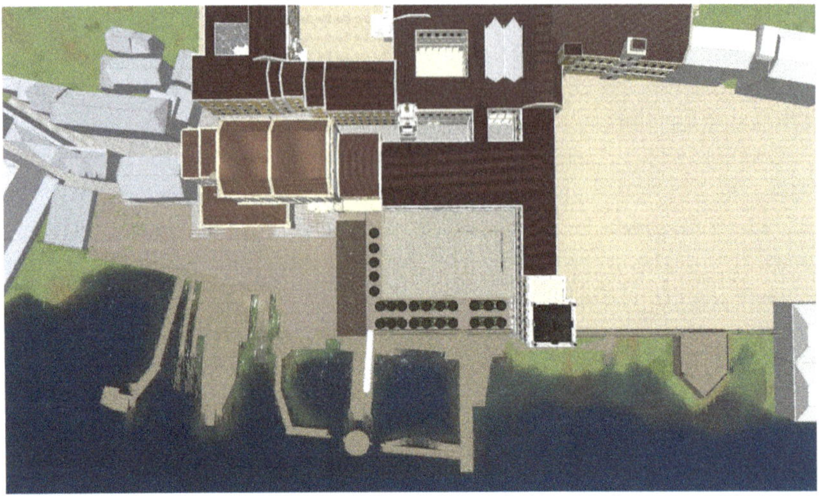

**Figure 11.8** Aerial view of the re-creation of the Ribeira Royal Palace complex. OpenSim version, November 2012.
*Credit:* project City and Spectacle: a vision of pre-earthquake Lisbon.

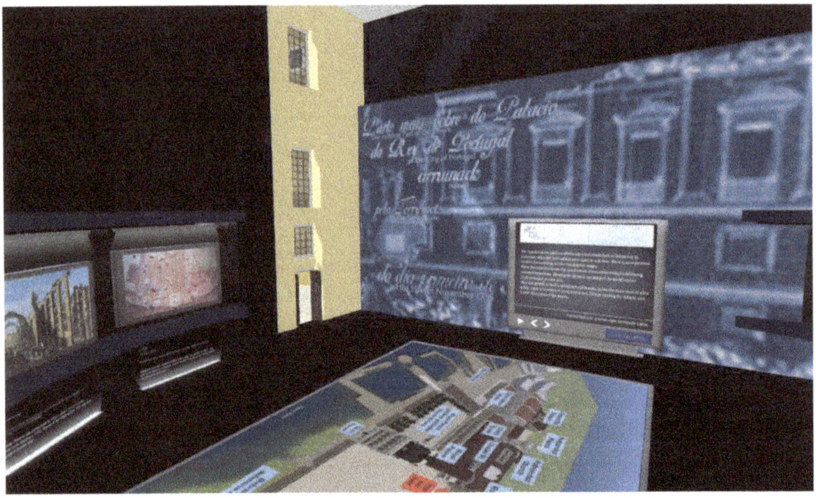

**Figure 11.9** Scene of the Virtual Museum. OpenSim version, August 2013.
*Credit:* project City and Spectacle: a vision of pre-earthquake Lisbon.

This simulation of pre-earthquake Lisbon also includes a first sketch of the interior of the Royal Opera House (figure 11.10).

The project is available to the public on the Kitely platform, and all the information about the project can be accessed on its website (https://lisbon-pre-1755-earthquake.org/en/about/). Videos of different stages of the project, one of which was part of the exhibition The Prodigious Commission.

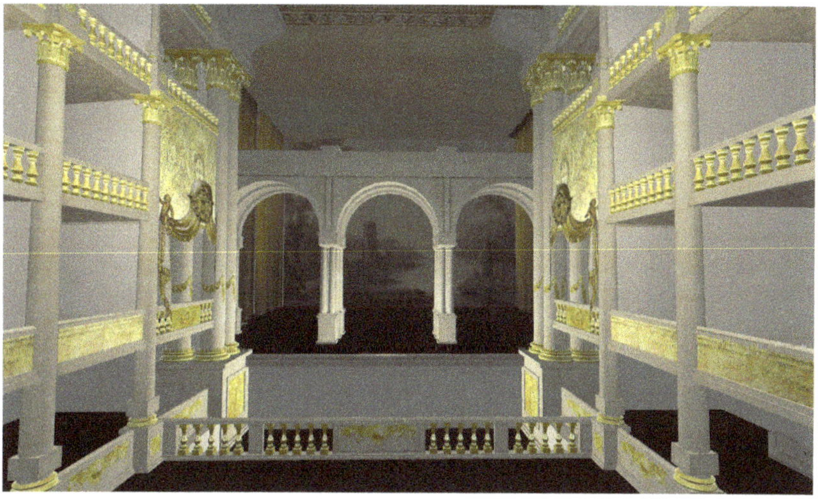

**Figure 11.10** View inside the Royal Opera House. OpenSim version, November 2014.
*Credit:* project City and Spectacle: a vision of pre-earthquake Lisbon.

From the Patriarchal Basilica to the Royal Chapel of St. John the Baptist (National Museum of Ancient Art, Lisbon, May 18 –October 23, 2013) are also available on the project's website and on YouTube and Facebook www.facebook.com/Lisbon-pre-1755-earthquake-372857996094063/), and Vimeo (https://vimeo.com/lisbonpre1755). To date, these videos have had more than 240,000 viewings.

There is still a long way to go to achieve the goals of the project—that is, extending the cityscape re-creation to the large urban area included in the rebuilding plan; re-creating several building interiors, of which there is scarce documentary evidence, and re-creating some aspects of Lisbon's social and cultural life at the time. Continuous digital development and funding options are key challenges for the project. It is essential to work with all these equations in order to build an encompassing visual outlook of a memory of Lisbon that has been for centuries dispersed in documentary sources.

The project was able so far to put forward a first model for the Palace Courtyard area of the city. This was achieved by testing in a 3D interactive and immersive framework the most relevant data on the subject. In so doing, the project created new data, of a digital nature that added to the information and knowledge of this period in Lisbon's history. It also made available to a wide and diversified audience the possibility of a sensory revisit of the memory of Lisbon in a context of social interaction. Concepts such as embodiment (Forte 2014) and cultural presence, and the notion of being "there and then" (Champion 2005, Tost and Champion 2007, Forte 2014), acquire here a precise meaning (particularly when the simulation of cultural and social events will be achieved), as the user is able to interact with the model, be part of the simulation, and share this experience with the project team and other users. As noted:

> In fact, visitors are not only able to take part in the simulation of the major events of baroque Lisbon, but they also become a part of it. Their experience of a past reality becomes part of the model itself and, as such, produces data which can be manipulated and reinvented both by the research team and by other users. Therefore, the memory of the past is experienced and changed by our own actions and experience. (Murteira and Rodrigues 2016)

The experience and knowledge gathered by the project team has also been made available in several papers and conference presentations. As such, the project has been a means of research and theoretical exercise on digital heritage. It needs to be re-examined at the light of the most recent technological developments for interactive and immersive digital heritage.

The use of digital tools and the acknowledgment of their potential for transforming knowledge creation and sharing had a late appearance in Portugal. These projects showed the impact that digital and virtual heritage have had on scientific investigation, education, tourism, and knowledge dissemination. Unfortunately, the tensions between funding sources and the constraints of technological development have not been favorable to their widespread use in Portugal. However, what was accomplished opened the way to perceiving the dialectic interconnections they allow between past and present and among data, interpretation, and public fruition.

## CONCLUSION

The past as perceived by the present is an abstract construction that derives from individual or collective memory. Historical research draws upon these memories through some precise viewpoints and through analyzing, interpreting, rearranging, and extending them. It exercises its approach to these memories by attempting to translate evidence into feasible working hypothesis.

Virtual representations/recreations/simulations of past urban realities enhance and strengthen our relationship with the past by allowing sensory experiences of historical memories and their historical interpretations to take place, often in a social significant context. The possibility of experiencing a simulacrum of a lost or profoundly changed urban reality in an interactive and immersive mode, either from a distance or in situ, enables the historical memory to act as a memoryscape open to continuous reconfiguration.

As a memoryscape, the idea or representation of the past can be redefined not only by the digital translation and re-equation of historical research but also by the input of the users' interactions and experiences. However, all the projects discussed here have, as an implicit or explicit feature, the 1755 earthquake as the reference event from which the past to be represented is defined. This fact, along with the continuous presence of the Palace Courtyard and the Ribeira Market, and of their relationship with the river as the key elements that give shape to the imaginability of Lisbon, restricts the redefinition of the representation of the past in the measure that they determine the initial point of view of the memoryscape.

Digital historic cities are thus a place for a close interchange between scientific exercise and contemporary fruition. Therefore, cultural presence ought to be deemed more than the "there and then" axiom as it becomes a present-day circumstance. The several case studies presented with regard to pre-earthquake Lisbon are relevant since they present themselves, from a 3D interactive point of view, as attempts to look at a lost city, from which little evidence has survived. Notwithstanding the scarce and fragmentary historical documentation available, there is a comprehensive and extensive historiography on the subject. It was crucial to be able to translate the latter into a coherent and visual model. The digital world provided the solution. Although all these projects relied on the state-of-the-art technology of their period, their advancement and survival depend on the role they ascribe to digital technology and on their ability to establish an interactive relationship with their users. The projects whose longevity was more limited were the ones that presented more rigid models in which interactivity was too restricted, as happened with the 3D model of the Lisbon Museum. Interaction with the public demands the project be conceived and developed within a sustainability logic according to the guidelines of the London Charter and the Seville Principles—in other words, predicting possibilities for its continuity when the technology used becomes outdated or obsolete. This was one of the priorities of the project City and Spectacle, allowing it to be currently under deep reassessment.

As more than a mere vehicle for research and dissemination, digital technology is extending and even reformulating epistemological boundaries in the study of the past, particularly in the case of urban phenomena. This revolution is relentless. There is a dialectic interchange between historical research and digital technology. Both strive to find a response to the demands and constraints of this novel field of knowledge. On the one hand, technology attempts to address the needs and goals of the ever-expanding character of cyber-archaeology and digital urban studies. In this context, as already mentioned, it is particularly relevant to safeguard the survival of digital data in the inexorable dynamic of technological advancement and expansion. On the other hand, historians and archaeologists need to ceaselessly address the methodology in use, achieve results, and disseminate their potential for transformation. The *memoryscape* is as able to be reconfigured as the human sensory experience is able to be expanded through the mediation of digital devices.

## REFERENCES


Almeida, Betâmio de. 2009. "The 1755 Lisbon Earthquake and the Genesis of the Risk Management Concept." In *The 1755 Lisbon Earthquake: Revisited, Geotechnical, Geological and Earthquake Engineering*, edited by L.A. Mendes-Victor et al., 7:147–165. Dordrecht: Springer.

André, Paula. 2012. "The Harmony, Symmetry and Programme Triad and the Modus Operandi of Intervening in the Portuguese Urbanism Territorium." In *Conferência Internacional PNUM 2012—Morfologia Urbana nos Países Lusófonos. Morfologia Urbana nos Países Lusófonos*, 280–282. Lisbon: ISCTE—Instituto Universitário de Lisboa.

Belting, Hans. 2014. *Antropologia da Imagem*. Lisbon: KYM.

Benjamin, Walter. 1992. "Pequena História da Fotografia." In *Sobre Arte, Técnica, Linguagem e Política*, 115–135. Lisbon: Relógio d'Água.

Câmara, Alexandra, Helena Murteira, and Paulo Rodrigues. 2016. "City and Spectacle: A Vision of Lisbon Before the 1755 Great Earthquake." In *Virtual Palaces, Part II Lost Palaces and their Afterlife Virtual Reconstruction between Science and Media*, edited by Stephan Hoppe and Stefan Breitling, 245–266. Munich: PALATIUM e-Publications.

Câmara, Alexandra, Helena Murteira, Paulo Rodrigues, and Luis Sequeira. 2009. "City and Spectacle: A Vision of Pre-earthquake Lisbon." In *15th International Conference on Virtual Systems and Multimedia (VSMM '09)*, 239–243. Vienna: IEEE. doi:10.1109/VSMM.2009.43.

Champion, Erik M. 2005. "Cultural Presence." In *Encyclopedia of Virtual Communities and Technologies*, edited by Subhasish Dasgupta, 95–101. Hershey, PA: IGI Publishing.

Collot, Michel. 2013. *Poética e Filosofia da Paisagem*. Rio de Janeiro: Oficina Raquel.

Forte, Maurizio. 2014. "Virtual Reality, Cyberarchaeology, Teleimmersive Archaeology." In *3D Recording and Moddelling in Archaeology and Cultural Heritage. Theory and Best Practices*, edited by Fabio Remondino and Stefano Campana, 113–127. British Archaeological Reports, International Series, No. 2598. Oxford: Archaeopress.

França, José-Augusto. 1987. *Lisboa Pombalina e o Iluminismo*. 3rd ed. Lisbon: Bertrand Editora.

Gautier, Théophile. 2015. "Exposição Fotográfica (1857)." In *O lápis mágico. Uma história da construção da fotografia*, edited by Carlos Sousa de Almeida and Carlos M. Fernandes, 125–130. Lisbon: IST Press.

Gschwend, Annemarie Jordan, and Kate Lowe, eds. 2015. *The Global City: On the Streets of Renaissance Lisbon*. London: Paul Holberton Publishing.

Gutenberg, Beno, and Charles Francis Richter. 1949. *Seismicity of the Earth and Associated Phenomena*. Princeton, NJ: Princeton University Press.

Halbwachs, Maurice. 1968. *La Mémoire Collective*. Paris: Presses Universitaires de France.

Leal, Joana da Cunha. 2004. "Legitimação artística e patrimonial da Baixa pombalina. Um percurso pela crítica e pela história da arte portuguesas," *Monumentos* 21: 6–17.

London Charter for the Computer-based Visualization of Cultural Heritage. 2009. http://www.londoncharter.org/.

Lynch, Kevin. 1999. *A Imagem da Cidade*. Lisbon: Edições 70.

Malraux, André. 2000. *O Museu Imaginário*. Lisbon: Edições 70.
Martinho, Bruno. 2010. "O Paço da Ribeira nas Vésperas do Terramoto." MA dissertation, FCSH da Universidade Nova de Lisboa.
Matos, Rute Sousa, and Desidério Sales Batista. 2013. "O Jardim Planetário. Uma Utopia para o Século XXI?." In *Arte & Utopia*, edited by Margarida Acciaiuoli, Paula André, Maria João Castro, Ana Duarte Rodrigues, and Paulo Simões Rodrigues, 300–313. Lisbon: CHAIA—Centro de História da Arte e Investigação Artística da Universidade de Évora, DINÂMIA'CET-IUL—Centro de Estudos sobre Mudança Socioeconómica e do Território do ISCTE-IUL, Faculdade de Ciências Sociais e Humanas da Universidade Nova de Lisboa.
Mendonça, Joaquim Moreira de. 1758. *História Universal dos Terremotos*. . . Lisbon: Off. de Antonio Vicente da Silva.
Murteira, Helena. 1999. *Lisboa, da Restauração às Luzes*. Lisbon: Editorial Presença.
Murteira, Helena. 2017. "Between Despair and Hope: The 1755 Earthquake in Lisbon" In *Catastrophe, Gender and Urban Experience*, edited by Deborah Leigh Simonton and Hannu Salmi, 42–63. New York: Routledge.
Murteira, Helena, Alexandra Câmara, Paulo Rodrigues, and Luis Sequeira. 2017. "Lost Cities as a Virtual Experience: The Example of Pre-Earthquake Lisbon" In *Memories of a City*, edited by Jonathan Westin and Ingrid Martins Holmberg, 57–88. Gothenburg: University of Gothenburg.
Murteira, Helena, and Paulo Rodrigues. 2016. "Lost Cities in the Digital Era: The Case of Pre-earthquake Lisbon." In *Space and Time Visualisation*, edited by Maria Bostenaru Dan and Cerasella Craciun, 151–172. Cham: Springer. doi:10.1007/978-3-319-24942-1.
Murteira, Helena, Paulo Rodrigues, Luis Sequeira, and Alexandra Câmara. 2015. "Virtual Lost Cities: The Case of Pre-earthquake Lisbon." In *2015 Digital Heritage*, 519–524. IEEE. doi:10.1109/DigitalHeritage.2015.7419437.
Nutall, Mark. 1991. "Memoryscape: A Sense of Locality in Northwest Greenland." *North Atlantic Studies* 1(2): 39–50.
Paice, Edward. 2008. *Wrath of God: The Great Lisbon Earthquake of 1755*. London: Quercus.
Parry, David. 2012. "The Digital Humanities or a Digital Humanism." In *Debates in Digital Humanities*. Minneapolis: University of Minnesota Press. http://dhdebates.gc.cuny.edu/debates/text/24.
Parry, Ross. 2007. *Recoding the Museum: Digital Heritage and the Technologies of Change*. London and New York: Routledge.
Pereira, Álvaro S. 2009. "The Opportunity of a Disaster: The Economic Impact of the 1755 Lisbon Earthquake." *Journal of Economic History* 69(2): 466–499.
Pimentel, António Filipe. 2012. "D. João V e a imagem do poder: o Terreiro do Paço ao revés." In *Do Terreiro do Paço à Praça do Comércio—História de um Espaço Urbano*, edited by Miguel Faria, 73–91. Lisbon: Imprensa Nacional Casa da Moeda, Universidade Autónoma de Lisboa.
Principles of Seville. 2011. International Principles of Virtual Archaeology. http://smartheritage.com/seville-principles/seville-principles
Rodrigues, Teresa. 2008. "Lisboa. Das longas permanências demográficas à diversidade social." In *Jornadas Demografia Histórica de Lisboa*. Lisbon: Gabinete de Estudos Olisiponenses (GEO). http://geo.cm-lisboa.pt/fileadmin/GEO/Imagens/GEO/Demografia_historica/Jornadas_demografia_2008/Demografia_doc1.pdf.
Senos, Nuno. 2002. *O Paço da Ribeira 1501-1581*. Lisbon: Editorial Notícias.
Stoichita, Victor. 2011. *O Efeito Pigmalião. Para uma Antropologia Histórica dos Simulacros*. Lisbon: KKYM.

Tost, Laia Pujol, and Erik Malcolm Champion. 2007. "A Critical Examination of Presence Applied to Cultural Heritage." In *PRESENCE 2007. International Society for Presence Research. The 10th Annual International Workshop on Presence*, 245–256. Barcelona: Starlab Barcelona S.L https://astro.temple.edu/~lombard/ISPR/Proceedings/2007/P2007ProceedingsIntro.pdf.

Tostões, Ana, and Walter Rossa. 2009. *O Plano da Baixa Hoje*. Lisbon: Câmara Municipal de Lisboa.

PART FOUR

# *Research, Planning, and Learning*

CHAPTER 12

# Spatial Representation of Vienna's Street-Level Environment

## *Urban Parterre Modeling*

ANGELIKA PSENNER

Translated by Michael Strand

### INTRODUCTION

Vienna's *Stadtparterre* (street-level environment) is in a precarious condition: on the one hand, ground floors are not being put to adequate use (problem of vacancies, underuse, use for storage and garages), on the other hand, the street space is basically monofunctional, being for the most part reserved for car parking and as a traffic surface. This fact appears all the more abstruse against the background of Vienna's rapid population growth[1] and the concomitant housing shortage. The need for additional housing cannot be met by drawing on the city's reserves of undeveloped land; therefore, new residential space will have to be provided in built-up neighborhoods in addition to the projected suburban development areas. To be able to do so, it is necessary to be aware of the nature, quality, and structure of use of the urban fabric and especially of the *Stadtparterre*, or urban ground floor.

The discussion of ground floor uses in inner urban areas is characterized by a certain hypocrisy. The general complaints about the more public-related uses of ground floors being increasingly discontinued are being mitigated by inactivity when it comes to launching initiatives to counteract the creeping privatization of this entire urban level. Everyday social exchange, whether

Angelika Psenner, *Spatial Representation of Vienna's Street-Level Environment* In: *Digital Cities*. Edited by: Maurizio Forte and Hele-na Murteira, Oxford University Press (2020). © Oxford University Press.
DOI: 10.1093/oso/9780190498900.003.0013

intentional or incidental, while navigating the urban space on foot, essentially depends on a varied use of the ground-floor level, which in spatial terms is the plane that physically connects *all* urban levels. However, such multiple spatial engagement is prevented by current building regulations.

> It is indeed an inherent quality of the city that all routes intersect and connect, converge and ramify on the ground floor level. This of course also includes the spatial layers directly below and above the ground floor [which is why in this text we speak of a street-level *environment*], provided that they are interrelated as one topological unit. But who is able to identify, to read and map this network? (Luchsinger 2014, 4)

This is where the study described in this chapter takes its starting point—namely, from giving visibility to the spatial connections of the base zone and their functional engagement. A differentiated 3D representation visualizes the actual spatial relations of ground floor, street, and courtyard spaces, as well as the "Souterrain" and "Hochparterre" (mezzanine floors). This is much more than a mere delineation in the sense of a quantitative or qualitative description; the issue here is the sensual and holistic perception of the street-level environment. "The quasi-archaeological representation of the spatial situation of the base zone prompts a cognition and design process that is entirely different from previous attempts at a functional and social re-engagement of spaces. The focus here is on the re-interpretation of the ground-floor situation on the basis of patterns that only become visible in the 3D representation, just like excavated ruins provide indications of past cultures, revealing patterns of everyday life that are as unexpected as they are topical and exemplary for the future" (Luchsinger 2014, 4).

## THE PROJECT'S BENCHMARKS

Creating residential space through rooftop conversions—which was done a lot over the past twenty years in Vienna—is often accompanied by a walling-up of ground floors (figure 12.1), as local building regulations require the provision of garages for private vehicles (cf. Psenner 2015a, 2015b, 2015c; 2014a, 2014b, 2014c; 2013; 2012a, 2012b).

The result is that actually no additional residential space is created; instead, the city merely moves one floor up, with detrimental effects such as disuse and depopulation on the public space at street level. If continued, this development will ultimately render an already precarious urban environment even more unsustainable.

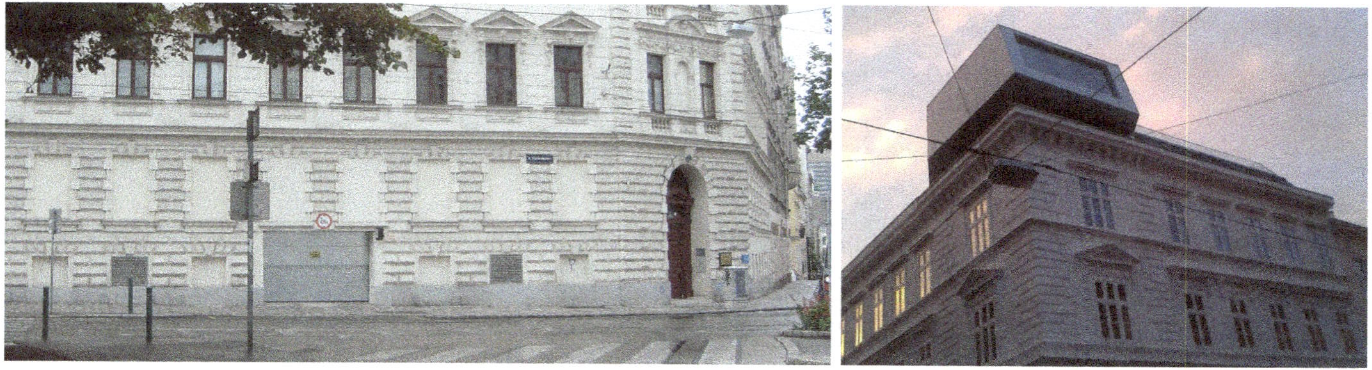

**Figure 12.1** Creating additional residential space through rooftop conversions is often accompanied by a walling-up of the ground floor, which renders an already precarious urban environment even more unsustainable.
*Credit:* Left: © A. Psenner; right: © S. Nöbauer.

The reasons why our cities have been made ever more car-friendly over time are well known.[2] However, historical photographs of our city streets speak to a different, highly diversified structure of uses. Before the street was turned into an exclusive traffic and parking area, it also served a number of other purposes, from meeting economic needs and providing necessities (trade, sales, transportation of goods), to cultural purposes, private self-representation, and the manifestations of public life. Motion and rest were considered as equally formative factors that defined the urban-design significance of the street (Walewa-Coen 1946, 97). As a matter of fact, street parking of cars in the nighttime was illegal in Vienna and was penalized with prison terms; things only changed in 1938, when the road-traffic regulations of Nazi Germany were adopted in Austria (Psenner 2013, 142ff; 2014a; 2015a; 2015c).

We also are familiar with images of megacities such as Tokyo, where residential streets—in spite of the tremendous mobility pressures of the Greater Tokyo Area with its population of 36 million—are primarily understood as spaces for living rather than traffic, and they have retained much of their public-space character (cf. Krusche and Vogt 2011, Krusche and Rost 2010, Krusche 2008). It is therefore necessary to put the issues concerning the street-level environment of Vienna in a broad systemic perspective and to identify ways in which architectural and urban research may contribute to understanding the existing problems. This also means relating to the street-level environment as a whole, rather than to certain parts of it.[3]

### *Stadtparterre:* The Street-Level Environment

There is a strong, direct relationship between urban street space and the structures and uses of ground floors of buildings.[4] In addressing urban-development issues, it cannot be emphasized enough that it is necessary to consider the totality of the *Stadtparterre*, or street-level environment, as a consistent urban zone of public, semi-public, and private spaces (Psenner 2017; Davis, 2012). The "spatial representation of Vienna's street-level environment," or Urban Parterre Model (UPM),[5] therefore includes both built-up and non-built-up areas, covering the street, as well as the adjacent houses and courtyards. Thus street, ground floor, and courtyard are treated as separate entities, so that their interrelations can come to light, as we are aware that the potentials of ground-floor use and the structure of the correlating public street space are directly related.[6]

### The Research Field

Buildings and tracts from the Gründerzeit period still make up a large portion of the Viennese cityscape. Despite busy construction activity, one in four apartments in the city is in a Gründerzeit building. This chapter therefore takes a closer look at street-level structures and environments from this period in an initial approach to the broader and more general subject of the *Stadtparterre*. Also, the project focuses on secondary streets because—unlike shopping streets or high-traffic roads—they have so far been mostly neglected by urbanist research and municipal administrations alike.

### The Research Questions

As the conventional cadastral map (figure 12.2, left) only shows building perimeters, it offers no information about the city's internal structure. Street, ground-floor, and courtyard uses are not documented with the necessary clarity and completeness, and can therefore not be objectively analyzed in a structural context. The Comprehensive Ground Plan Map, or—as we call it—the Urban Parterre Model (UPM; figure 12.2, right) provides precisely the kind of information that is prerequisite for an authentic morphological analysis.[7]

In this context, the study presented here addresses the following questions:

- How was the Viennese ground level originally used? What urban functions were located there?
- What are the (historical) interrelations between public space and life inside buildings?

The causes of the current *Stadtparterre* crisis are thus analyzed from a historical and systemic perspective.

## THE PROJECT'S METHODOLOGY

### The 2D Comprehensive Ground Plan Survey

Originally, the two-dimensional Comprehensive Ground Plan Survey (CGPS) was derived from studies on the relationship between urban morphology and building typology, similar to the ones used by Saverio Muratori in Venice and by Gianfranco Caniggia in Florence and Como (cf. Peters 1999, 137).

Several such morphological studies were subsequently conducted by Swiss architects and historians. In the 1960s, architects in the canton of Tessin initiated an inventory that was continued at the Swiss Federal Institute of Technology (ETH) in Zurich in the 1970s; further progress in the field was made with a comprehensive survey of Zurich's urban center under the direction of architectural researcher Margareta Peters.[8]

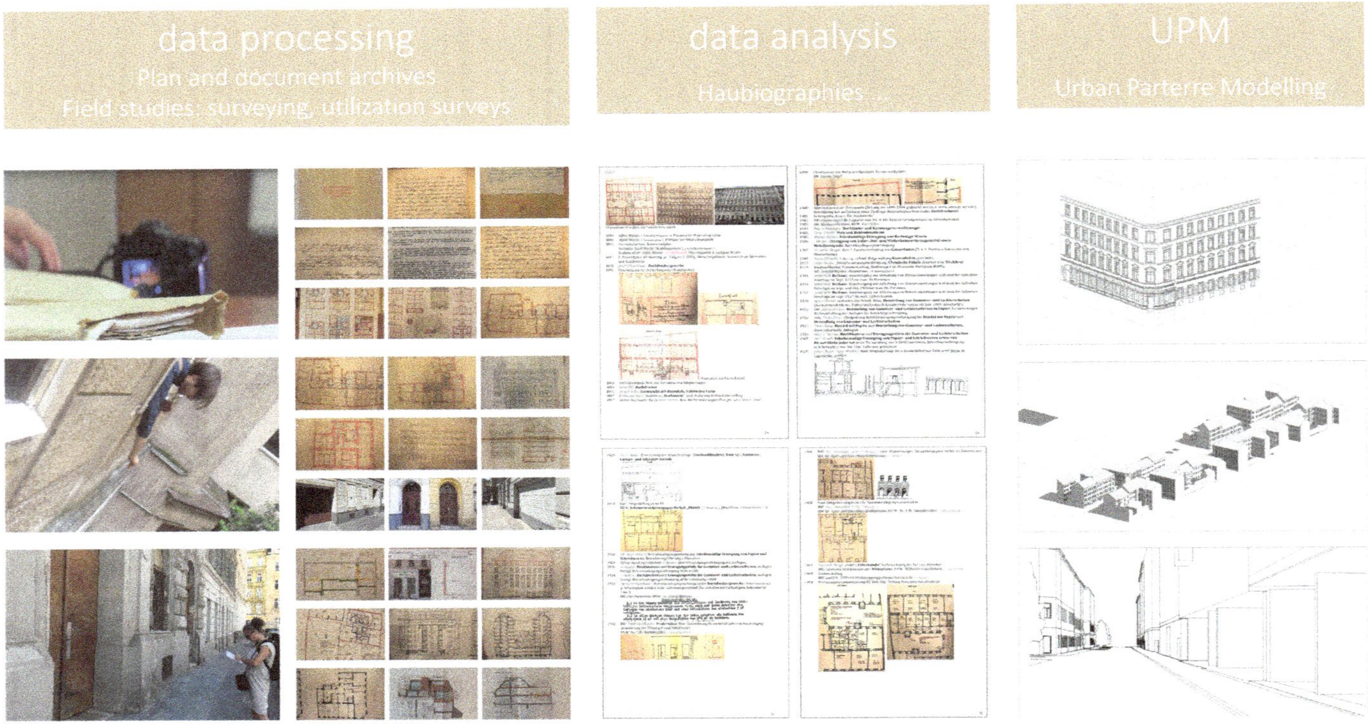


**Figure 12.2** Visualization of the Urban Parterre Modelling UPM-procedure: data processing, data analysing and 3D-modelling of the data.
*Credit:* © A. Psenner.

Comprehensive ground plan surveys are available for the following Swiss towns and cities:

| | | | | *RBTC*[9] |
|---|---|---|---|---|
| Bellinzona | Luigi Snozzi, Tita Carloni, (Livio Vacchini) | 1965–1969 | basement (bm), ground floor, all upper floors (uf) | – |
| Zurich (Old Town[a] 1955) | Supervisor Prof. Aldo Rossi, ETH Zurich | 1973 | ground floor (gf) | – |
| Lucerne | Andy Raeber | 1974–1984 | ground floor | – |
| Bern (Old Town) | Supervisor Prof. Dolf Schnebli, ETH Zurich | 1975 | ground floor | – |
| Solothurn (state of 1900) | Supervisor Prof. Paul Hofer, ETH Zurich | 1978 | bm, gf, typical upper floor | – |
| Bern (Old Town 1200) | Supervisor Prof. Paul Hofer, ETH; Janine Martinez | 1978–1981 | bm[b] | – |
| Biel | Supervisor Prof. Paul Hofer ETH[c] (A. Corboz) | 1979 | bm, gf, typical upper floor | X |
| Tessin[d] | Supervisor Prof. Aldo Rossi, ETH[e] | 1979 | no specification | – |
| Zurzach | Prof. Hans Rudolf Sennhauser | since 1980 | no specification | X |
| Le Landeron[f] | Doz. Marie Claude Betrix, Technikum Biel | 1986–1992 | ground floor | X |
| Baden | Baden Municipal Construction Dept. | 1989–1990 | ground floor[g] | – |
| Wil SG (Old Town) | Margareta Peters | 1994 | bm, gf, typical upper floor | – |
| Zurich (Old Town) | Supervisor Prof. V. Lampugnani, ETH; M. Peters | 1995–1999 | bm, gf, first floor | X |
| Zurich (industry quarter) | Supervisor Prof. V. Lampugnani, ETH; M. Peters | 1997–1999 | ground floor | X |
| Zurich (Stadelhofen district) | Supervisor Prof. V. Lampugnani, ETH; M. Peters | 1999 | ground floor | X |

Notes: [a] Covering the area inside the Baroque town walls; [b] "Kellerplan" (cellar plan) of the medieval city center; [c] Project leaders: Margareta Peters, Jürg Reber and Christian Sumi; [d] Comprehensive survey of the Tessin Canton; [e] In collaboration with Max Bosshard, Eraldo Consolatio, Bruno Reichlin, Fabio Reichart, and Christian Sumi; [f] The CGPS Le Landron, the typological studies of the buildings and the students' design proposal finally led to the passing of new building regulations for the Old Town; [g] The plan is also entitled "Brandmauernsystem" (map of the fire walls).

After years of absolute negation of this approach - only in isolated cases did researchers resort to this methodology (Fortier, 1989) - the CGPS has recently attracted renewed attention again (Lampugnani et al. 2018, Ebbing 2018).

Conventional cadastral maps (called *Mehrzweckkarten* in Vienna) provide no sufficient information about ground floor plans and the structure of urban interiors (see figure 12.2, left), whereas a CGPS sets buildings in relation to public space and their topographic environment, *visualizing the interrelation between the interior life of individual buildings and the public street space surrounding them.* Comprehensive Ground Plan Surveys therefore are useful to examine the relationships between buildings, streets, and courtyards.[10]

## The 3D Urban Parterre Model, UPM

A baseline set of data is provided by the existing digital cadastral map, which details land use for the entire municipal area of Vienna. It is a fundamental tool, a matrix that is filled in with ground-level plans of individual buildings, covering both historical floor plans and documented building uses, as well as most recent conversion records,[11] so as to reflect the morphological evolution of the *Stadtparterre*. All plans are then verified on site and, if necessary, adapted and amended, in particular with regard to actual building uses.

Given the importance of ceiling and building heights,[12] one of the study's primary objectives is to provide a three-dimensional model of the urban street-level environment. To do so, the method was further developed into the three-dimensional Urban Parterre Model (UPM). This three-dimensional representation of the data—with modeling done with Autodesk Revit[13]—helps in the analysis of links between buildings' interior spaces: Where is, or was, the basement or courtyard directly connected to ground-floor use? And how do these interior connections affect the entire parterre function? The introduction of 3D is the major difference from the Comprehensive Ground Plan Surveys done in the 1960s in many Italian and Swiss cities.[14]

In the Urban Parterre Model, the ground floor is rendered in every detail, while for the standard upper floor and basement, a set of simple structural data, such as weight-bearing elements (main supporting walls), access points (stairways and corridors), and the façade (opening axes), provides sufficient information for a rough-and-ready outline. Where the adjoining levels (basement and first upper floor) are directly connected to the ground floor by stairs or doorways, they are treated as extensions of the ground floor and therefore represented in greater detail. Normally, however, the areas above and below the ground floor are broadly outlined. Included in this are the street profile and daylighting conditions on the ground floor and street level. The façade is interpreted as a permeable interface between building and public space, with information on the formal façade design available as an option.

The UPM is considered a particularly sound representation, which includes basic information on ground floor use (by means of color codes). Thus, it produces precisely the information needed for analysing and defining the building's relationship to public space and the internal coherence in the *Stadtparterre*.

In this sense, the UPM represents a special type of 3D urban model, with precise detail data and offering a ground-level perspective. Usual 3D urban models mostly provide a bird's-eye view and do not contain the information needed for a ground-level investigation.

## THE PILOT STUDY: ANALYSIS

So far, the Vienna street-level environment has been explored in terms of an exemplary street in an inner district, an archetypal Gründerzeit street totally rebuilt in the late nineteenth century. Preliminary research took an in-depth historical approach: (1) a theoretical study of the evolution of the unique architectural type of the Viennese Gründerzeit-Stadthaus, which covered topics like the original historical use of that type of architecture (which was by no means exclusively residential, as one might presume from today's perspective); and (2) consideration of the impact of relevant legislation (building regulations, commercial code, public health care, road traffic regulations, tax incentives, urban planning, etc.).[15]

In order to better understand the role that the street played within the larger urban fabric, the biography of the street was extensively researched, including detailed information on layout (geography, spatial and urban planning), architectural development (historical and current building plans on the selected lots), and development over time of the uses of adjacent ground floors and basements (trading documents, business licenses and tax regulations).

The Urban Parterre Model uses color codes—for example, green for semi-public spaces and spaces with high user frequency, such as shops, coffee-houses, studios; orange for work spaces, such as offices, workshops, factories; yellow for housing; dark gray for garages and storage spaces; light gray for vacancies—to identify the type and intensity of use (figure 12.3).

The study actually confirms the hypothesis that, from a historical point of view, ground floor space is semi-public space, with the façade functioning as a permeable interface allowing constant interchange between public and semi-public areas. Photographs of the time also substantiate this assumption: ground-floor façades were permeable, and private ground-floor uses extended into public street space; conversely, the street-level interior rooms were open to the public.[16]

The façades were analyzed in terms of their permeability with regard to the actual engagement of the street-level environment; the question was: What

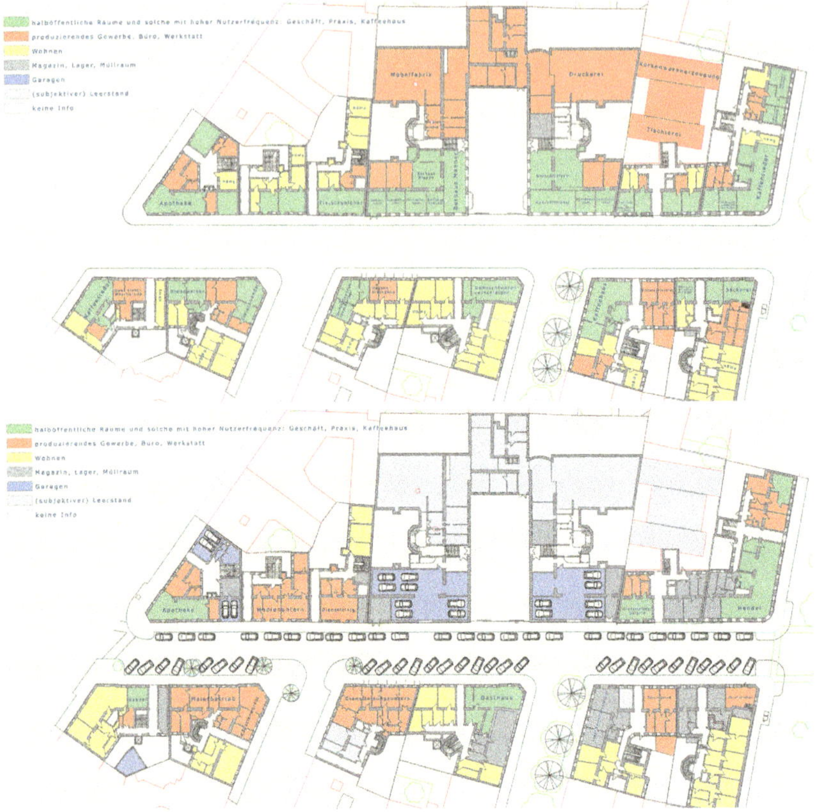


**Figure 12.3** Above: Uses of the *Stadtparterre* around 1910. Below: Today's uses of the *Stadtparterre* (UPM).
*Credit:* © A. Psenner.

goes on behind them and how does it affects the public space of the street, both during the day and in the night, so as to determine the effects of ambient lighting from house entrances and ground-floor residential and shop windows and night-time use frequency (figure 12.4).

Research also involved an insight study of actual workplaces on the ground floor, which involved a resident and visitor frequency analysis.[17] Also analyzed in this context was the internet presence of businesses situated on the ground floor, in order to evaluate the quality of the interlacing of Euclidean urban and ephemeral web space.

## RESULTS AND FINDINGS

Including all additional relevant ground-floor data, the spatial representation of Vienna's street-level environment facilitated a conclusive analysis of the (use)

(a)

(b)

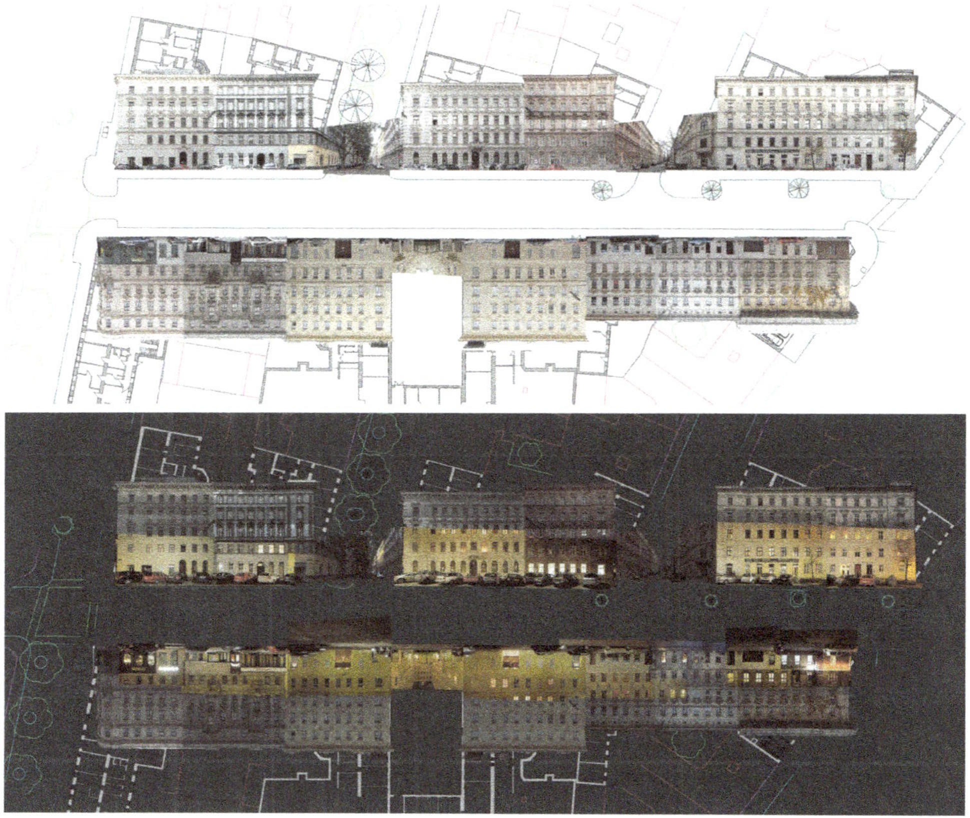

**Figure 12.4** Pilot test showing daytime permeability of façades (a) and lighting/permeability during the night (b).

*Credit:* © A. Psenner.

structure and potential of the ground-floor zone and related it to the uses and functions of the street space. Thus, interrelations could be identified and problematic situations considered and resolved in a broader context (figure 12.5).

## The Sidewalk

The holistic approach embodied in the term *Stadtparterre* shows the primary function of the sidewalk in its true light: easily and directly accessible, the sidewalk is the most important area of interaction in the public realm, a space where people can meet and move among, or simply observe each other relatively freely. Not least, it is the site of integrative interaction between minorities and majorities in a diversified society.[18] Clearly, it should be possible to engage this multifunctional urban space in all sorts of ways, provided it is adequately dimensioned.[19] At present, however, the average 1.7 to 2.2 meters between building fronts and parked cars are reserved for moving (pedestrian) traffic; section 78 of the Austrian Road Traffic Regulations even expressly prohibits "hindering pedestrian traffic by stopping for no reason," while failing to provide any information as to what might constitute an admissible reason to do so (Psenner 2004, 133; 2011a, 2011b; 2012b; 2013; 2014a, 137). Also, using the sidewalk "for other purposes than those of street traffic" is still conditional on administrative approval.[20]

This problem of lack of space also has to be seen as a consequence of a deliberate shifting of the rights of use: while more and more street space was assigned to cars, pedestrians found themselves reduced to increasingly narrow walking lanes. In the years after World War II, sidewalks were built with a minimum width of 1.25 meters (according to ÖFS building standards, 1956), whereas, in the Gründerzeit period, the *trottoirs—trotter* meaning to "roam about" in French—on both sides of the street had been assigned no less than one third (!) of the total street width (Kortz et al. 1905, 174; Drexel 2000, 207 and 209). On the initiative of the city administration, these narrow postwar sidewalks have been increasingly widened in recent years, with additional bulges added at street crossings so that the pavements now offer a little more space to walk and stand, but this still is a far cry from the comfortable 4 meter sidewalks common in New York City. To make things worse, the clear width (unobstructed passageway) achieved at great cost has in many places been drastically reduced again by signposts.

## Structures of Historical Ground-Floor Use

One important aspect of Gründerzeit ground floors is their generous ceiling heights. The large amount of space thus offered is a special asset, not least

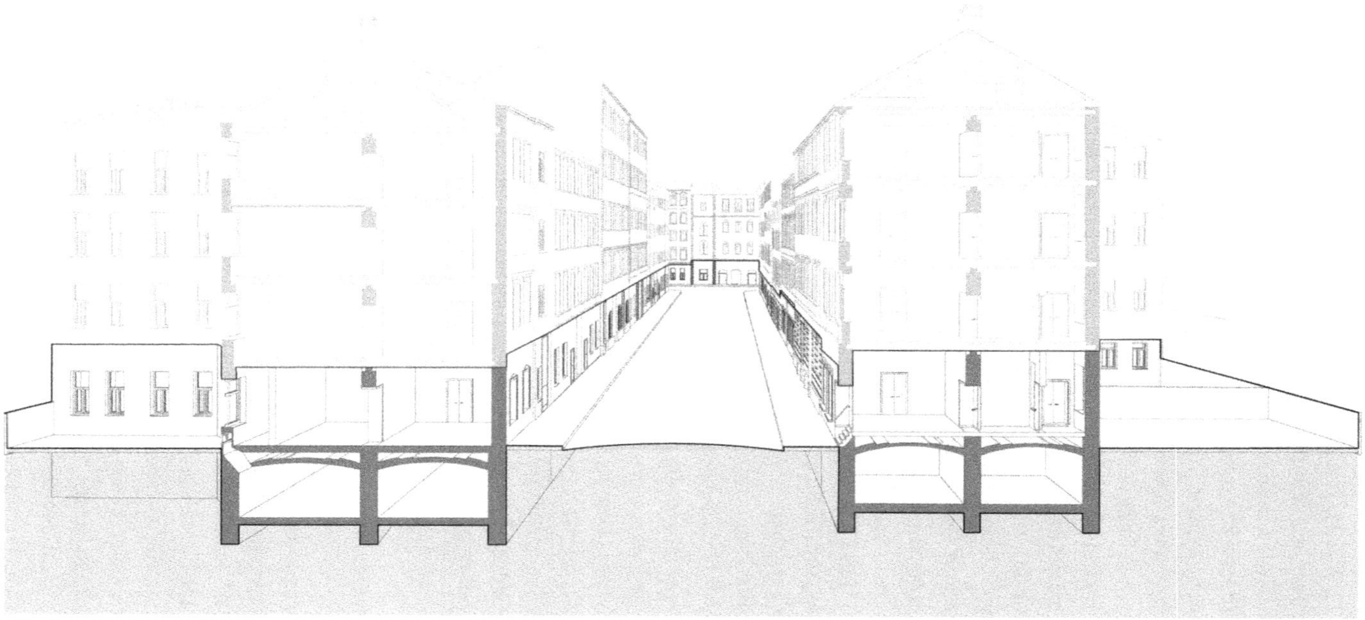

**Figure 12.5** UPM: a 3D model of the urban parterre (Stadtparterre) and its coherent microstructure.
*Credit:* © A. Psenner and K. Kodydek.

because it opens up possibilities for a variety of uses, lending itself not to a specific market segment but, rather, to a broad range of prospective occupants (Psenner 2011c, 2012a, 2012b).[21]

At that earlier time, Vienna's economic, social, and urban fabric was characterized by a large number of small businesses, many of them workshops or small manufacturing operations, preferably located on the ground floors of Gründerzeit buildings. Of the businesses registered in the Establishment Census of 1869, only a few had more than ten workers; enterprises that employed a larger workforce were more or less all located in the suburbs—a specific structural characteristic which also continued to prevail throughout and after the economic crisis that came in the wake of the stock market crash of 1873.

In these conditions, Vienna had an incredible range of registered businesses, among them, for example, the *Pfaidlergewerbe* (shirt maker), *Paramentenerzeuger* (vestments manufacturer), *Banderzeuger* (ribbon maker), *Bettwarenerzeuger* (bed linen maker), *Naturblumenbinder* (florist), *konzessionierter Spirituosenschänker* (licensed liquor dealer), *Brunnenmeister und Brunnengräber* (well digger and builder), *Büchsenmacher und Schwertfeger* (gun and blade smith), *Bürsten- und Pinselmacher* (brush maker), *Deichgräber* (ditch digger/trencher), *Nadler Webkammmacher und Drahtwarenerzeuger* (needle, weaving comb, and wire products maker), *Federnschmücker* (plume feather decorator), *Flaschenbierfüller* (beer bottler), *Fragner und Greisler* (grocer), *Hafner* (stove fitter), *Kamm- und Fächermacher* (fan and comb maker), *Gemischtwarenverschleißer* (general store operator), *Kanal- und Senkgrubenraeumer* (sewer and cesspit cleaner), *Kostgeber* (landlord or landlady), *Lohnfuhrwerker* (haulage contractor), *Sauerkraut- und Saure Rübenverschleißer* (sauerkraut and pickles merchant), *Seiden- Schön- und Schwarzfärber* (dyer), *Tuchscherer* (cloth shearer), and *Wirkwarenerzeuger* (knitware manufacturer).

### Historical Development of Ground-Floor Use in the Street Study

Despite its proximity to a parallel higher-level shopping street, the street examined has the characteristics of a supply road, so that the ground floors of the Gründerzeit buildings mainly accommodated workshops, small manufacturing, and trade businesses, while in the side streets residential use predominated in the majority of cases. This structure of ground-floor uses has partially subsisted to this day, but recently was disrupted by massive breaches of use patterns (e.g., large garage built into the ground floor of house nos. 5–7).

A morphological-use structure analysis of the stretch of street examined here showed that, historically, the typical Viennese ground floor was a semi-public space. As the apartments in most buildings were really small,[22] and also were mostly used by several occupants in shifts, residential functions were often moved to the outside, out of the private units. Also, only very few apartments had fully equipped kitchens, or pantries and other food storage facilities, so that shopping for perishable goods was an everyday chore. These circumstances also account for the relatively large number of eateries, coffee houses, and liquor stores (seven on a stretch of only 190 meters), as well as grocers and other small retailers.

Over the past century, the following businesses, or business premises, were accommodated on the street-side ground floor of buildings along the 190 meter street stretch examined:

- 7 restaurants/coffee houses/wine spirit shops, of which:
  - 1 restaurant; after 1909: smoked meat production; after 1921: breakfast room; after 1960: patisserie
  - 1 restaurant; after 1904: coffee roaster; after 1933: restaurant; after 1970: cafe; after 1981: shop
  - 1 coffee roaster; after 1967: Cafe; after 1978 shop
  - 1 restaurant; after 1970: cafe
  - 1 wine spirit shop; after 1993: paint shop
  - 1 restaurant
  - 1 coffee house
- 5 not specifically defined salesrooms (called *Gassenlokale* and occasionally *Gewölbe*)
- 4 general stores (*Gemischtwarenverschleißer*)
- 3 bakeries/pastry shops
- 2 underwear and linen makers; after 2000: offices
- 2 shirt makers
- 1 pharmacy
- 1 dairy
- 1 barbershop
- 1 printing house (of the Sonn- und Montagszeitung)
- 1 furniture maker /carpenter
- 1 synagogue (after 1990: furniture store)
- 1 pharmaceutical company (*Pharmakon*); after 1964 advertising agency
- 1 necktie sewing school (*Privatlehranstalt für Krawattennaehen*)
- 1 producer of cork goods; after 1912: carpentry; after 1971: printing house
- 1 frame and molding maker; after 1928: glass-cutting workshop
- 1 shop for celluloid articles; after 1970: installer
- 1 stationer's shop (*Papierverschleißer*); after 1911: locksmith; after 1970: installer and plumber
- 1 dry cleaners; after 1967: grocery store
- 1 plumber's showroom
- 1 butcher shop; after 1999: video store
- 1 carpenter; after 1942: butcher shop and smoked meat production

### Structural Characteristics of Use of the Historical Viennese Ground Floor

As mentioned earlier, the historical Viennese ground floor was a semi-public space, with no clear-cut boundary between inside and out: rather, doors and windows were left open most of the time so that there were many points giving access to the ground-floor premises. Photographs from the period attest to this: the ground-floor façades were permeable; semi-public, or even private uses of the ground floor extended to the street and, conversely, the premises were easily accessible to the "flow of the public."

Many of the ground-floor premises in the studied street (the study underlying this text had to be anonymized for data-protection reasons) were connected with basement floors or cellars underneath, which meant an extension of the ground floor zone. The (commercial) use of the street-facing premises in most cases also included the interior courtyard.[23]

The historical *Stadtparterre* was a ramified, varied, much-used, and hence engaging space. Permeable ground-floor façades provided a flexible interface between public and semi-public spaces that were intensely interacting with one another.

## CONCLUSION

Currently, the Vienna municipality has no oversight or control over the ongoing urban modification described at the beginning of this text, as there are no data being collected on the actual ground-floor situation, nor are there numbers available about store vacancies or recently built-in private garages. Unlike for a number of Italian and Swiss cities and towns, no comprehensive ground plan survey has ever been drawn up for Vienna. This is the first major innovation that the study brings.

From an international research perspective, another innovative asset of the project is the three-dimensionality of the comprehensive model. Successful regulatory measures in urban planning, administration, and economic policies depend on sound and detailed knowledge of the actual architectural structure, as well as the current actual—and potential—use of the street-level environment. The UPM provides this information in an easily accessible and locally contextualized form. Thus, the development potential of the street-level environment can be clearly identified to inform urban planning.

The goal of this study was to map and outline the potential of the street-level environment, among other things, through an in-depth analysis of its past development. The current state of affairs is not satisfactory: moving and stationary traffic make the street an unattractive space to be in, without much amenity value and usability for the pedestrian public. Ground-floor vacancies are on the rise, and consequently even more space is being converted into indoor parking. Thus, the city moves upward—and when attics are converted into penthouses, the existing parking space regulation leads to the installation of further private ground floor garages and hence to further loss of potentially semi-public urban space. This precarious tendency is also reflected in new construction types: here, the ground floor usually accommodates secondary uses, garages or self storages (Psenner 2019), with residential use starting only on the first upper floor. The ground floor façades are closed and impermeable and therefore the building does not communicate with the surrounding urban space.

The complex micro-analytical possibilities of the UPM will allow the documentation of vacancies and other issues relating to street use in various neighborhoods and the analysis of contributing economic, traffic, and social factors. Systematic 3D modeling of the built-up environment coupled with an inventory of historical, current, and potential ground-floor uses will provide a basis for putting Vienna's street-level environment into a long-term development plan that can serve as a practical guide for future interventions in various neighborhoods and for the re-design of individual street complexes.[24]

This street-level environment study is a groundbreaking approach that will provide data that can be highly usable for urban landscape research and planning, including in particular computer-based simulations. Both the UPM and the historical analysis of the urban street-level environment, transforming as it has over time between the built-up and the non-built-up in Vienna's Gründerzeit neighborhoods, will be a sound basis for further in-depth analysis (Psenner 2017). The benefits that may be expected from this approach include:

- Ability to visualize the historical interrelation between street space and ground-floor use over time and to identify existing discrepancies between private use of the street space and the public interest.
- Transparent and objective visualization of interrelated functions so as to provide a sound argument action on public administration and/or business.
- Reliable planning and implementation guidelines for concrete improvements in the urban street-level environment.

In any case, the application possibilities of the model go far beyond the situation delineated in this chapter, as the UPM will be usable not only in an urban context but also in suburban or rural settings—for example, to study and address in regional planning terms certain architecture-related aspects and conditions of the problem of vacancies in rural areas (small towns, rural communes, and villages). Once fully developed, the analytical UPM approach may also be applied to any other critical ground-floor situation in urban peripheries or "shrinking city" situations.

First and foremost, though, the point here is to acknowledge the significance of ground floors for the functioning of a city—a fact that has somewhat fallen from view in the twenty-first century, ever since the emergence of 3D city modeling. The reason for this may be that conventional 3D city models cannot really represent the intricate, small-scale, multilayered, and ramified ground-floor structures as well as the UPM, and thus prevent us from perceiving them in a broader functional perspective.

## REFERENCES


Campagnol, Gabriela / Caffey, Stephen / Clayton, Mark J. / Kang, Julian / Booth, Geoffrey / Klein, Nancy and Glowacki, Kevin. 2014. "Unreal Projects: Using Immersive Visualization to Learn about Distant and Historical Locales" In *12th International Conference on Urban History: Cities in Europe, Cities in the World. European Association for Urban History (EAUH), (Proceedings).* Lisbon.

Caniggia, Gianfranco. 1963. *Lettura di una citta': Como.* Rome: Centro Studi di Storia Urbanistica.

Caniggia, Gianfranco. 1986. *Die Morphologsiche Betrachtungsweise von Stadt und Territorium.* Eidgenössische Technische Hochschule Zürich. Architekturabteilu ng: Geschichte des Städtebaus.

Davis, Howard. 2012. *Living Over the Store: Architecture and Urban Life.* London and New York: Routledge.

Drexel, Anita. 2000. *Pflaster auf städtischen Fußböden: Bauhandwerkliche und freiraumplanerische Qualitäten von Gehwegen in Wien und anderen Städten.* Vienna: Österreichischer Kunst- u. Kulturverlag.

Ebbing, Georg, ed. 2018. Christoph Mäckler. Lehre 1998-2018. *Von der Rematerialisierung der Architektur zur Rekultivierung des städtischen Raumes.* Berlin: DOM publishers

Fortier, Bruno. 1989. *La Métropole Imaginaire: Un atlas de Paris.* Brussels: Peirre Mardaga.

Kortz, Paul, and Österreichischer Ingenieur- und Architekten-Verein. 1905. *Wien am Anfang des XX. Jahrhunderts: Ein Führer in technischer und künstlerischer Richtung.* Vol I: *Charakteristik und Entwicklung der Stadt: Ingenieurbauten.* Vienna: Gerlach & Wiedling.

Krusche, Jürgen, ed. 2008. *Der Raum der Stadt: Raumtheorien zwischen Architektur, Soziologie, Kunst und Philosophie in Japan und im Westen.* Marburg: Jonas-Verlag.

Krusche, Jürgen, and Frank Roost, with Department Architektur ETH Zurich. 2010. *Tokyo. Die Straße als gelebter Raum.* Baden: Lars Müller.

Krusche, Jürgen, and Günther Vogt. 2011. *Strassenräume Berlin, Shanghai, Tokyo, Zürich: Eine foto-ethnografische Untersuchung.* Baden: Lars Müller.

Lampugnani, Vittorio M., Stühlinger, Harald R. and Tubbesing, Markus. 2018. *Atlas zum Städtebau. Band 1 und 2.* Munich: Hirmer.

Lewis, Robert. 2004. *Manufacturing Suburbs: Building Work and Home on the Metropolitan Fringe.* Philadelphia, PA: Temple University Press.

Luchsinger, Christoph. 2014. Preface. In *Das Wiener Gründerzeit-Parterre* by Angelika Psenner. Vienna: TU Vienna, Department of Urban Design.

Malfroy, Sylvain, and Gianfranco Caniggia. 1986. *Die morphologische Betrachtungsweise von Stadt und Territorium.* Zurich: ETH, Architekturabt., Geschichte des Städtebaus.

Muratori, Saverio. 1960. *Studi per un operante storia urbana di Venezia.* Rome: Istituto Poligrafico e Zecca dello Stato, Liberaria dello Stato.

Österreichische Forschungsgesellschaft für das Straßenwesen im österr (ÖFS). 1956. *Normblätter: Richtlinien für den Ausbau von Strassen; Bearbeitet von der Arbeitsgruppe "Planung und Verkehr."* Vienna: ÖFS.

Peters, Margereta. 1990. "Stadtgrundriss als Arbeitsinstrument: dem Mittelalter auf der Spur." *Hochparterre: Zeitschrift für Architektur und Design* 4: 30–31. http://retro.seals.ch/cntmng?pid=hoc-001:1990:3::1295.

Peters, Margereta. 1999. "Elektronische Erfassung eines Industriequartiers: zusammenhängende Grundrissaufnahme in Zürich; ein Experiment." *Schweizer Ingenieur und Architekt* 117: 779–784. http://retro.seals.ch/cntmng?pid=sbz-003:1999:117::587

Plunz, Richard. 1990. *A History of Housing in New York City.* New York: Columbia University Press.

Psenner, Angelika. 2004. *Wahrnehmung im urbanen öffentlichen Raum.* Vienna: Turia und Kant.

Psenner, Angelika. 2005. "Parterre. Wechselwirkung zwischen Erdgeschoss und Straßenraum in Wien." *Derive* 18: 8–11.

Psenner, Angelika. 2011a. "Integrative Diversität zu ebener Erd'? Das Entwicklungspotenzial der stadträumlichen Struktur des Wiener Gründerzeit-Parterre." *SWS Rundschau* 2: 195–218.

Psenner, Angelika. 2011b. "Vienna's Historic Ground Floor and Its Socio-Urban Potential—Integrative Diversity Parterre?" In *REAL CORP 2011. Change for Stability. Lifecycles of Cities and Regions,* edited by Manfred Schrenk, Vasily V. Popovich, and Peter Zeile, 1119–1126. Schwechat: Competence Center of Urban and Regional Planning. www.corp.at/archive/CORP2011_11.pdf.

Psenner, Angelika. 2011c. "The Price of Generous Ceiling Heights: The Influence of Historic Building Value on Vienna's Gründerzeit Architecture." In *Theory: For the Sake of the Theory,* edited by Efe Duyan, 395–409. Istanbul: Dakam.

Psenner, Angelika. 2012a. "Mixed Building Use Promotes Mixed Urbanity: Insights from Historical Use-Neutral Architecture." In *REAL CORP 2012: Re-Mixing the City: Toward Sustainability and Resilience?,* edited by Manfred Schrenk et al., 463–473. Schwechat: Competence Center of Urban and Regional Planning. www.corp.at/archive/CORP2012_18.pdf.

Psenner, Angelika. 2012b. "Wie wollen wir das Wiener Gründerzeit-Parterre nützen?" In *Perspektive Erdgeschoss, Werkstattbericht 121,* edited by R. Scheuvens and T. Schütz, 18–23. Vienna: Magistratsabteilung 18, Stadtentwicklung und Stadtplanung.

Psenner, Angelika. 2013. "Wem gehört die Straße? Genealogie der Nutzerrechte in Wiens Straßen." *SWS Rundschau* 2: 131–159.

Psenner, Angelika. 2014a. "Don't Even Think of Parking Here. Wiener Straßenraum: Verhandlung von Nutzungsrechten und Nutzungsansprüchen." In *Place-Making in urbanen Diskursen: Interdisziplinäre Beiträge zur Stadtforschung,* edited by Ingo. H. Warnke and Beatrix Busse. Diskursmuster—Discourse Patterns 7, 121-147. Boston and Berlin: de Gruyter.

Psenner, Angelika. 2014b. "A Smart Researching and Planning Tool for the Neuralgic Urban Zone: 3D-ZPA." In *REAL CORP 2014: Plan It Smart? Clever Solutions for Smart Cities,* edited by Manfred Schrenk et al., 35–44. Schwechat: Competence Center of Urban and Regional Planning. https://repository.corp.at/297/1/CORP2014_28.pdf .

Psenner, Angelika. 2014c. *Das Wiener Gründerzeit-Parterre: Eine analytische Bestandsaufnahme. Pilotstudie (Final Report of the Study).* Vienna: TU Vienna, Department of Urban Design.

Psenner, Angelika. 2015a. "Historischer Überblick über die Entwicklung der Straße & des öffentlichen Raums." In *Mission Mikrourbanismus*, edited by Erich Raith, 102–109. Vienna: TU Vienna, Department of Urban Design.

Psenner, Angelika. 2015b. "Fakten zum Wiener Gründerzeit-Parterre und seinen Nutzungschancen." In *Mission Mikrourbanismus*, edited by Erich Raith, 110–118. Vienna: TU Vienna, Department of Urban Design.

Psenner, Angelika. 2015c. "Street Life! It's the Only Life I Know. Street Life, And There's a Thousand Parts to Play." In *REAL CORP 2015: Plan Together—Right Now—Overall. From Vision to Reality for Vibrant Cities and Regions*, edited by Manfred Schrenk et al., 579–590. Schwechat: Competence Center of Urban and Regional Planning. https://repository.corp.at/22/1/CORP2015_30.pdf.

Psenner, Angelika. 2017. "Funktionen des ‚Ebenerds'—StadtParterre reloaded." In *Umbau*, edited by Österreichische Gesellschaft für Architektur ÖGFA, 70–83. Basel: Birkhäuser.

Psenner, Angelika. 2019. „Wenn der Laden zum Lager wird: Kritische Anmerkungen zum Boom von Selfstorages im urbanen Erdgeschoß." In *Wo die Dinge wohnen. Das Phänomen Selfstorage*, edited by Martina Nußbaumer and Klaus Pichler and Wien Museum, 68–73. Zürich: Park Books.

Walewa-Coen, Slawa Blagowesta. 1946. "Die architektonische Bedeutung der Straße mit besonderer Berücksichtigung der Laubenstraßen." Doctoral disseration, Vienna University of Technology.

CHAPTER 13

# Unreal Projects

## *Using Immersive Visualization to Learn About Distant and Historical Locales*

GABRIELA CAMPAGNOL, STEPHEN CAFFEY, MARK J. CLAYTON, KEVIN GLOWACKI, NANCY KLEIN, JULIAN KANG, AND GEOFFREY BOOTH

### INTRODUCTION

The technology of immersive visualization (IV), sometimes also known as virtual reality (VR), has progressed such that it can be employed to deliver experiences of places and environments that are otherwise too difficult or even impossible to reach in the real world in real time.* Since the Rome Reborn initiative led by the UCLA Virtual Reality Lab in the 1990s and 2000s, digital heritage platforms and virtual reality environments have served as laboratories within which scholars from across the disciplines can speculatively re-engage with some of history's most expansive gestures of architectural expression (Frischer et al. 2000, 2003; Frischer 2003, 2008; Frischer and Stinson 2007). Building upon this tradition, the Unreal Projects endeavor brings together faculty and students from architecture, construction science, and landscape architecture and urban planning at Texas A&M University to conduct research

* The authors wish to acknowledge the support of the Texas A&M University's Core Curriculum Technology Enhancement Grant Program, the Center for Heritage Conservation, and the following individuals for their contributions in the completion of this chapter: Kendall Clarke, David Creamer, Jong Bum Kim, Fatemeh Pariafsai, Duygu Yenerim, and Shermeen Yousif.

Gabriela Campagnol, Stephen Caffey, Mark J. Clayton, Kevin Glowacki, Nancy Klein, Julian Kang, and Geoffrey Booth, *Unreal Projects* In: *Digital Cities*. Edited by: Maurizio Forte and Helena Murteira, Oxford University Press (2020). © Oxford University Press.
DOI: 10.1093/oso/9780190498900.003.0014

on the construction of 3D digital models of places too distant—or from too far in the past—for most people to visit easily, if at all. One goal is to enable students in core curriculum university courses to enhance their spatial and sensory understanding of place by accessing the virtual sites, structures, and spaces generated for the project. For this project, we use a building information modeling cave automatic visualization environment (or BIM CAVE), a computing and display system developed by Dr. Julian Kang, that enables small groups of students to construct and navigate digital models of some of the most important, influential, and distinctive monuments of the world's architectural heritage. The BIM CAVE is already in use in courses supporting the degrees of Bachelor of Science in Construction Science and Bachelor of Environmental Design, and in graduate studies at Texas A&M University (figure 13.1). In these programs, students build and explore BIM CAVE models to aid in development of a construction schedule, systems installations, or architectural design. One of our objectives was to determine the feasibility and effectiveness of the BIM CAVE environment for a wider range of student learners.

While numerous platforms can be used for the IV experience, the BIM CAVE used in our study provides particular advantages that include specific types of practicality and accessibility that differs from similar initiatives at other

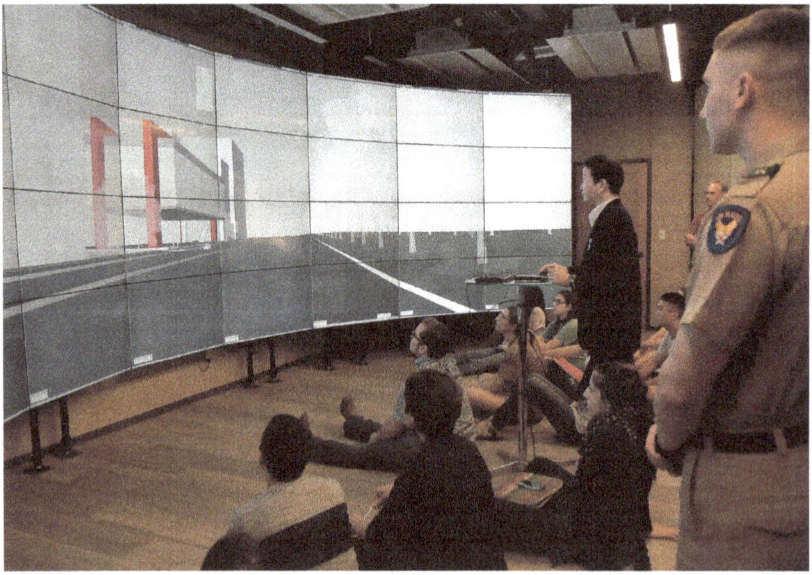

**Figure 13.1** Students in a History and Theory of Modern and Contemporary Architecture class exploring MASP in the BIM CAVE, fall 2015.

*Credit:* Photo taken by Professor Geoffrey Booth of students from Dr. Gabriela Campagnol's History and Theory of Architecture class, experiencing Texas A&M University's BIM CAVE with the guidance of Dr. Jong Bum Kim.

educational institutions and research laboratories. The BIM CAVE takes advantage of commercially available products, such as Autodesk Navisworks, that are used widely in architecture and construction industries. Similarly, Autodesk Revit Building Information Modeling (BIM) software enables the quick and complete 3D modeling of architectural sites, structures, and urban spaces with the added capability of embedding nongraphic information (or metadata) into the model to better inform the user and guide the visualization. A student may watch a building being constructed or inspect it at different phases of its history, view details of a heritage conservation intervention or highlight the structure's and systems' technical characteristics. The modeling and navigation software programs are available free of charge to students and faculty through university-funded licensing agreements, while the BIM CAVE hardware configuration and display array are proprietary to Texas A&M University. The skills acquired in generating content for and using the BIM CAVE to conduct formal, technical, and historical analyses are readily transferable to courses in other disciplines (i.e., for students who choose to pursue careers other than architecture, construction, or urban planning) and to activities undertaken in professional practice (e.g., for students who wish to go on to become licensed architects). The Unreal Projects platform, content-development exercises, and content-delivery mechanisms supplement existing curricula intended to improve students' spatial literacy, as well as expand their understanding of and engagement with "place" as a theoretical, critical, and practical concept.

## PROJECT RATIONALE

The potential to enrich the understanding of history and culture by allowing students to navigate freely through the simulated, "unreal" world of an IV environment is essentially unlimited. The value of such an approach derives from two important factors. First, the generation of students who produce content for the BIM CAVE system in their studies of architectural and urban histories are acculturated to the digital: they have only ever lived in a world dominated by computer technology and, in many instances, seem to prefer the virtual to in-real-life (IRL) experiences. For both content producers and system users/consumers, the Unreal Projects initiative teaches architectural history, urban history, construction processes, and heritage conservation in the modality most appropriate for millennials who plan to pursue careers as educators and practitioners. Those whose interests and professional plans lie outside the disciplines of architecture and construction science will also benefit from the increased sensitivity to the designed and built world derived from their encounters with the Unreal Projects BIM CAVE experience. Second, and perhaps more important, the BIM CAVE spatializes knowledge, presenting site, structure, life span, and technical and contextual information on a scale sufficient to engage more fully the perceptual and cognitive experiences of the content developer and the end user. Inhabiting this unreal world has the potential to convey a greater understanding and insight about the spatial and sensory qualities of the environments than can be obtained through other media, while at the same time enhancing the understanding and appreciation of real-world planned, designed, and built environments. The use of an IV facility such as the BIM CAVE can become a standard for excellence in teaching historic, contemporary, and speculative environments in architectural studies and other disciplines.

Based on preliminary analyses conducted by architecture and construction science faculty who have successfully deployed the combination of BIM and 3D model-navigation software, the team predicted the following advantages and outcomes:

1. Visualization of construction of existing and no-longer-extant sites, structures, spaces, and complexes.
2. Visualization of modification and other forms of change through time.
3. Visualization of diagrams or abstractions of the design (e.g., regulating lines, symmetry and asymmetry, structural systems, mechanical systems, circulation patterns).
4. Hyperlinks to additional information/metadata (e.g., material and their sources, explanation of historical styles, precedents and quotations, and historical events and figures).
5. Access to privileged vantage points (e.g., the alignment of a special view, view sheds, etc.).
6. Alternative schemes proposed or considered before the actual construction.
7. Visualization of projects never completed or never begun.

Building on earlier research by one of our team members toward the development of a Virtual Vauxhall Pleasure Gardens from eighteenth-century London (Caffey et al. 2010), we compiled a preliminary list of environments selected for modeling and navigating in the BIM CAVE. The bases for selection included availability of documentation sufficient to produce a complete model, viability of a currently extant structure to enable comparisons of the real world to the unreal world, and a range of suitable study situations, such as reconstruction of monuments through various phases over the life span of the structure, the demonstration of sequence of construction, or the analysis of orientation and solar path and the resulting shadow and light patterns (using the sunlight simulation feature of commercially available BIM software).

Unreal Projects has focused, at various levels of investigation and modeling, on the following environments:

- The Villa of the Bronzes, Olynthus, Greece, fourth century BCE (available for visit, but in a ruined state) (figure 13.2).

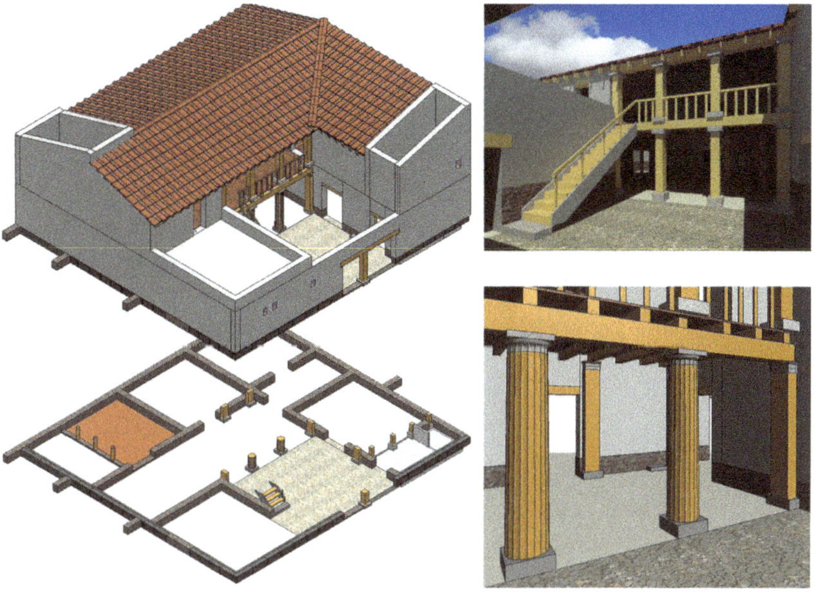

**Figure 13.2** Revit model of the Villa of the Bronzes at Olynthus.
*Credit:* Revit model prepared by Dr. Kevin Glowacki and Master of Architecture Student Kendall Clarke at Texas A&M University.

- Villa Capra (La Rotonda) by Andrea Palladio, Vicenza, Italy, sixteenth century (limited visits permitted) (figure 13.3).
- Château de Maisons by François Mansart, Maisons-Laffitt (northwest of Paris), France, seventeenth century (available for visit).
- Vauxhall Pleasure Gardens, London, England, eighteenth century (no longer extant but with extensive visual, textual and physical model documentation).
- Imperial Sugar Company, Sugar Land, Texas, early twentieth century (partially available for visit; largely abandoned, but with an adaptive use/rehabilitation initiative undergoing).
- Museu de Arte de São Paulo (MASP), São Paulo, Brazil, by Lina Bo Bardi, mid-twentieth century (available for visit; still functioning as an art museum, but with periods of extensive modifications of interior).
- Rachofsky House, Dallas, Texas, by Richard Meier, late twentieth century (previously available for visit, but currently not available).

In its earliest stages, the Unreal Projects initiative aimed to (1) serve as an opportunity for undergraduate and graduate students who are fluent in BIM software to develop content for the BIM CAVE display and navigation environment; (2) create a pedagogical tool that could enhance the classroom experiences of advanced undergraduate students in architectural history

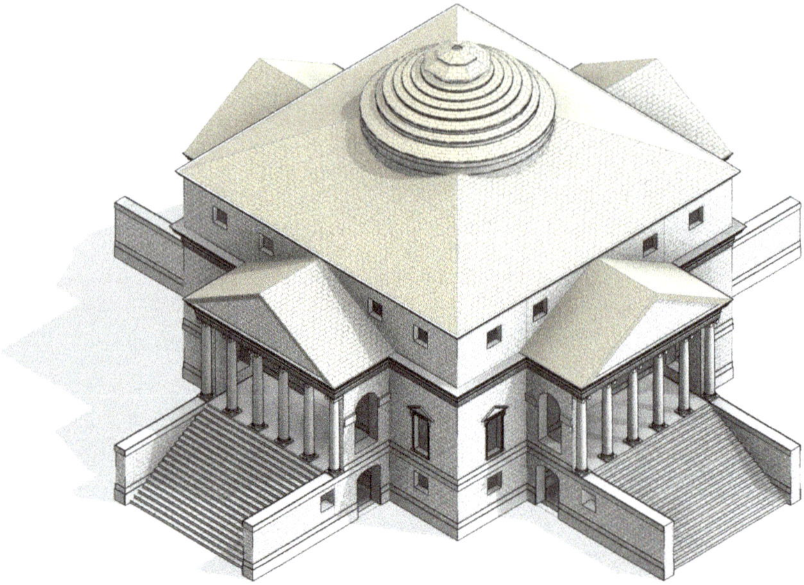

**Figure 13.3** Revit model of Palladio's Villa Rotonda.
*Credit:* Revit model prepared by Dr. Jong Bum Kim.

courses; (3) function as a research archive for advanced undergraduate and graduate students pursuing degrees in architecture, environmental design, construction science, and landscape architecture and urban planning, as well as those graduate students pursuing a special certificate in heritage conservation through the Texas A&M University Center for Heritage Conservation; and (4) provide a rich resource for research into high-impact, technology-based learning experiences in the context of higher education more broadly, and in the contexts of design education and visual and spatial literacies more specifically.

The Unreal Projects pilot study discussed here focuses on one example: the Museu de Arte de São Paulo (MASP) (São Paulo Art Museum). A PhD student, whose area of expertise was in modeling settlements by utilizing BIM technology, developed the initial model using Autodesk Revit 2014. Drawing from the original construction documents, the building also served as a case study for a graduate-level class in BIM. Our study compared and evaluated four modes of representation and presentation to students in an undergraduate survey class of modern and contemporary architectural history: a text-only description, images only (including photographs, plans, sections, and elevations), a navigable IV model delivered via Google Cardboard or a head-mounted display (HMD), and the same navigable model experienced in the BIM CAVE. The objectives of the pilot study included the assessment of student responses to the various forms of representation.

## PILOT TEST: MUSEU DE ARTE DE SÃO PAULO

Designed in 1957 by the Italian-born, Brazilian architect Lina Bo Bardi (1914–1992), and inaugurated in 1969, the Museu de Arte de São Paulo (MASP) hovers above a 2100-square-meter belvedere on Avenida Paulista in the heart of the city of São Paulo, Brazil. The building was registered as historic building at the State of São Paulo level in 1982, and it received a designation as a National Heritage monument in 2003. However, these designations did not prevent later modifications to the interior design of the building, including the 1996 removal of the original museum displays designed by the architect herself. Thus, one of the Unreal Projects team's first objectives was to re-create the original museum display and offer the virtual experience of a space that, until late 2015, no longer existed.

Motivations for the selection of this heritage monument as the first case study included: (1) the building's simple geometry and precursor structural system; (2) the availability of original architectural design and construction drawings; (3) the opportunity to reproduce the original interior with glass easels, allowing viewers to experience the art collection through IV; (4) the museum's urban context and integration with the city fabric; (5) recent protests against the possibility of closing the open public space below the gallery; (6) the commemoration of Bo Bardi's 100th birthday in the year of 2014; and (7) previous research on the building by some of our team members (Campagnol and Caffey 2011, Caffey and Campagnol 2015, Campagnol and Caffey 2015, Kang 2012, Kang et al. 2012, 2015).

### The Original Museum Display and Later Modifications

According to Bo Bardi's original design, the works of art were to be hung on the glass easels at eye level but were arranged in a staggered grid; the viewer's periphery was full of the other paintings in the gallery, even while focusing on one work of art. This simplicity allowed for a unification of all works from all time periods and styles to be experienced together. Bo Bardi thought, "seeing a painting hanging on a glass panel is like seeing the work as the painter saw it when he created it" (Bo Bardi 1997; Miyoshi 2011). The back of each painting was to include a composite of descriptions, diagrams, and pictures, as well as revealing the back of the frame. This would allow the viewer a deeper understanding of the artist, but only after an encounter with the artwork itself. Thus, in Bo Bardi's design, a viewer experienced a work of art first and foremost as the artist saw it during the process of painting.

There were several problems, however, that led the museum to change the displays to a more traditional museum-gallery style of display in the 1990s. For example, the size of paintings that could be displayed was limited by the size of the easels designed by Bo Bardi, with the result that the MASP curator could not exhibit all of the works in their collection. In addition, the angles at which a work of art could be viewed were limited by the physical location of the works and the available space of the gallery floor. Moreover, the glass easels themselves were found to be damaging to some works of art. Finally, despite the architect's design intent, some people still preferred to see the name and information on the work of art at the same time as viewing the art. Some critics even claimed that a solid display wall served as an important component of the experience of a painting (Campagnol and Caffey 2011; Caffey and Campagnol 2015). Immersive visualization allows the viewer to explore all these options, experience the effects, and compare the results (figure 13.4).

## The Revit Model

Drawing upon the various visual and textual descriptions of MASP, the project team created virtual construction materials and libraries of features (called "families") for MASP in Revit. The model was developed by a graduate student in the PhD program of the Department of Architecture.

The modeling process comprised five steps: (1) drawing the toposurface; (2) creating conceptual mass objects by setting up regulating lines; (3) drawing the structural elements (foundation, columns, beams, slabs, cables, etc.); (4) assigning nonstructural objects such as walls, floors, doors, windows, roofs, staircases, elevators, glass easels and art objects; and (5) adding landscape elements.

Modeling in Revit can be fast and accurate. Most important, the use of conceptual mass objects, grids, and reference planes highlights the formal qualities and ideas expressed by the architect. An axis of symmetry or line of alignment or array of repeated elements is modeled explicitly. Lock constraints enable the modeler to express the logic of the design. The creation of custom families for the building also provided a toolkit that a designer could then also use to make a MASP-like building. The use of constraints and libraries of objects begins to reveal the architectural language of the designer. A key insight in modeling a historic structure is that one must interpret existing drawings and measurements, and then model the structure as if one was a designer. The person doing the modeling must often think, *If I was designing this project, how would I do it?* Extensive education and professional experience help lead to a good model.

Because of the interoperability capabilities of BIM, it is also possible to conduct several different types of analyses of the MASP. Energy simulation, quantity takeoffs, structural analysis, code analysis, and other analyses can be easily conducted to better understand how the building performs. These are

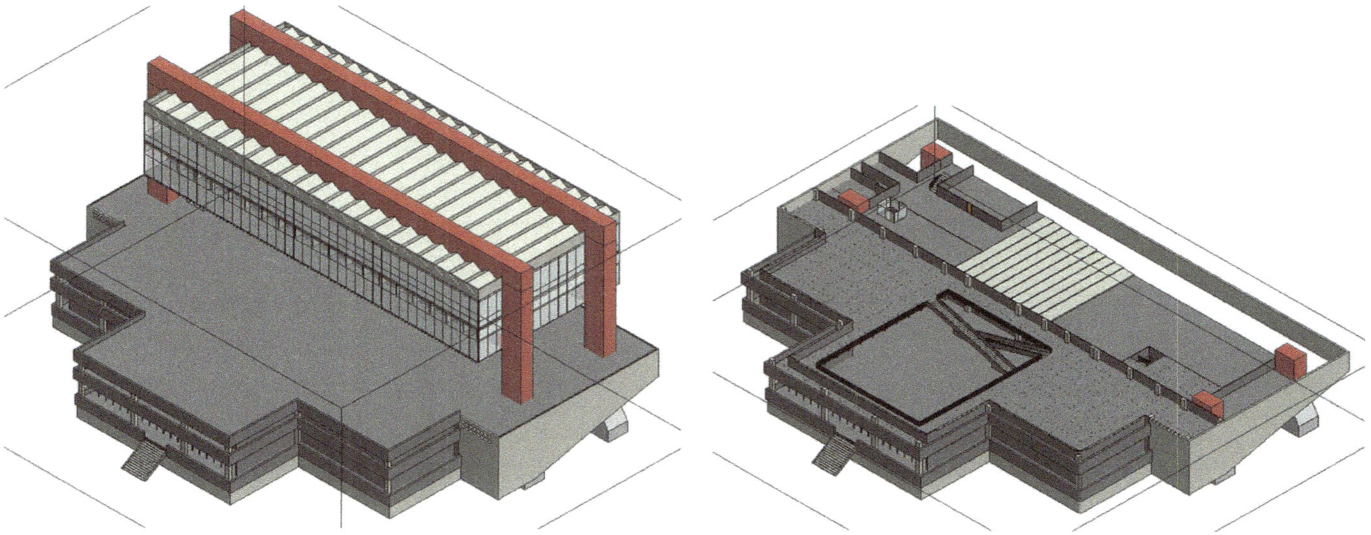

**Figure 13.4** Revit model of MASP under development by students in Dr. Julian Kang's class.
*Credit:* Revit model prepared by students from Dr. Julian Kang's course on BIM at Texas A&M University.

special kinds of information that are not often used in historical analysis but can potentially enrich the critical discussion.

### MASP in the BIM CAVE

The MASP Revit model was subsequently imported into the BIM CAVE, enabling groups of students to navigate through the MASP Revit model. Our BIM CAVE provides multiple displays arranged in a wide angle of vision (170 degrees) to fill the visual field. Using Autodesk Navisworks, the viewer can navigate in real time through the model in an immersive experience. A particular advantage of the BIM CAVE is the ease by which the Revit model can be imported and displayed, since the setup is based on commercial software.

### MASP in 3D Head-Mounted Display

Through virtual reality (VR), our team is able to take BIM to the next level of aesthetic immersion. Our goal was to display the originally intended design of the museum's top floor as Bo Bardi had envisioned. By using head-mounted 3D displays, a research assistant was able to place a viewer in a highly realistic experience that enabled the derivation of insights into the character of the original context (figure 13.5).

In the Unity 3D gaming platform, the process began with the establishment of a character model that was capable of receiving motion input from a Bluetooth controller and head-tracking gyroscopic input from an Android smartphone. The Revit model was adapted using 3D Studio Max and prepared for export to Unity format. With the character and Revit models both in Unity, the next step was to add materials to the museum and form the landscape. From there the Unity file was made into an application that was able to run on an Android smartphone. The phone was inserted into a head-mounted display based on the designs of Google Cardboard (figures 13.6 and 13.7).

The MASP model in the 3D head-mounted display faced two major challenges: (1) how to deal with a brand-new medium that has not yet established standard conventions; and (2) how to compensate for limitations by the hardware and software to enable the model to run smoothly on a mobile device. In order to achieve an acceptable level of smoothness and speed within these limitations, we utilized a newer head-mounted display and updated the software to accommodate the upgrade. This consisted of reconstructing and modifying the character model and camera scripts. These new scripts allowed for adjustment to the picture displayed, distorting the view in such a way that it would look appropriate when viewed through the lenses. This anticipatory

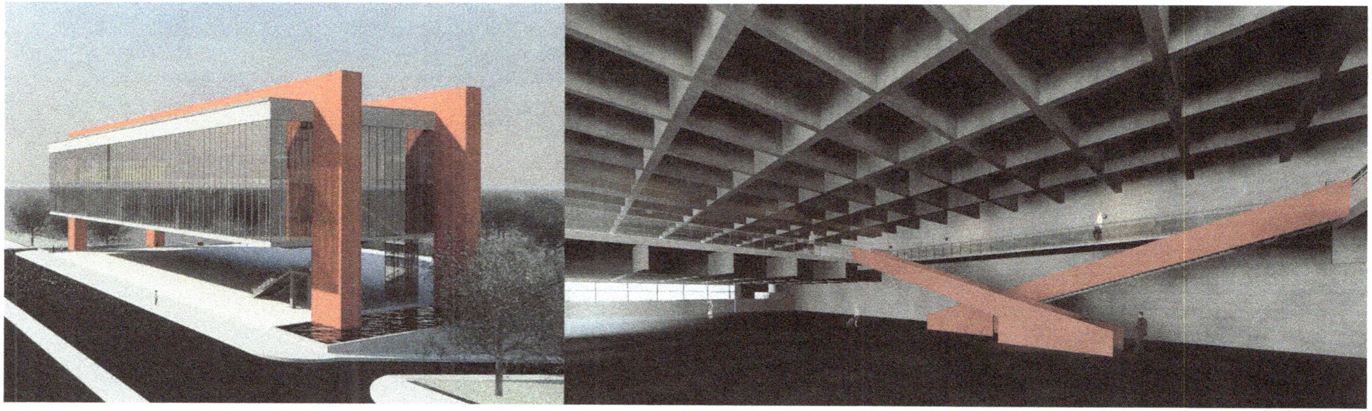

**Figure 13.5** Revit model of MASP for the BIM CAVE.
*Credit:* Revit model prepared by Dr. Campagnol and PhD student Duygu Yenerim at Texas A&M University.

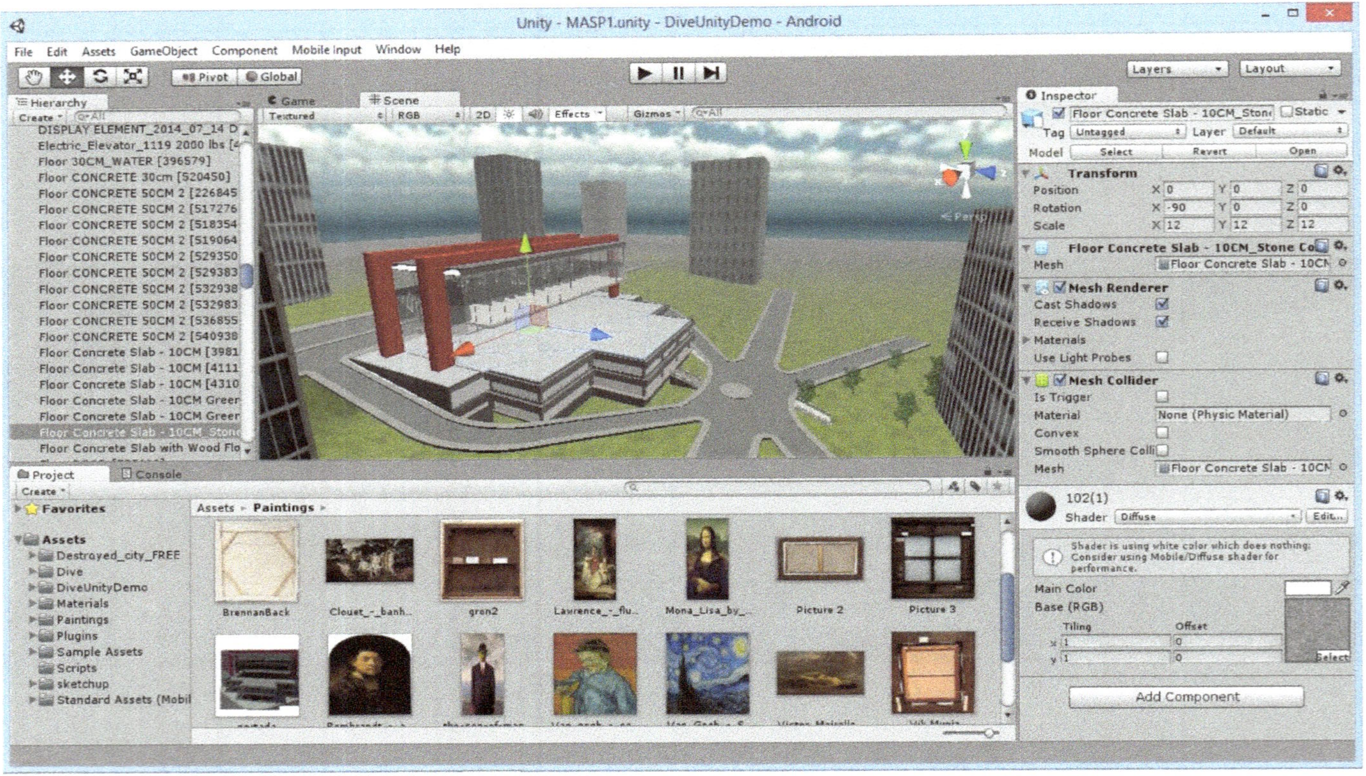


**Figure 13.6** Screenshot of the Unity environment and MASP model.
*Credit:* Revit model prepared by Dr. Campagnol and Master of Architecture student David Creamer at Texas A&M University.

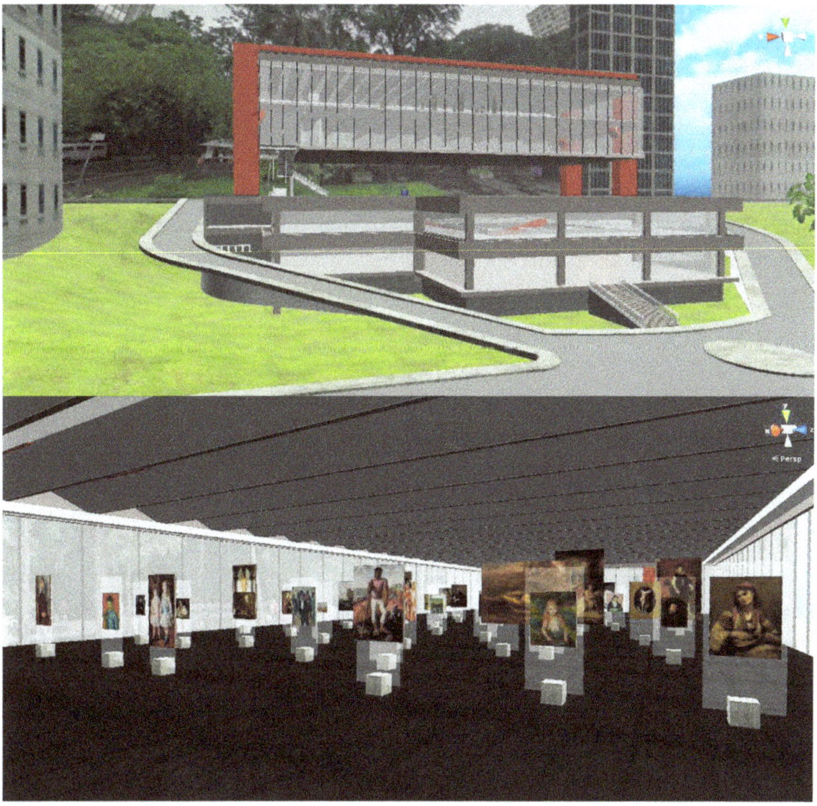

**Figure 13.7** Exterior and interior views of MASP model in Unity 3D.
*Credit:* Model prepared by Dr. Campagnol and student David Creamer at Texas A&M University.

warping also included chromatic aberration that accommodated for the red and blue shifting of the individual pixels that appears near the edge of the lenses. Another feature that allowed an overall better experience for the viewer is called "asynchronous time warp," which helps smooth out the experience and reduce head-tracking latency. In summary, this process allows the head-tracking to maintain functionality even when the rendered picture experiences freezing.

The similarity between the head-mounted display and BIM CAVE is that both are experiences of moving through a three-dimensional virtual world. The difference arises when it comes to the interactions with the model itself. In the head-mounted display, the experience is individual and more intuitive to the user, partially due to the developer's efforts on "polishing and simplifying" the experience. Along with this, the interactions with the model are only the ones that the developer designed to be available. On the other hand, the BIM CAVE allows more collaborative/group experience and functionality when manipulating the digital model on the spot.

## ASSESSMENT OF STUDENT LEARNING

Based on anecdotal evidence gathered in pilot testing, students and faculty in architectural design and heritage conservation found the experience of the BIM CAVE both novel and stimulating. As the project moved forward, data gathering on perception and performance were formalized through a research protocol approved by the university's Institutional Review Board. Assessment has been incorporated into special surveys and regular assessments of the courses. Classroom lectures presented the necessary background information and course examinations included appropriate questions for gathering data. Assessing student learning was accomplished through comparative analyses between the various conventional pedagogical methods and content delivered through the VR headset and the BIM CAVE.

### Methodology

We conducted the pilot experiment in one of our core curriculum classes in the history and theory of modern and contemporary architecture. We subdivided the class into four groups: Group 1 received only textual information describing MASP; Group 2 received only static, two-dimensional images to view; Group 3 was taken to the BIM cave for a "walk through" the building; and Group 4 experienced the head-mounted display. Before taking the pilot test, students in each group were also asked to answer an online questionnaire on their learning styles: Visual, Aural, Read/Write, or Kinesthetic (VARK, n.d.). A random mixture of learning styles was represented in each group. Group 1, which received only text information about the building, included 31.3% of students. In Group 2, which received only images, 30.1% of students participated. The cumulative percentage of these two groups indicates that the traditional tools were used for 61.4% of the class. The rest of the class (38.6%) received VR tools either BIM CAVE or head-mounted display (HMD), 24.1% and 14.5%, respectively. Regarding the VARK test highest score, a majority of the respondents (73%) received the highest score as either Visual or Kinesthetic, 41.9% and 31.1%, respectively. The rest got the highest score as either Aural or Read/Write, 14.9% and 12.1%, respectively.

The aim of dividing the students into these four groups was to test whether those who experienced the BIM CAVE and the HMD would develop better comprehension and improved perception of the building compared to the other conventional media of text and illustrations. In addition, it was hoped that the comparison of the results might inform us about the most effective aids for teaching courses on the history and theory of architecture across a variety of learning styles, and the impact of technology in the education and learning process. The working hypothesis was that students in the immersive visualization and head-mounted display groups would record higher scores regarding the overall comprehension of the building design elements, characteristics, and main features.

The survey questions were developed to test three main learning outcomes: (1) comprehension of building layout, arrangement, structure, circulation, urban context etc.; (2) recognition/identification of main design features and architectural concepts; and (3) and the ability to explain and communicate the building characteristics to others. The responses offered five options for the students in each group to reflect on how confident they were about the material they had just studied in terms of the learning outcomes: strongly disagree; disagree; undecided; agree; and strongly agree.

## Interpretation of Results

Our explorations permitted us to reconstruct the experience of the MASP as originally conceived and constructed. We have understood that different new types of technologies can affect positively some aspects of learning about the history and theory of modern architecture.

The collected data were analyzed using descriptive and analytical statistics. Quantitative analysis was carried out using IBM SPSS Statistics software, version 23, and statistical significance was set at $p < 0.05$. Descriptive statistics were used to summarize the results of the questionnaires. In addition, rank-based nonparametric tests, including Mann-Whitney U test and Kruskal-Wallis H test, were conducted at the 95% significance level to examine any statistically significant difference between responses of different subgroups (Text, Image, BIM CAVE, and HMD) based on learning tool (1: Traditional Tools, 2: VR Tools), VARK test high score (1: Visual, 2: Aural, 3: Read/Write, 4: Kinesthetic), and learning method (1: Text, 2: Image, 3: BIM CAVE, 4: HMD). For ordinal dependent variables (Likert scales), when the independent variable consisted of two categories, for instance, "traditional tools" and "VR tools," a nonparametric Mann-Whitney U test was conducted. When the independent variable consisted of more than two categories, for instance, Text, Image, BIM CAVE, and HMD, a nonparametric Kruskal-Wallis H test was conducted.

The differences in the responses of four subgroups (Text, Image, BIM CAVE, and HMD) indicate that the tools used for teaching/learning affected the participants' answers to half of the questions. In another words, four out of eight multiple choice-type of questions showed statistically significant responses. When the four groups were combined into two subgroups (the new media and the traditional media), five out of eight questions showed significance level. These results indicate that VR or IV tools positively affected the participants' understanding of the subject. Based on the information provided, the students' level of confidence in describing the building to others scored higher among the students who experienced the BIM CAVE and HMD than those who received information via conventional media.

## FINAL COMMENTS AND OBSERVATIONS ON THE PILOT TEST

Among the major benefits of creating a BIM model include the ability to visualize, experience, and analyze design intent; the revelation of functional and decorative aspects; the representation of construction processes, including structure and systems; and the explanation of precedents and influences. In the case of MASP, the BIM model also allows users to experience both the architect's original curatorial scheme of 1968, the subsequent and more conventional approach that was instituted in 1996, and the reinstatement of the original scheme (completed in 2015). Additionally, BIM offers MASP model users the opportunity to expand visual, spatial, and formal analyses to simulate and test lighting quality, as well as the ability to calculate and estimate costs of current and future renovations and operations of the museum. When paired with emerging IV technologies, BIM enables students of architectural design, history, and theory to spatialize knowledge of MASP and its radical museographical and museological strategies.

As in most museums, a challenge to the MASP curators was the lack of space to exhibit all of the works in their collection. Bo Bardi's solution was to mount the artwork on glass easels that were distributed throughout the entire gallery floor space. Traditional captions, names, and information about each painting could only be viewed by walking around to the other side of the display. Thus, in Bo Bardi's design, a viewer experienced a work of art first and foremost as the artist saw it during the process of painting. There were several problems that led the museum to change the displays to a more traditional style in the 1990s. For example, we know that the size of paintings that could be displayed was limited by the size of the easels designed by Bo Bardi. In addition, the angles at which a work of art could be viewed were limited by the physical location of the works and the available space of the gallery floor. Moreover, the glass easels themselves were found to be damaging to some works of art. Despite Bo Bardi's design intent, some people still preferred to see the name and information on the work of art at the same time as viewing the art. Some critics even claimed that a solid display wall served as an important component of the experience of a painting (Campagnol and Caffey 2011, Caffey and Campagnol 2015). Immersive visualization allows the viewer to explore all of these options, experience the effects, and compare the results.

In its applications for architecture, engineering, and construction (AEC), BIM technology generates and replicates both geometric and non-geometric features for sites, structures, and spaces virtually. The creation of a detailed BIM model of the Museu de Arte de São Paulo in Brazil provides a case in point. One of the reasons for selecting MASP for the pilot test was the possibility of exploring modern architecture and theory through a case study utilizing BIM technology as a tool of interrogation and analysis. Generated with off-the-shelf software programs (Autodesk Revit and Navisworks) and Android smartphone applications, a digital reconstruction and 2D and 3D walkthroughs make architectural knowledge both explicit and actionable in the virtual realm, which in turn aids in the production and retention of knowledge and its incorporation into behavior in the experiential realm. The MASP model for this project derives from photographs, original construction drawings, personal accounts, and visual and textual descriptions of the museum.

In addition to offering keen insight into MASP's structural and material innovations, BIM and virtual reality serve as a mechanism through which to explore the formal expression of architect Lina Bo Bardi's social, cultural, and political programs for the site, its structure, and its space. In particular, the combination of BIM and IV could reveal MASP's pioneering construction system, the magnitude of Bo Bardi's intervention in the urban fabric, in the collective consciousness of the local population, and in the formal language of architectural modernism in the twentieth century. We found the use of BIM and immersive visualization technology to be relatively easy and accessible for supporting these types of analyses. Furthermore, the technology allowed the researchers and students to obtain new insights about the case study. Future work will explore the impact of these techniques to enhance education and reinforce lessons in the participant's memory.

## INSIGHTS AND FUTURE WORK

Our case study suggests a new way of curating an exhibit derived from Bo Bardi's innovative design, but using advanced 3D viewing technology instead of a physical museum gallery. Just as Bo Bardi displayed art in a way that contrasted with contemporary practice in the 1960s, IV environments have the potential to allow art to be displayed in completely new modalities, thereby opening up new ways that future generations might experience art, education, and entertainment. Virtual art cannot be damaged, and space and size are no longer problems. An installation of an exhibit using 3D glasses could restore the original vision of the architect of MASP and extend it to accommodate more art and more artistic experiences. Changes to the design of the museum can not only respect its architectural and artistic heritage but still reflect the technology and culture of present times.

Related explorations of IV to support examination of architectural historical examples suggest other benefits of the technology. At smaller scales, one project participant focused on a 3D model of the Villa of the Bronzes at Olynthus, Greece, using previously published plans and photographs (Robinson 1946, 235–258; Cahill 2002, 97–103). Destroyed along with the rest of the city during a military attack in 348 BCE, the house was excavated by archaeologists from the American School of Classical Studies in Athens in the first half of the twentieth century. While only the lowest courses of the walls on the ground floor are preserved today, the information recovered from the site was sufficient to allow the "interpretive leaps" necessary for a hypothetical recreation of this two-storied, courtyard house (see figure 13.3). The poor preservation of the largely ephemeral materials (rubble masonry, sun-dried mudbrick/adobe, wood) commonly used in ancient Greek domestic architecture often prevents much detailed discussion in surveys of architectural history. Through the model, presented either as a sequence of static images or as an immersive visualization, students are able to experience different spaces and view sheds and explore questions that range from construction techniques to the integration or segregation of different parts of the house.

Another project participant produced a 3D model of François Mansart's Château de Maisons in Yvelines, France, first using photographs as the basis for generating the structure and then comparing the BIM model to design documents published in the early twentieth century (Guérinet 1908). Although this model is still in its earliest stages of development, the value of using BIM software has already proven useful in revealing tectonic complexities—a vital element that is missing from 2D design documents and from photographs. For example, in the Unreal Projects model of Château de Maisons, difficulties in reconciling some of the dimensions, proportions, and relationships between building materials revealed a substantial structural correction on the exterior of the building; this alteration was undetectable in plans, sections, elevations, and photographs.

The original impetus for this research was to explore ways of using advanced digital technology for enhancing the teaching and learning of art and architectural history. Our study has shown that readily available commercial software can provide a rich and powerful environment for the exploration of our case study building. The modeling process with BIM encourages a person to understand the thought processes of the original designer and reveals some of the fundamental challenges of design, construction, and operation. The use of the CAVE environment shows that BIM can support an immersive experience, enabling students to encounter structures that are distant in time and space. The use of 3D head-mounted displays provides another powerful delivery channel and may lead to new modes of museum display.

Future work will focus on further identifying and quantifying the benefits and liabilities of these methods. We hope to employ the model in large format courses in the near future. The veracity and verisimilitude of immersive visualization environments in comparison to real environments can be more fully explored once more models and metadata have been developed and tested in the array. The use of the BIM CAVE as a virtual heritage tool to reconstruct important and influential sites, structures, and spaces can be further explored.

## CONCLUSION

The incorporation of the BIM CAVE IV experience as a complement to the production and review of conventional 2D design documents, the reading of architectural history texts and journals, attending lectures and participating in class discussion will substantially enhance student experience, academic performance and, ultimately, professional practice by approximating real-space and real-time experiences of the monuments under investigation. Thus, one might reasonably predict that environments not accessible to students because of limitations imposed by realities of time, space, politics, religion or war can be studied with increased richness and more productive insight using the IV approach.

To sum up, it can be inferred that IV tools have enough potential to be used at least as a supplementary pedagogical tool in teaching about distant and historical locations.

## REFERENCES

Bo Bardi, Lina, and Aldo van Eyck. 1997. *Museu de Arte de São Paulo/São Paulo Art Museum*. São Paulo: Blau/ Instituto Lina Bo e P.M. Bardi.

Caffey, Stephen, and Gabriela Campagnol. 2015. "Dis/Solution: Lina Bo Bardi's Museu de Arte de São Paulo." *Journal of Conservation and Museum Studies* 13(1): Article 5, 1–13.

Caffey, Stephen, Robert Graf, Charles Culp, Wei Yan, Ehsan Barekati, and Matthew Marshall. 2010. "Reconstructing the Music Hall Rotunda Annex at Vauxhall Pleasure Gardens c. 1764." In *Electronic Visualisation and the Arts: Proceedings of EVA London 2010, BCS London, 5 Southampton Street, London, 5–7 July 2010*, edited by Alan Seal, 332–338. Swindon, British Computer Society.

Cahill, Nicholas. 2002. *Household and City Organization at Olynthus*. New Haven, CT, and London: Yale University Press.

Campagnol, Gabriela, and Stephen Caffey. 2011. "Pepper the Walls with Bullets: Lina Bo Bardi's Museu de Arte de São Paulo." In *Creating_Making Forum*, 145–151. Norman: University of Oklahoma College of Architecture.

Campagnol, Gabriela, and Stephen Caffey. 2015. "Construction of the Museu de Arte de São Paulo (MASP)." In *5ICCH Proceedings* (5th International Congress on Construction History), Vol. 1, edited by Brian Bowen, Donald Friedman et al. Chicago: Construction History Society of America (electronically).

Frischer, Bernard. 2003. "Mission and Recent Projects of the UCLA Cultural Virtual Reality Laboratory." In *Virtual Retrospect 2003: Proceedings of the Conference, Biarritz (France), November 6th-7th 2003*, edited by Robert Vergnieux and Caroline Delevoie, 65–74. Bordeaux: Ausonius.

Frischer, Bernard. 2008. "Introduction: From Digital Illustration to Digital Heuristics." In *Beyond Illustration: 2D and 3D Digital Technologies as Tools for Discovery in Archaeology*, edited by Bernard Frischer and A. Dakouri-Hild, v–xxiv. Oxford: Archaeopress.

Frischer, Bernard, Diane Favro, Dean Abernathy, and Monica De Simone. 2003. "The Digital Roman Forum Project of the UCLA Cultural Virtual Reality Laboratory." *International Archives of the Photogrammetry, Remote Sensing and Spatial Information Sciences* 34: 5/W10.

Frischer, Bernard, Diane Favro, Paolo Liverani, Sible De Blaauw, and Dean Abernathy. 2000. "Virtual Reality and Ancient Rome: The UCLA Cultural VR Lab's Santa Maria Maggiore Project." In *Virtual Reality in Archaeology*, edited by J. A. Barcelo, M. Forte, and D. H. Sanders, 155–162. British Archaeological Reports International Series 843. Oxford: ArchaeoPress.

Frischer, Bernard, and Philip Stinson. 2007. "The Importance of Scientific Authentication and a Formal Visual Language in Virtual Models of Archeological Sites: The Case of The House of Augustus and Villa of the Mysteries." In *Interpreting the Past: Heritage, New Technologies and Local Development. Proceedings of the Conference on Authenticity, Intellectual Integrity and Sustainable Development of the Public Presentation of Archaeological and Historical Sites and Landscapes, Ghent, East-Flanders, 11–13 September 2002*, edited by Neil Asher Silberman and Dirk Callebaut, 49–83. Brussels: Flemish Heritage Institute, Provincial Archaeological Museum Ename, Province of East-Flanders, Ename Center for Public Archaeology and Heritage Presentation.

Guérinet, Armand. 1908. *Monographie du château de Maisons (Seine-et-Oise)*. Paris: Guérinet.

Kang, Julian. 2012. "BIM/CAVE: Immersive BIM Reviewing System." *Journal of Building Information Modeling* (Spring): 23–24.

Kang, Julian, Gabriela Campagnol, Stephen Caffey, and Mark Clayton. 2015. "4D Construction Simulation Model of MASP." In *The Next City—New Technologies and the Future of the Built Environment: Proceedings of the 16th International Conference* (CAAD Futures 2015), São Paulo, July 8–10, 2015, edited by Gabriela Celani, David M. Sperling, and Juarez M. S. Franco, 370–379. São Paulo: Universidade Estadual de Campinas–Biblioteca Central César Lattes.

Kang, Julian, Adithya Ganapathi, and Hussam Nseir. 2012. "Immersive Virtual Environment for BIM." In CONVR 2012. *Proceedings of the 12th International Conference on Construction Applications of Virtual Reality*, November 1–2, 2012, Taipei, Taiwan, edited by Yu-Cheng Lin and Shih-Chung Jessy Kang, 183–186. Taiwan: National Taiwan University Press.

Miyoshi, Alex. 2011. *Arquitetura em Suspensão: O Edifício do Museu de Arte de São Paulo*. Campinas: Armazém do Ipê.

Robinson, David M. 1946. *Excavations at Olynthus, Part 12. Domestic and Public Architecture*. Johns Hopkins University Studies in Archaeology 36. Baltimore, MD: Johns Hopkins University Press.

VARK. n.d. "The VARK Questionnaire: How Do I Learn Best?" http://vark-learn.com/the-vark-questionnaire/.

CHAPTER 14

# At the Interface: Multimodal Sensing and Intelligent Learning Systems

## *The Dynamic Transformation of the Cityscape and Its Ongoing Study*

BILL SEAMAN

### INTRODUCTION

Differing forms, scales, and uses of computational media are changing the face of the city in multiple ways. The ubiquitousness of ever-changing media-based facades, pre-programmed and/or driven by different forms of interface, present a changing environment that has no historic parallel. Thus where buildings previously had a beginning date and aged over time, our new situation explores a relation between the physicality of buildings and new forms of computational media that can be continuously updated in terms of both hardware and software. The use of screens, connectivity, and multiple forms of interfacing enable one to enter conceptual portals to gain information; explore multiple computational functions; and gain access to a vast compendium of digital sites. Thus the city is "every city" in some aspects; particularly in terms of its connective media processes, it becomes an *everyplace* of digital relations that are linked to all those who are connected via the internet throughout the world.

#### Big Data and the Omnipresence of Screens

Media researchers, media historians, cyber-archaeologists, and archaeologists are all tasked with creating new modes of interdisciplinary and transdisciplinary

Bill Seaman, *At the Interface: Multimodal Sensing and Intelligent Learning Systems* In: *Digital Cities*. Edited by: Maurizio Forte and Helena Murteira, Oxford University Press (2020). © Oxford University Press.
DOI: 10.1093/oso/9780190498900.003.0015

research to map and explore the quixotic nature of the city in terms of both its physical and its informational domains. Here we must point to the differing time scales that the computational media bring to bear on our relation to physical space, the environment, and architecture. We can use the computer as a tool for the study of that situation, the creation of new forms of interfacing with Big Data (the collection of multimodal data of varying types), the archiving of differing forms of media objects including time-based streams of multimodal sense data, the development of textual meta tags and meta-meta tags (or a hierarchy of meta tags that extends into regions of further categorical articulation), the generation of new forms of annotation and hyper-linkage to other relevant media objects, the employment of image-recognition systems, the discernment of data patterns from many differing conceptual perspectives, the employment of language analysis and new forms of search modalities, and the exploration of multiple forms of crowd sourcing as a means of articulating new forms of relationality. All these explore the digital ↔ digital, digital ↔ analogue, and digital ↔ environment relations we face today. In terms of the city, we must also consider the meta-Big Data—that is, a Big Data approach to approaches to Big Data.

Perhaps one of the biggest changes to our perceptual field of fields is the proliferation of screens, from the smallest of small wearable screens on bracelets and digital watches (Ryan 2014), to building-size screens (Seaman and Todd 2012), and to entire city blocks of conglomerate media access (figure 14.1).

We might say there is a kind of taxonomy of screens, which includes use of multiple screens employed on architecture and in foyers for both information and advertising; the use of architecture itself as an impromptu screen; the employment of mobile screens driven by moving vehicles; the adoption of large external lighting systems as a means of aesthetic layering, of changing color as well as providing information—in particular changing texts; the exploration of moving screens on blimps; the inclusion of screens in trucks, SUVs, and cars; and the use of small mobile screens by pedestrians standing, walking, and commuting on trains and buses, or the use of built-in media in trains, trams, taxis, and buses or by drivers and others for text messaging. All these examples contribute to our changing world of computational media proliferation.

Along with screens, there are also mobile devices such as mp3 players, iPhones, iPods, iPads, Android smartphones, touch devices, and related mobile devices as other portals to multimodal information—in short, we experience visual, haptic, and/or sonic layers of image, music, and/or text. These devices sprawl into and across the city.

## Multimodal Interfaces

It is interesting to watch the developing trends in the creation and/or abstraction of biomimetic interfaces—touch, camera (pattern recognition),

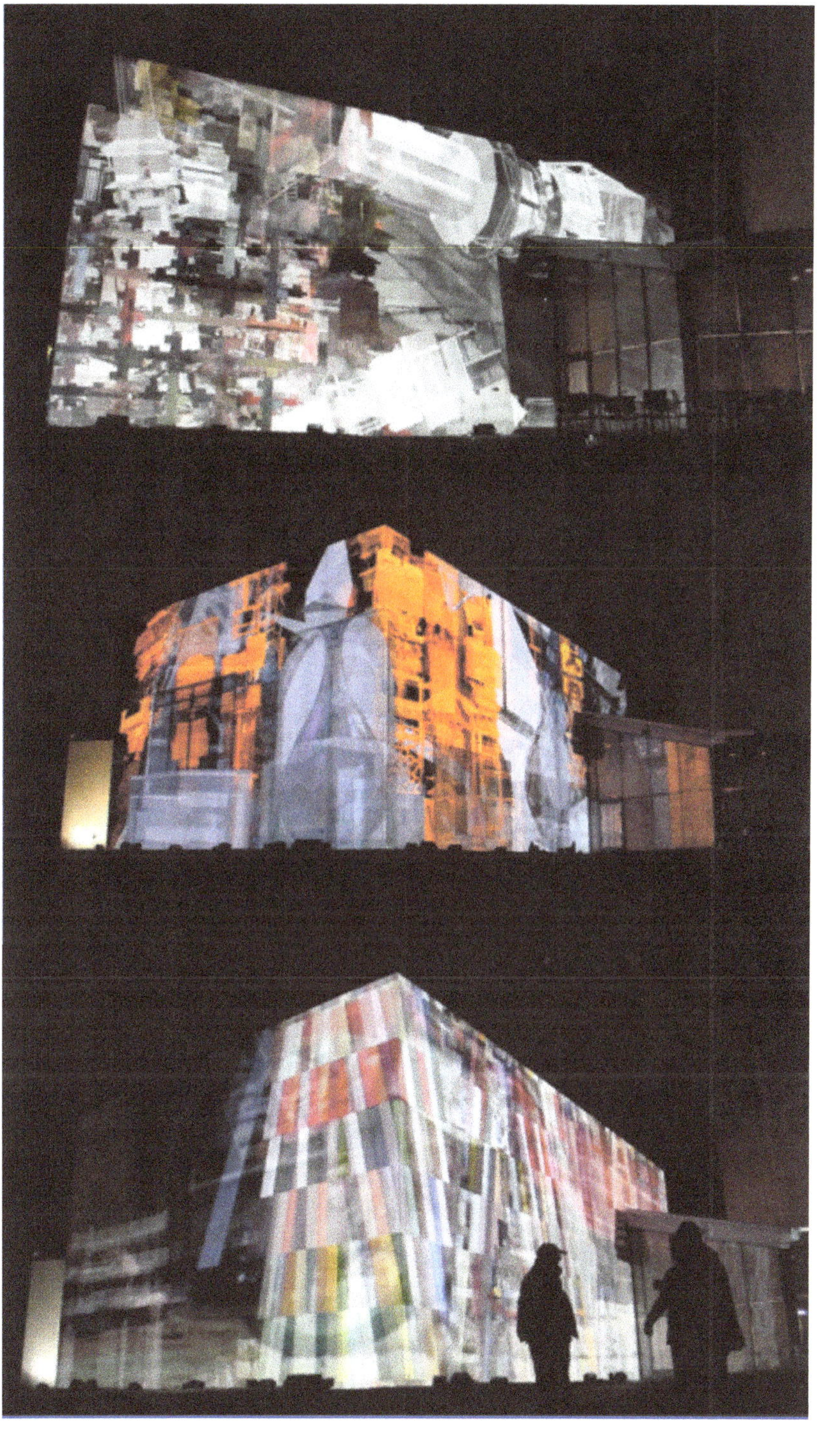

**Figure 14.1** A China of Many Senses (Nasher version) Chat Festival 2012.
*Credit:* Photo details of installation by Todd Berreth.

microphone and voice recognition, haptic recognition, spatial alignment and balance, and chemical sensors, not to mention biometric interfaces as a new form of "lock"—the biometric login. Differing "reading systems," like the Kindle, are also changing the function of libraries—for example, one can be reading *Ulysses* by James Joyce at a local pub, or negotiating via many differing devices new forms of interactive and other contemporary poetic works via online spaces like *UbuWeb* (www.ubuweb.com). Such widespread systems can provide intellectual stimulus as an overlay to any environment. It is also clear that these experiences change how we understand our world.

Locative media, both aesthetic and information-based, enable information to be accessed, triggered by particular "locative" relations (Brucker-Cohen 2014). GPS units provide another layer, or shall we say give a computational "voice" to ongoing navigation. Machinic sensing systems become commonplace, both built-in and mobile, and function as different forms of interface with multiple aesthetic assemblages and intellectual potentials. This is particularly true of the Apple iPhone, which houses many forms of multimodal sensing. It goes without saying that video cameras, used for both pleasure and surveillance, are found everywhere across the cityscape.

The smartphone, for example, enables a multiplicity of distributed built-in machinic sensing devices to be employed that form a digital connective tissue—routing images and information in real time via "facetime," Skype, and other connective applications. It employs small portable camera devices, internet access, and media streaming, with related texting and email, as communication streams.

Social media are everywhere, providing access to reproductions of works of art, among other wonderous imagery. Artists have taken advantage of these forms to use as a new space for showing and/or distributing their works, in part moving them out of the gallery and into and across the city as a new site for aesthetic activity. Galleries of digital reproductions, as those available on Snapchat, Tumblr, and Pinterest are huge image storehouses. Of course, Vimeo and YouTube hold vast video repositories that are also instantly accessible, enabling the intersection of physical and psychological spaces. As the "internet of things" grows ever larger, so does our digital connectivity to the physical world and its physical processes.

## AUGMENTED AND VIRTUAL REALITY: SCIENCE FICTION BECOMES REALITY

The employment of augmented reality enables multiple forms of aesthetic overlay and/or textual interplay with a relational environment (Bonsor, n.d.). Virtual reality (VR) presents many approaches to the design and study of the city, as well as creative visualizations of imaginary landscapes including cityscapes, nested in the actual space. Such ubiquitous systems as Oculus Rift (a stereo immersive VR interface) and repurposed game engines provide spaces for varying forms of content (including aspects of cyber-archeological materials) with the potential to access vast relational multimodal databases, calling in media objects on demand and facilitating 3D navigation of these generative spaces. All in all, the city as portrayed in many forms of science fiction—for example, the movie *Blade Runner*—is now the city as it (almost) is. How do all these developments extend the potentials of cyber-archaeology?

## The Changing Cityscape

Two authors early on pointed to the new nature of the city—the late William J. Mitchell, in *City of Bits: Space, Place and the Infobahn* (1995), wrote eloquently on how the digital environment was bringing about vast changes: "Electronic interaction will become increasingly multimodal, as when videoconferencing combines sound and vision. Robotic effectors combined with audio and video sensors will provide telepresence" (Mitchell 1995) lending to the complexity of the intersecting spaces that weave the fabric of the contemporary city. We are still on the cusp of developing new forms of physical interfacing, empowering physical interactivity at a distance. Another author who has greatly contributed to the field is Nicholas Negroponte, in his book *Being Digital* (1995), in which he points to the many differences between bits and atoms. Negroponte was optimistic about the future of our digital world and the Media Lab at MIT (which he headed), and he became a midwife to many changes in the digital/physical landscape. Negroponte stated: "we are not waiting on any invention. It is here. It is now. It is almost genetic in nature in that each generation will become more digital" (231). Negroponte also collaborated with cyberneticist Gordon Pask on *The Architecture Machine* (Negroponte 1970) in which he expressed interest in intelligent generative systems where "The design process, considered as evolutionary, can be presented to a machine, also considered as evolutionary, and a mutual training, resilience, and growth can be developed" (6) and in his *Soft Architecture Machine* (Negroponte 1975).

Given that the physical environment is not in perfect harmony with every person's lifestyle, and given that architecture is not faultless in meeting human needs, and given that the architect is not the consummate manager of physical environments, I view the physical environment as an evolving organism, as opposed to a designed artifact. In particular, I consider this an evolution aided by a specific class of machines. Warren McCulloch (1956) calls them "ethical robots"; in the context of architecture, I call them "architecture machines" (Negroponte 1970).

## Interflow Architectures

In terms of the contemporary city, can we begin to make machines that learn and have intelligent functionality (Seaman 2015)—that both augment building and keep track of systems but that also help us study the history and flow of change related to the construction of cities. In the past, I have called this "interflow architectures."

The body is a flow-architecture in continuous interactive motion, embedded in constructed architectures and the environment. It is a semi-mutable and mobile environment nested in a larger set of environmental flows. A series of architectural flows pivot, bridge, become layered, and intermingle, enabling a set of physical/conceptual domains to arise in the flow of thought—human observation. All communication needs a physics of instantiation with varying qualities of change to derive bodily affective (sensual) properties. The articulation of these physics of flow constitute a layered set of architectures that function in dynamic interactive relationships.

Thought arises as the sum of all change, producing flows through interactions with this enfolded set of architectural flows. The intermingling fields of perception form a continuous flow of state changes in the "thoughtbody." The patterns of sensual perceptions with internalizations and introspection related to those perceptions intermingle with this ongoing enfolded space of flows, as follows:

> The architecture of constructed spatial flows.
> The architecture of material flows.
> The architecture of sense flows.
> The architecture of a physics of flow.
> The architecture of body flows.
> The architecture of electro-chemical flows.
> The architecture of linguistic flows.
> The architecture of semiotic flows.
> The architecture of spatiotemporal flows.
> The architecture of thought flows.
> The architecture of biological flows.
> The architecture of chemical flows.
> The architecture of functional flows.
> The architecture of mathematical flows.
> The architecture of gravitational flows.
> The architecture of liquid flows.
> The architecture of waste flows.
> The architecture of energy flows.
> The architecture of information flows.

The architecture of media flows.
The architecture of electronic flows.
The architecture of optical flows.
The architecture of digital flows.
The architecture of manufactured nano-flows.
The architecture of code flows.
The architecture of aesthetic flows.
The architecture of sexual flows.
The architecture of political flows.
The architecture of social flows.
The architecture of virtual flows.
The architecture of the flow of time.
The architecture of flows of feeling.
The architecture of desire flows.

The architecture of flows outside of our perceptual range are:

The architecture of intermingling fields of flow.
The qualities of differing viscosities.

Negotiating the force of these flows, the intersubject ↔ object of these flows, is what I call "interflow architectures" (Seaman, 2002). Our thoughts are composed, to paraphrase Ranulph Glanville.

## A Relational Multiperspective Approach to Knowledge Production

My intention has been to explore what I call a multi-perspective approach to knowledge production. In terms of information, one could take multiples of the lines shown earlier and map out a relationality between them. One might even parse an individual focus and observe aspects of its operative qualities over time relevant to its individual nature. The question then becomes: Can a multimodal machinic observation system contribute to this ongoing sense of human learning and understanding, intelligently registering particular qualities of flow? There is a sublime vastness to space both physical and cyberspace, as well as to the histories of its understandings—its various articulations.

Intelligent systems can in part contribute to how we can come to explore the multiple perspectives that function as part of our complex contemporary technological milieu. The question becomes: How can we learn from the driving problems this suggests for the next generation of thinking and acting in the world in terms of authoring new intelligent systems of storage, retrieval, and observation?

### Digital ↔ Physical Relationalities

Our contemporary city is, just as Negroponte discusses, the intermingling of bits and atoms on vast scales. Here, we could potentially begin to define a compendium of digital ↔ physical relationalities where we intermingle the internet with the internet of things, and then expand that into actual environmental space, as well as move further outward into the space of all apps and computational functionalities. In fact, my long-term goal is to create a compendium of relationalities that we might draw upon in terms of machinic intelligence.

In my class on experimental interface design, I discuss interface systems in terms of the following scheme: input + punctionality = output. The viewer/user of the system falls into a cybernetic loop with that system, contributing to its steering and/or observation. Yet now we must nest this simplified system of classification in the larger environment of the cityscape. We must also realize that when intelligent systems are set in motion, they form a seamless continuum of interaction/intra-action with the interactant/participant.

## ARCHITECTURES: PHYSICAL AND CONCEPTUAL

In 1991, my essay entitled "Hybrid ArchitecturesMedia Information Environments, was included in the book *Intelligent Environments: Spatial Aspect of the Information Revolution* (1997). This text suggested many speculations that have since come true.

Witnessing the ever-quickening flow of technological invention from the shifting perspective of the present, one can project potential hybrid architecture/media information environments involving state-of-the-art communication systems, advanced virtual reality networks, speed-of-light transfer and retrieval of vast selections of information, and new forms of art and entertainment. The city presents a concentration of nodes where users are informed and transformed through the identity and functionality of a series of layered networks, interfacing the physical world with the conceptual space of electronic interchange. The information architectures that facilitate this exchange are becoming increasingly palpable, both in terms of the "physical" world change and in relation to potential sensual feedback systems that serve to substantiate the illusionistic/metaphorical levels of potential interaction (Seaman 1997).

It is easy to forget just how much our world has changed since the 1980s and '90s. In terms of cyber-archaeology, the quixotic nature of the city becomes highly problematic in respect to contemporary historical study. In order to best study the cityscape, then, we must both create new forms of multimodal machinic "sensing" and empower new approaches to media relationality as it pertains to historical materials. Can we create new interfaces and flexible software that enables many different kinds of contexts to be generated exploring relationality over time, as viewed via multimodal media apparatus? What are the most helpful digital micro-processes to contribute to this ongoing research?

### A Universal Format Translator

In the 1990s, I discussed the potential for creating a "universal format translator" for digital information. This robotic system would size up the storage media and seek to emulate all functionalities necessary to access particular forms of digital historical media. This hypothetical technology would be authored in terms of both hardware to read the digital media from the storage medium and software to emulate their original functionalities—potentially including interactive structures and would enable access to material that would usually be limited to differing operating systems. This might also mean accessing historical fragment-modalities of the internet via the Internet Archive and the Wayback Machine (https://archive.org/web/).

Great strides have been made in certain forms of individual document translation, especially in terms of the pdf and the format translation and character recognition (OCR) systems that facilitate analogue-to-digital conversion. Speech-to-text translation systems are also improving, but still need work. Creation of a universal format translation device would be an expensive proposition in the beginning, but if we consider the amount of important digital information that has been lost because of changes in storage format and operating system, it is clear that such a device would be critical to the future of both cyber-archology and media archaeology.

### The Realm of Cyber-Archaeology

A definition of *cyber-archaeology* has been offered by Maurizio Forte and Nicolò Dell'Unto, introduced at a conference on the subject:

> The ontology of archaeological information, or the cybernetics of archaeology, refers to all the interconnective relationships which the datum produces, the code of transmission, and its transmittability. Because it depends on interrelationships, by its very nature information cannot be neutral with respect to how it is processed and perceived. It follows that the process of knowledge and communication have to be unified and represented by a single vector. 3D information is regarded as the core of the knowledge process, because it creates feedback, then cybernetic difference, among the interactor, the scientist and the

> ecosystem. It is argued that Virtual Reality (both offline and online) represents a possible ecosystem, which is able to host top-down and bottom-up processes of knowledge and communication. In these terms, the past is generated and coded by "a simulation process." Thus, from the first phases of data acquisition in the field, the technical methodologies and technologies that we use, influence in a decisive way all the subsequent phases of interpretation and communication. Thus, the questions which we pose in a phase of bottom-up knowledge (for example, in an archaeological excavation) will influence the top-down phases of interpretation, or the mental patterns (for example, a comparative analysis and reconstruction of models). If we peremptorily separate knowledge and communication, we risk losing information along the way, reducing the relationships that are constructed between acquisition/input and transmission/output. (Forte and Dell'Unto 2009)

This definition in some ways outlines the birth of a new field. The second-order cybernetic (von Forester 1979) relationship of the researcher to the system, one in which they can potentially navigate and contribute, is central. Second-order cybernetics is the cybernetics of cybernetics, or the cybernetics of observing systems. Here, instead of separating the researcher from the system, the researcher actually "negotiates" it, participating through differing levels and qualities of interaction—potentially contributing new materials, annotations, and commentary to the database or archive in an ongoing manner.

## Open-Order Cybernetics

Gaugusch and Seaman (2004) have discussed what they call "open-order cybernetics." As intelligent systems become ubiquitous, we begin to have a new form of observer, a machinic observer that will have its own characteristics and it will lend new perspectives to any field of potential study, including cyber-archaeology. Also, as we define new methods and new languages in an ongoing manner, this suggests the notion of a cybernetics that is continuously expanding and redefining itself via the creation of new terms and new forms of "intelligent" systems and processes—in particular, new varieties of learning systems.

One can imagine autonomous robotic learning systems as discussed by people like Ray Kurzweil, in *How to Create a Mind* (2012), or the very early texts of Ross Ashby and others, like *Design for a Brain* (Ashby 1960), as well as systems with self-writing code that will drastically change the way we understand the observer—both human and machinic. Joscha Bach's book *Principles of Synthetic Intelligence* (2009) also outlines an approach to an intelligent robotic system, as do Seaman and Rossler's *Neosentience: The Benevolence Engine* (2011).

## From Recombinant Poetics to Recombinant Informatics

New forms of search mechanisms enable the generation of complex mixed-media virtual environments as intelligently drawn from vast media databases. I argued in my PhD thesis, "Recombinant Poetics" (1999), that we are on the cusp of a non-logocentric linguistics that such interactive multimodal computational media enable. I discussed the notion of "fields of meaning" that are brought into proximity, each with its own meaning force.

Since generative virtual environments are constantly changing through interaction, there is a summing of the meaning of these conceptual forces, as understood by the individual exploring the system. Thus I built a virtual-world generating system that was user friendly, working with Gideon May as programmer. One could explore many different media elements and processes in real time, and literally experience meanings arising and changing based on the perception of neighboring media, as well as via negotiation and interaction with the system. It was this early work that led to building the generative computational system I am now exploring; I call this new area of research "recombinant informatics." As such, I am interested in generative systems that are both combinatoric and have life-like characteristics as part of their operation.

I currently have an operative system called the Insight Engine.[1] It was funded by both DIBS (Duke Institute for Brain Sciences; http://athanasius.trinity.duke.edu/projects/I_E/i_e.html) and BASS connections at Duke. The idea behind this system was to reverse-engineer different research practices across multiple disciplines. I assume that many researchers undertake similar practices— reading papers, viewing diagrams, exploring data sets, creating and viewing visualizations, watching videos, and partaking in discussions among other activities. Interdisciplinary and transdisciplinary research means crossing such "linguistic" domains in framing that research. Here, the generation of shared language (a bridging of languages) is essential. I am still working on this aspect of the system. In regard to this, I am considering a gaming, playful approach to articulating shared definitions for particular jargon, inspired in part by Luis von Ahn's "human computation," where players work the *ESP Game* to provide captions for pictures. Von Ahn explains, "I build systems that combine humans and computers to solve large-scale problems that neither can solve alone" (von Ahn, n.d.).

The Insight Engine brings multiple computational functionalities to individuals and/or teams of users through a new interface design and connectivity paradigm (figure 14.2). The system keeps tracks of its ongoing use and computationally discerns dynamic relationalities between different aspects of the research in an ongoing manner, as driven by users. One functionality generates an intelligent network of associated media objects that can be navigated, annotated, shared, and added to in an iterative manner. The goal

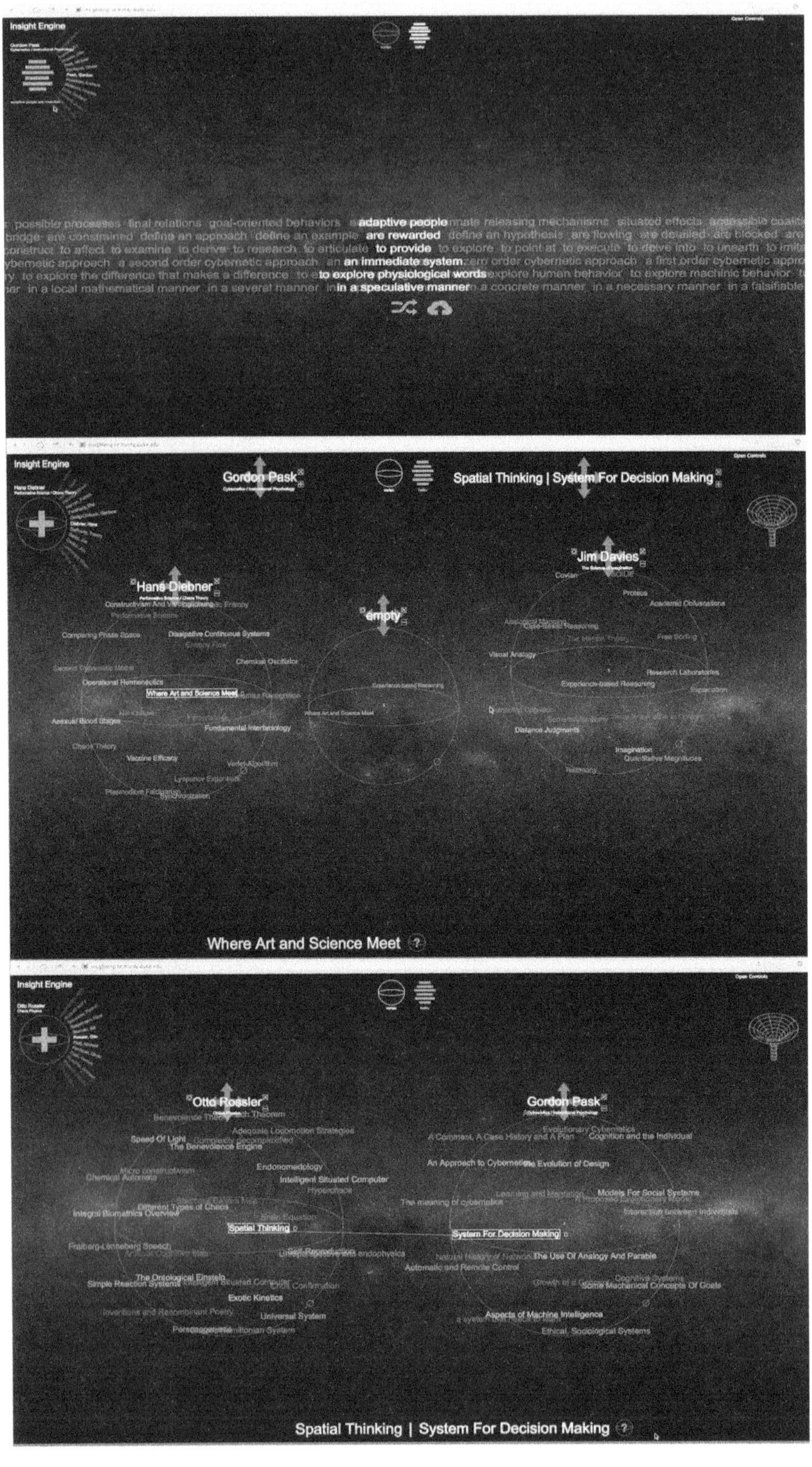


**Figure 14.2** Interface details of Insight Engine, 2014 version.
*Credit:* Bill Seaman, with Todd Berreth, Olivier Perriquet.

is to computationally enable potential bi-sociational processes (see Koestler 1964) through human/human and human/computer interactions.

The Insight Engine, as one of its multiple functions, seeks to empower new forms of intelligent searches across disciplinary domains. Key functions enable a juxtaposition of sources of "relevant" multimodal data, as drawn from a relational database; in the future, the system will draw from distributed sources from across the internet. This function points to a multiperspective approach to research and knowledge production, as drawn from different disciplinary foci.

## The Neosentient Architecture Generator

Currently, I am authoring a new system, with Matthew Kenney as the programmer, to be used, in part, as a vehicle for cyber-archeological research. The system, named the Neosentient Architecture Generator,[2] draws on many different interpretations of the word *architecture*: the architecture of meaning production, the architecture of thought, digital architecture generation, and code architecture. This system (Seaman 2015) will embody many of the ideas presented earlier in the chapter concerning interflow architectures. Thus the system will aim for many different potential uses.

The Neosentient Architecture Generator, or N_Arch.Gen, seeks to combine the functionality of my World Generator system (embodying the Recombinant Poetic processes mentioned earlier) with the Insight Engine (recombinant informatics), and will service cyber-archeological research. Thus, one could load the database with many different kinds of media objects, including entire virtual worlds or pointers to particular media objects that are not currently computationally compatible; for example, one might have a still photograph and want to load a scanned example of virtual reality derived from the site under study). The Neosentient Architecture Generator will serve an open-order cybernetic function, enabling the user to bring media objects into virtual proximity. The system will use textual analysis to empower "intelligent" queries, and hypothetically will empower additional open-ended textual annotation to the given media elements.

We seek, over time, to include additional intelligence and learning capabilities. As you can imagine, one gradually "chips away" at the creative act for such a system, yet this slow process differentiates me from many other computer scientists. Being a media artist and media researcher, I give myself permission to explore such long-term goals, always with functional short-term goals in mind.

In *Neosentience: The Benevolence Engine* (2011), Otto Rössler and I wrote a book with many micro-chapters that were concerned about the future of AI and learning systems (figure 14.3). In a post-Turing Test manner,

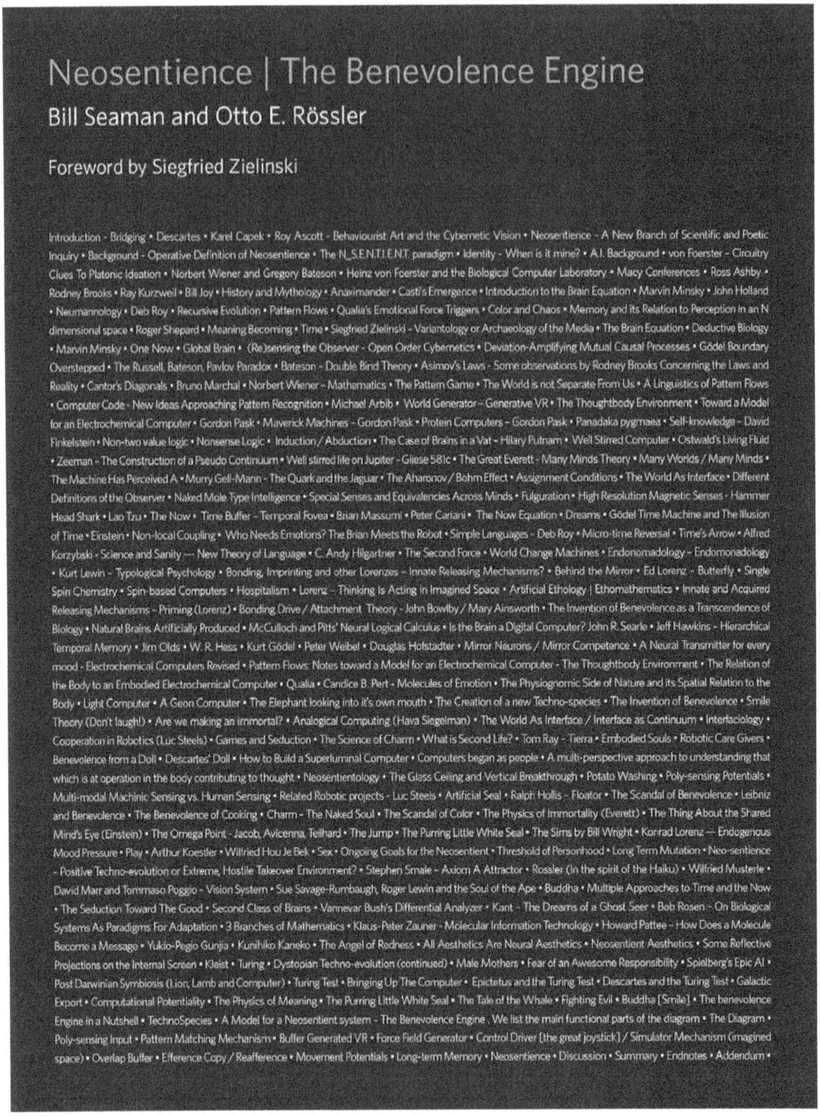

Neosentience | The Benevolence Engine

Bill Seaman and Otto E. Rössler

Foreword by Siegfried Zielinski

Introduction - Bridging • Descartes • Karel Capek • Roy Ascott - Behaviourist Art and the Cybernetic Vision • Neosentience - A New Branch of Scientific and Poetic Inquiry • Background - Operative Definition of Neosentience • The N_S.E.N.T.I.E.N.T. paradigm • Identity - When is it mine? • A.I. Background • von Foerster - Circuitry Clues To Platonic Ideation • Norbert Wiener and Gregory Bateson • Heinz von Foerster and the Biological Computer Laboratory • Macy Conferences • Ross Ashby • Rodney Brooks • Ray Kurzweil • Bill Joy • History and Mythology • Anaximander • Casti's Emergence • Introduction to the Brain Equation • Marvin Minsky • John Holland • Neumannology • Deb Roy • Recursive Evolution • Pattern Flows • Qualia's Emotional Force Triggers • Color and Chaos • Memory and its Relation to Perception in an N dimensional space • Roger Shepard • Meaning Becoming • Time • Siegfried Zielinski – Variantology or Archaeology of the Media • The Brain Equation • Deductive Biology • Marvin Minsky • One Now • Global Brain • (Re)sensing the Observer - Open Order Cybernetics • Deviation-Amplifying Mutual Causal Processes • Gödel Boundary Overstepped • The Russell, Bateson, Pavlov Paradox • Bateson - Double Bind Theory • Asimov's Laws - Some observations by Rodney Brooks Concerning the Laws and Reality • Cantor's Diagonals • Bruno Marchal • Norbert Wiener - Mathematics • The Pattern Game • The World is not Separate From Us • A Linguistics of Pattern Flows • Computer Code - New Ideas Approaching Pattern Recognition • Michael Arbib • World Generator – Generative VR • The Thoughtbody Environment • Toward a Model for an Electrochemical Computer • Gordon Pask • Maverick Machines - Gordon Pask • Protein Computers – Gordon Pask • Panadaka pygmaea • Self-knowledge - David Finkelstein • Non-two value logic • Nonsense Logic • Induction / Abduction • The Case of Brains in a Vat – Hilary Putnam • Well Stirred Computer • Ostwald's Living Fluid • Zeeman - The Construction of a Pseudo Continuum • Well stirred life on Jupiter - Gliese 581c • The Great Everett - Many Minds Theory • Many Worlds / Many Minds • The Machine Has Perceived A • Murry Gell-Mann - The Quark and the Jaguar • The Aharonov / Bohm Effect • Assignment Conditions • The World As Interface • Different Definitions of the Observer • Naked Mole Type Intelligence • Special Senses and Equivalencies Across Minds • Fulguration • High Resolution Magnetic Senses - Hammer Head Shark • Lao Tzu • The Now • Time Buffer – Temporal Fovea • Brian Massumi • Peter Cariani • The Now Equation • Dreams • Gödel Time Machine and The Illusion of Time • Einstein • Non-local Coupling • Who Needs Emotions? The Brian Meets the Robot • Simple Languages – Deb Roy • Micro-time Reversal • Time's Arrow • Alfred Korzybski - Science and Sanity --- New Theory of Language • C. Andy Hilgartner • The Second Force • World Change Machines • Endonomadology - Endomonadology • Kurt Lewin – Typological Psychology • Bonding, Imprinting and other Lorenzes – Innate Releasing Mechanisms? • Behind the Mirror • Ed Lorenz – Butterfly • Single Spin Chemistry • Spin-based Computers • Hospitalism • Lorenz – Thinking Is Acting In Imagined Space • Artificial Ethology | Ethomathematics • Innate and Acquired Releasing Mechanisms – Priming (Lorenz) • Bonding Drive / Attachment Theory - John Bowlby / Mary Ainsworth • The Invention of Benevolence as a Transcendence of Biology • Natural Brains Artificially Produced • McCulloch and Pitts' Neural Logical Calculus • Is the Brain a Digital Computer? John R. Searle • Jeff Hawkins – Hierarchical Temporal Memory • Jim Olds • W. R. Hess • Kurt Gödel • Peter Weibel • Douglas Hofstadter • Mirror Neurons / Mirror Competence • A Neural Transmitter for every mood - Electrochemical Computers Revised • Pattern Flows: Notes toward a Model for an Electrochemical Computer - The Thoughtbody Environment • The Relation of the Body to an Embodied Electrochemical Computer • Qualia • Candice B. Pert - Molecules of Emotion • The Physiognomic Side of Nature and its Spatial Relation to the Body • Light Computer • A Geon Computer • The Elephant looking into it's own mouth • The Creation of a new Techno-species • The Invention of Benevolence • Smile Theory (Don't laugh!) • Are we making an immortal? • Analogical Computing (Hava Siegelman) • The World As Interface / Interface as Continuum • Interfaciology • Cooperation in Robotics (Luc Steels) • Games and Seduction • The Science of Charm • What is Second Life? • Tom Ray - Tierra • Embodied Souls • Robotic Care Givers • Benevolence from a Doll • Descartes' Doll • How to Build a Superluminal Computer • Computers began as people • A multi-perspective approach to understanding that which is at operation in the body contributing to thought • Neosentientology • The Glass Ceiling and Vertical Breakthrough • Potato Washing • Poly-sensing Potentials • Multi-modal Machinic Sensing vs. Human Sensing • Related Robotic projects - Luc Steels • Artificial Seal • Ralph Hollis – Floator • The Scandal of Benevolence • Leibniz and Benevolence • The Benevolence of Cooking • Charm - The Naked Soul • The Scandal of Color • The Physics of Immortality (Everett) • The Thing About the Shared Mind's Eye (Einstein) • The Omega Point - Jacob, Avicenna, Teilhard • The Jump • The Purring Little White Seal • The Sims by Bill Wright • Konrad Lorenz — Endogenous Mood Pressure • Play • Arthur Koestler • Wilfried Hou Je Bek • Sex • Ongoing Goals for the Neosentient • Threshold of Personhood • Long Term Mutation • Neo-sentience - Positive Techno-evolution or Extreme, Hostile Takeover Environment? • Stephen Smale - Axiom A Attractor • Rossler (In the spirit of the Haiku) • Wilfried Musterle • David Marr and Tommaso Poggio - Vision System • Sue Savage-Rumbaugh, Roger Lewin and the Soul of the Ape • Buddha • Multiple Approaches to Time and the Now • The Seduction Toward The Good • Second Class of Brains • Vannevar Bush's Differential Analyzer • Kant – The Dreams of a Ghost Seer • Bob Rosen – On Biological Systems As Paradigms For Adaptation • 3 Branches of Mathematics • Klaus-Peter Zauner - Molecular Information Technology • Howard Pattee – How Does a Molecule Become a Message • Yukio-Pegio Gunjia • Kunihiko Kaneko • The Angel of Redness • All Aesthetics Are Neural Aesthetics • Neosentient Aesthetics • Some Reflective Projections on the Internal Screen • Kleist • Turing • Dystopian Techno-evolution (continued) • Male Mothers • Fear of an Awesome Responsibility • Spielberg's Epic AI • Post Darwinian Symbiosis (Lion, Lamb and Computer) • Turing Test • Bringing Up The Computer • Epictetus and the Turing Test • Descartes and the Turing Test • Galactic Export • Computational Potentiality • The Physics of Meaning • The Purring Little White Seal • The Tale of the Whale • Fighting Evil • Buddha [Smile] • The benevolence Engine in a Nutshell • TechnoSpecies • A Model for a Neosentient system - The Benevolence Engine . We list the main functional parts of the diagram • The Diagram • Poly-sensing Input • Pattern Matching Mechanism • Buffer Generated VR • Force Field Generator • Control Driver [the great joystick] / Simulator Mechanism (imagined space) • Overlap Buffer • Efference Copy / Reafference • Movement Potentials • Long-term Memory • Neosentience • Discussion • Summary • Endnotes • Addendum •

**Figure 14.3** Back cover copy for the book *Neosentience | The Benevolence Engine.*
*Credit:* Cover copy by Bill Seaman.

we suggest a set of practical benchmarks to articulate the components of neosentience, it:

- learns
- intelligently navigates
- interacts via natural language
- generates simulations of behavior (it "thinks" about potential behaviors) before acting in physical space

- is creative in some manner
- comes to have a deep situated knowledge of context through multi-modal sensing
- displays mirror competence.

Of course, this neosentient system is a long way off in terms of its fully "intelligent" functionality. But it is interesting to ponder how these benchmarks might begin to generate new fields of study. In a class a few years ago, my students articulated ideas on neosentient design, neosentient architecture, and neosentient gaming. Here, I imagine bringing artificial cognition (Vernon 2014) to bear on the functionality of this new system over time.

### The Neosentient System for Cyber-Archeological Study

Imagine that in my digital archive I have historical texts, videos, photographs, oral histories, data visualizations, and so on—any form of collectable digital data. Using annotation, intelligent textual search mechanisms, pattern matching, image recognition, and voice recognition and transcribing, I can explore media elements that may have never been considered before in relation to one another. By bringing these media objects into virtual proximity, I have the system potentially serving as an engine driving the construction of new knowledge. Using meta tags and what I call meta-meta tags, I can use new search methodologies—that is, semantic, syntactic, and statistical algorithms—to bring disparate forms of media research into visual, multimodal, and conceptual proximity. This is "neighboring," as Gaston Bachelard calls it (1951, 6) brought about in virtual space; but in this space the virtual-world models of actual sites can potentially be nested within each other as part of a more extensive virtual environment (or set of linked environments). Maurizio Forte, in discussing cyber-archaeology suggests:

> The rapid and incremental use of 3D digital technologies in archaeology in the field and labs is able to design diverse and uninvestigated workflows in the generation, representation and communication of data. This digital migration of data and models in such diverse domains creates unexpected results and more advanced knowledge. According to Bateson's thought . . . we can define this process as the map-code of the cybernetic cycle. The study of the code is mandatory for analyzing the interpretation process in the light of a cybernetic perspective: the feedback created by different interactors operating in the same environment/ecosystem generates further feedback and not pre-determined inter-connections. (Forte and Dell'Unto 2009)

### Relational Archaeological Media

A goal for the contemporary media researcher in terms of the city, its history and archaeology, is to create new interactive systems that contribute to the understanding of that history and archaeology. It is clear that 3D models enable new kinds of understandings. The next set of goals for me has to do with relationality. Here, we seek an ongoing, deep understanding of context from many multimodal perspectives, both human and machinic. In his book *Entangled* (2012), Ian Hodder points to many such forms of relationality. Again, the notion of generating a crowd-sourced compendium of relationalities, inspired by the research of Luis von Ahn into "human computation" (von Ahn, n.d.), is a deeply compelling concept.

## CYBER-ARCHAEOLOGY ↔ MEDIA ARCHAEOLOGY

We must begin to take stock of a broad range or series of completely different kinds of technological artifacts. That is, the city can be studied in terms of its relationships between its architectures, transportation systems, lighting systems, and physical infrastructures, streets, and parks, as well as in terms of viewing mechanisms, surveillance systems, the acquisition of still and moving images, and multimodal digital delivery systems. All the city's physical features, above and below ground, can be studied as they fall into relationship with digital artifacts over time, be they related to the given space (like a digital timetable accessed at the airport or subway system) or superimposed on the space (like streaming a movie). Additionally we need to be aware of the archaeology of the media. Media archaeology is a different field from that of cyber-archaeology, yet it is valuable to intermingle the study of these two fields, to bring them into dynamic relation. Media archaeology is itself a quixotic field. In their edited book *Media Archaeology: Approaches, Applications and Implications*, Erkki Huhtamo and Jussi Parikka, point out in the introduction ("An Archeology of media Archaeology") that the field is still one of becoming.

### New Media and New Systems

The advent of "new media" (in common parlance, a loose conglomeration of phenomena such as the internet, digital television, interactive multimedia, virtual reality, mobile communication, and video games) has challenged many scholars to investigate this culture of media in late modernity. Research agendas have varied from network analysis to software studies, from mappings of the new network economies to analysis of "ways of seeing" (or hearing, reading, and touching). Efforts have been made to pinpoint where the "newness" of social networking, interactive gaming, or data mining lays the foundations for "philosophies" and "languages" about the new media. For some researchers, the main concerns are social or psychological, while for others they are economical and ideological, or are motivated by the search for technological determinates that hide behind myriad manifestations of media (Forte and Dell'Unto 2009).

I seek to create new systems that will empower us to study and come to better understand the networks of new media elements and processes that are becoming ever more pervasive. We can explore a mix of human and computer intelligence to negotiate the quixotic technological artifacts that form the intermingled backgrounds of the cityscape. The individual cannot be subtracted out of this; the new media are constantly relational, forming the continuum of our lived experience.

If we make a dynamic timeline as part of this visualization processes, we suddenly have a new series of technological reference points, as embodied in the dramatic changes in technologies and multimodal media delivery systems that have become part of the fabric of the city. These technologies are the purveyors of new forms of culture, education, business, and leisure. As such, our definition of the historical artifact as it relates to the archeological study of cities must now be completely revised. The digital artifact as well has unique qualities that are distinct from traditional "physical" artifacts. So, we ask: What are the commonalities and what are the differences between the two? How can we create new categorizations to help us intellectually and culturally frame these artifacts and "hybrid" spaces of bits and atoms?

Anyone living in or near a city in the past fifty years knows that drastic changes have taken place: a proliferation of computational media and new infrastructures that empower its linkages and ubiquitous nature. Here, media archaeology and cyber-archaeology intersect and begin to spawn a new field—one that enables us to reflect upon the cyber-artifacts, cyber-relationalities, and cyber-histories of this new world. Indeed, we must not underestimate this paradigm shift that is technologically and computationally driven. It is important that we understand the digital/physical relationality that is centered on the human–computer interface design, as well as these new forms of multimodal perception methodologies, both human and machinic. Here, we can experience across myriad scales, moving from the smallest of personal screens, to architectural screens that occupy the façades of entire buildings, to the conglomerate media environments that seem to be overtaking entire regions of the city—city blocks of media, advertisements, information, news, and so on. We understand, read, and inhabit the city differently in relation to these new technologies. We are also mapped, our behaviors traced, as our cyberbodies play in this dramatic physical/media environment.

## Symmetrical Archaeology

One contemporary approach to archaeology has been described as "symmetrical archaeology" by a lab of researchers at Stanford:

> We aim to re-characterize our relationships with what remains of the material past and contemporary things in ways that are not based upon the dualistic schemes characteristic of modernist thought. Therefore symmetrical archaeology avoids modernist divides (subject | object, structure | agency, nature | culture, etc.) altogether. In place of these "Great Divides," we articulate mixtures, imbroglios, hybrids of humans and things. In this way, symmetrical archaeology is "founded on the premise that things, all those physical entities we refer to as material culture, are beings in the world alongside other beings, such as humans, plants and animals." . . . Within this posthuman movement human beings are not detached and singular "intentional agents, but rather always are implicated in complex socio-technical assemblages." . . . This recognition requires redefining what we do as a more entangled activity—as practices such as mapwork that are comprised of multiple fields. (Stanford University Humanities Lab, http://humanitieslab.stanford.edu/23/9)

These themes are not meant to be prescriptive—indeed, they are suggestive of what we regard as some of the core issues for a symmetrical archaeology:

> archaeological time; *belles noiseuses*; beyond interpretation; chiasmatic culture; (nonembedded) collaboration; companion species; conceptual therapy; cyborg ontology; digital ontologies; machinic assemblages; mapwork; media; mediation; the modernist predicament; modes of engagement; multiple fields; long term and short term; network ontologies; post-human movement; pragmatology; proprioception; sociotechnical complexities; sociotechnical genealogy; structural reflexivity; symmetry. (http://humanitieslab.stanford.edu/23/27)

Each of these terms and approaches is unpacked at length on the Stanford website. Further, in the text *Deep Mapping the Media City*, Mattern (2013) discusses this group. He states:

> Olsen and his colleagues regard these various disciplinary infrastructures — technologies, instruments, protocols, and standards—as modes of engaging with and manifesting the past. Different tools for record keeping and representation manifest different aspects of the past, including those ineffable qualities that don't really lend themselves to "accumulate," standardized formats and representations. Whitmore argues that using a mixture of media—field recording, photographs, maps and so on—can allow archeologists to "translate something of the sensory, physical presence of the material past." (Mattern 2013)

### Interface Study and Design

In terms of my own teaching about interface study and design, we can unpack differing information systems by looking at them as input + functionality = output, and interactant, as one particular methodology involving nested loops. Yet when we use these systems, they function as a cybernetic unity with the interactant. This notion is closely tied to the "symmetrical archeological" approach discussed earlier. The systems can react in a biologically inspired manner that is coded—employing the cybernetic vocabulary, I use the term "circular causal feedback system." A more contemporary term would be "the study of complex systems."

When it comes to code, we must also think in terms of a unity that comes about through physical change over time. This is subtle; code functions simultaneously on a physical level. The physics level of the hardware—system connectivity, technologies of connectivity, and distributed technologies—while also propagating the digital information level; both these levels function together to form a cybernetic loop with the interactant. Thus the city is a magnificent series of these systems working both in parallel and asymmetrically, simultaneously in a personalized manner like that of the internet via a series of local sites of connectivity that enable information flows, as well as played out through city-wide technologies, building-wide technologies (all those system that enable a building to run on every level), and landscape wide technologies (lighting systems, stoplights, warning signals, etc.) that enfold the physical with the digital.

## MULTIMODAL SENSING—BOTH HUMAN AND MACHINIC

We extend our senses with a series of machinic senses that are housed in a landscape of new functionalities—in our smartphones, GPS devices, home computers, car functionalities, and audio devices, as well as in mixed arenas of computational ecologies, constantly being focused on different pragmatic tasks and leisure potentials. The computer functions as a pivoting hub enabling our attention to shift quickly in new ways. Because this sensing is distributed and can be collected by machines, we might say that every city becomes one large sprawling distributed city-related system spanning the connected globe. This includes viewing information from any manner of situation to radically altering our biological functioning and human understanding via the creation of new sensing systems that work in tandem with human sensing.

Ironically, one can be on a mountain in Maine and yet be connected to the city via a smartphone or laptop computer. Of course, there are spaces of non-connectivity ("Can you hear me now?") So this chapter's title is bi-directional in a sense: it points at multimodal sensing technologies and interfacing multimodal sensed data that can extend our historical understandings of the city, while also pointing at multimodal human sensing in the viewing/perception of these materials. My essay "Pattern Flows | Hybrid Accretive Processes Informing Identity Production" (2005) points to how we construct identity through re-enforced flows of similar but also different repetitions of multimodal sense perturbations.

I often tell the story of the child whose mother holds out a red ball, smilingly offering it to the child. This action presents a series of sense-perturbation flows: visual, tactile, smell, and touch. Even haptic flows are explored, as the child bounces the ball or tries to catch it. This image gets reinforced over time by similar embodied experiences. Written language comes much later as a new pattern flow—one that builds on this history of embodied patterning over time. Similarly, we can build machinic polysensing systems such that computers begin to learn and understand aspects of context in a similar manner as our own learning, as linked to multimodal sensing systems. Early on, I explored this notion of polysensing, winning an academic border-crossing grant to work with Ingrid Verbauwhede (engineer), Mark Hansen (statistical computer scientist), and grad students Fabian Winkler (Design | Media Arts) and Shenlin Yang (engineering).[3]

This simultaneously suggests that polysensing technologies and their visualization/sonification/hapticization (when linked to appropriate hardware) can be housed along with other digital materials in the relational database and can potentially help us better understand the fleeting artifacts of the contemporary city—both atoms and bits—while providing an accretive cyber-archeological environment to study history through new modes of search and juxtaposition commensurate with these new technological potentials. I am especially interested in how AI systems can contribute to this understanding of the city through new sensing systems, learning systems, as new artificial cognitive systems, not to mention the employment of distributed and mobile systems.

## CONCLUSION

How can we view and understand historical cities in a contemporary technological manner? And how can this impact how we now come to re-understand historical societies— their social relations and cultures? Perhaps this new approach toward the relationality of new media, technologies, spaces, people, things, and environments will enable new technological methodologies that can empower us to understand historical processes in new "accretive" ways. It is clear that the contemporary cityscape is quickly changing in keeping with the quixotic nature of contemporary computational media. In this chapter, I have discussed different ways we might begin to create new technological methodologies to aid in the study of the cityscape over time.

In terms of the cityscape, the digital media and their presence in differing forms, are becoming more ubiquitous in our lives. Here, my approach has been to combine two different areas of study—cyber-archology and media archaeology in the service of future research. The notion is to create intelligent open-order cybernetic systems for archiving and perusing multimodal media. New computationally intelligent approaches to generative virtual environments and relational databases then come to the fore. Also, one points to the nature of context and the limits of current computing in terms of discerning that context. The exploration of polysensing offers a new methodology to help us approach computational contextual learning and understanding. As humans we can quickly size up situations, shifting contexts without a problem. The goal now is to build new polysensing systems to help augment both human and machinic understanding of context. We can also work toward defining a vast compendium of relationalities, as well as develop new multimodal search methodologies, as examined and explored in generative virtual environments. Such systems can help us create contemplative media contexts that lend insight into our ongoing learning in terms of research and cultural understanding of the present, enfolding multiple chosen perspectives. These ideas are central to the intersection of cyber-archaeology and media archaeology. It is here that we are at the interface of multimodal sensing and intelligent learning systems, as explored through circular causal interactions.

## REFERENCES


Ashby, W. Ross. 1960. *Design for a Brain: The Origin of Adaptive Behavior*. London: Chapman & Hall.

Bach, Joscha. 2009. *Principles of Synthetic Intelligence*. New York: Oxford University Press.

Bachelard, Gaston. 1951. *L'Activité Rationaliste de la Physique Contemporaine*. Paris: Presses Universitaires de France.

Bonsor, Kevin. "Augmented Reality." http://computer.howstuffworks.com/augmented-reality.htm.

Brucker-Cohen, Jonah. 2014. "Locative Media Revisited." http://rhizome.org/editorial/2014/mar/26/locative-media-revisited/.

Forte, Maurizio, and Nicolò Dell'Unto. 2009. "Cyber-archaeology." Paper given at conference at Stanford University, May, 1–3, workshop on cyber-archaeology. http://web.stanford.edu/dept/archaeology/cgi-bin/TAG/drupal/?q=content/cyber-archaeology-0.

Gaugusch, Andrea, and Bill Seaman. 2004. "(Re)Sensing the Observer: Offering an Open Order Cybernetics." http://projects.visualstudies.duke.edu/billseaman/pdf/resensing.pdf.

Hodder, Ian. 2012. *Entangled: An Archaeology of the Relationships Between Humans and Things*. West Sussex: Wiley Blackwell.

Huhtamo, Erkki, and Jussi Parikka, eds. 2011. *Media Archaeology: Approaches, Applications and Implications*. Berkeley: University of California Press.

Koestler, Arthur. 1964. *The Act of Creation*. New York: Macmillan.
Kurzweil, Ray. 2012. *How to Create a Mind: The Secret of Human Thought Revealed*. New York: Viking.
Mattern, Shannon. 2013. *Deep Mapping the Media City*. Minneapolis: University of Minnesota Press.
McCulloch, Warren. 1956. "Towards Some Circuitry of Ethical Robots or An Observational Science of the Genesis of Social Evaluation in the Mind-Like Behavior of Artifacts." *Acta Biotheoretica* 11: 147–156. www.vordenker.de/ggphilosophy/mcc_ethical.pdf.
Mitchell. William. 1995. *City of Bits: Space, Place and the Infobahn*. Cambridge, MA: MIT Press.
Negroponte, Nicholas. 1970. *The Architecture Machine*. Cambridge, MA, and London: MIT Press.
Negroponte, Nicholas. 1975. *Soft Architecture Machines*. Cambridge, MA, and London: MIT Press.
Negroponte, Nicholas. 1995. *Being Digital*. New York: Alfred A. Knopf.
Verbauwhede, Ingrid, Bill Seaman, Mark Hansen, et al. 2002. "The Poly-sensing Environment and Object Based Emergent Intention Matrix: Toward an integrated Physical / Augmented Reality Space." users.design.ucla.edu/~fwinkler/PSE/descr.html and http://students.dma.ucla.edu/~fwinkler/PSE/.
Ryan, Susan. 2014. *Garments of Paradise: Wearable Discourse in the Digital Age*. Cambridge, MA: MIT Press.
Seaman, William. 1997. "Hybrid Architectures/Media Information Environments." In *Intelligent Environments: Spatial Aspect of the Information Revolution*, edited by Peter Droege, 641-661 . New York: Elsevier.
Seaman, Bill. 1999. "Recombinant Poetics: Emergent Meaning Examined and Explored Within a Specific Virtual Environment." PhD dissertation, University of Wales.
Seaman, Bill. 2005. "Pattern Flows | Hybrid Accretive Processes Informing Identity Construction." *Convergence Magazine* 7.2, special Issue on Intelligent Environments
Seaman, Bill. 2010. *Recombinant Poetics: Emergent Meaning Examined and Explored Within a Specific Virtual Environment*. Berlin: VDM Press.
Seaman, William. 2014. "A Multi-Perspective Approach to Knowledge Production." *Kybernetes* 43(9/10): 1412–1424.
Seaman, Bill. 2015. "Neosentient Architecture Generator, N_arch.gen." Talk given at Cyberworlds 2015 Conference Proceedings, 2015. www.siggraph.org/attend/events/cyberworlds-2015.
Seaman, Bill. n.d. "Interflow Architectures." Poetic text presented at The Theatre of Interflow Architecture: An Exploration into a Poetics of Machinic Sensing, Multimodal Searching and Thought Augmentation Approaches within Networked Architectural Spaces. http://projects.visualstudies.duke.edu/billseaman/pdf/interflowArchitectures.pdf.
Seaman, Bill, and Todd Berreth. 2012. "A China of Many Senses." https://vimeo.com/36984212.
Seaman, William, and Otto E. Rössler. 2011. *Neosentience: The Benevolence Engine*.: Intellect Press, Bristol England.
Vernon, David. 2014. *Artificial Cognitive Systems*. Cambridge, MA: MIT Press.
von Ahn, Luis. n.d. Author website. www.cs.cmu.edu/~biglou/.
von Forester, Heinz. 1979. "Cybernetics of Cybernetics." www.alice.id.tue.nl/references/foerster-2003.pdf.

# NOTES

## CHAPTER 2

1. Several of Diderot's meditations explore this insight in particular: "The Salon of 1767: mediation 105," in Goodman 1995. I am indebted to Odai Johnson's forthcoming book, *Ruins: Classical Theatre and the Archeology of Memory*, for bringing this to my attention.
2. Lucretius, for example, suggests in Book 4 of *De Rerum Natura* that images we experience through our imagination are films thrown off from the surface of things. When they enter our eyes, we see; when they enter our minds, we imagine things or dream them.
3. This is a topic which my research team and I explored extensively in a project, Skenographia, sponsored by the Leverhulme Trust (http://www.skenographia.cch.kcl.ac.uk/).
4. Diderot famously discussed the "paradox of the actor"—the simultaneous coexistence of the performed character and the enactor of that performance. "One is oneself by nature; one is another by imitation; the heart you imagine for yourself is not the heart you have" (Diderot 1994, 13).
5. Von Ranke 1874, vii, the thing itself as it actually was.
6. Body, Mask and Space: http://www.kvl.cch.kcl.ac.uk/masks/. In this project, sponsored by the British Arts and Humanities Research Council, we worked with professional actors, experienced in masked performance, to prepare an extensive repertoire of scenes from ancient drama, which we recorded in both ChromaKey and motion capture. We then "inserted" these scenes into a great variety of ancient places of performance (including triclinium G) for which we had researched and prepared detailed 3D models.

## CHAPTER 3

1. An example is the impact of the insertion of train stations into urban environments; in Venice, this entailed an 180 degree rotation of the main means of axis to the city and the introduction of pedestrian walkways and bridges, which in turn transformed the primary modes of access into many buildings from the "water side" on a canal to the "street side."
2. See Kennedy 2011.
3. See most recently on this topic, Kallis 2014.
4. Indeed, it can be assumed that the tomb niches in the completed lower part of the façade helped to finance the building; at the upper levels, construction was halted perhaps in part because this funding option no longer existed. See Bruzelius 2014, 152–154, 160–162.

5. See the moving obituary of this great scholar by Florence Day (1954).
6. The walls were often funded by taxation of local citizens; by the early to mid-thirteenth and fourteenth centuries, however, the walls of some cities were partially supported by income derived from Inquisitorial activities (Bruzelius 2014, 178–179).
7. See Bruzelius 1987.
8. Burns 1995.
9. Bruzelius 1987.
10. Seymour 1939, 106–118.
11. Trachtenberg 2013.
12. Trachtenberg 2013.
13. See Huffman et al. 2016 for articles that examine all parts of this enterprise, from initial concept to exhibitions, teaching, and research.
14. In 2010, the team expanded to include Andrea Giordano and his department at the University of Padua, and numerous collaborators at Duke and in the partner institutions.
15. See http://www.dukewired.org/.
16. See http://www.dukewired.org/. A recent publication on the project is Ferrighi 2014, 59–66; also see Huffman et al. 2016.
17. The workshops have been supported by generous funding from the Gladys Krieble Delmas Foundation and, starting in 2015, the Getty Foundation.
18. Guidarelli and Tolîc 2016.

## CHAPTER 6

1. "In-world" refers to an activity inside the virtual world—i.e., while logged in.
2. The number of avatars per area is not "hard-coded" in OpenSimulator; it can be changed through configuration parameters and could theoretically be raised to thousands or millions. However, because OpenSimulator relies on the standard Second Life Viewer application to display content, and Second Life was designed having the limit of 100 avatars per region as being fixed, there are limitations to how many avatars can effectively be displayed realistically before the application crashes. As a consequence, users of OpenSimulator tend to configure the platform to mimic Second Life's own restrictions and limitations for what is experienced as the "best" performance, even though the ability to change such parameters allows experimentation and research.

## CHAPTER 7

1. Skarin 2011.

## CHAPTER 8

1. We should also mention the year 2001, in which Oporto was the European Capital of Culture, an event that helped to leverage this growing interest in Oporto and its Historical Center, in terms of both its diffusion (exhibitions, issues, etc.) and its own urban intervention.
2. About this topic, see also, among others, Real and Tavares 1993.
3. For example, see Costa 2002. Or search online using the word "Porto" in the Open Access Scientific Repository of Portugal—http://projeto.rcaap.pt/—or, in a more targeted manner, in the various repositories of Oporto's university (open, thematic, and of data), http://repositorio.up.pt/repos.php.

4. Just to name a few, *The City of Oporto in 1789*, drawing by the Architect Teodoro de Sousa Maldonado and engraving by Manuel da Silva Godinho; the drawing made to illustrate the work of P$^{e}$. Agostinho Rebelo da Costa, in *Edição Comemorativa da Inauguração da Ponte da Arrábida* (1844); the *Material Iconográfico do Arquivo Histórico Municipal do Porto* engraving; or *View of the Oporto City and Douro River* (1855), by F. P.Graça, in *Edição Comemorativa da Inauguração da Ponte da Arrábida* (1844).
5. For example, Arquivo Histórico Municipal do Porto, 1983. From Oporto's plans, see Cidade do Porto-Planta Redonda. See also Balk 1813; and Arquivo Histórico Municipal do Porto., *Álbum de Cartografia Portuense*, cota 4/B'-1(M).
6. For example, Passos 1935 and Ferreira and Abreu 1928.
7. See Braga 2001.
8. The Foto Beleza fund today includes the Portuguese Photographic Estate; see http://www.espoliofotograficoportugues.pt/.
9. Arquivo Histórico Municipal do Porto. Processo: Maqueta Porto Medieval (s/ cota).
10. In addition to the works already cited, see Ferreira 2007.
11. The author emphasizes, in this context, the Rua da Judiaria Nova do Olival (today, São Bento da Vitória), the Rua Nova (today, Rua do Infante D. Henrique), and the Rua das Flores.
12. The scale model is opened within a window that, besides protecting it, is endowed with a mechanism that allows interactive queries ranging over the medieval city, generated from a control panel, which allows a diverse range of questions whose answers are provided through light beams and laser, sometimes accompanied by spoken comments available in several languages. See Real 1999, 1 and 27.
13. At the Lisbon City Museum, this model shows the area that was destroyed in the great earthquake of November 1, 1755. See Museu de Lisboa 2015.
14. See "A Scale Model of the Medieval Oporto"; "The Documentary Sources"; "The Research Programme."
15. See "The City's Image"; "The Urban Relief"; "The Village's River"; "The Walls"; "The Estiva's Quay"; "The Miragaia's Shipyards"; "The Sé's Neighborhood"; "*O Corpo da Guarda*"; "The Congostas' Streets"; "The Nova Street"; "The Ribeira's Square"; "The Jewry"; "The City Block of Banhos"; "Cathedral and Bishop's Palace"; "The Mendicant Convents"; "Custons House"; "The Hospitals, Boarding Houses and Roadhouses"; and "The Urban Towers."
16. See "The Information Management"; "The Urban Space Construction: "The Finishes' Application"; and "The Museum Framing."
17. See AHMP, "Processo: Maquete Porto Medieval," S/cota.
18. Notice, for example, the care taken in the faithful restoration of the walled fences of the Medieval Oporto model, the topographic deployment level, perimeter, altimetry, towers, walls' gates, and shutters.
19. Porto Virtual no Séc. XVI (796/2.2/C/NRE), financed by the Operational Programme Information Society (POSI), is a project of the Research Centre in Science and Arts Technology (CITAR) of the Universidade Católica Portuguesa. See Porto Virtual Século XVI 2005.

## CHAPTER 9

1. See Oettermann 1997, Hyde 1988, Comment 2000, Altick, 1978, Avery 1995, and Huhtamo 2004 and 2012.

2. Research on presence is an established and key body of inquiry for virtual environments and VR that focuses on the degree and qualities of user immersion in virtual space. See, for example, the International Society for Presence Research (ISPR, http://www.temple.edu/ispr/) and Presence and Interaction in Mixed-Reality Environments (Presence II, http://cordis.europa.eu/ist/fet/pr.htm).
3. See https://www.google.com/culturalinstitute/project/world-wonders.
4. For details and imagery/footage of these works, see http://www.naimark.net/projects.html.
5. For footage, see https://youtu.be/D8LmzZrq14o.
6. For details and footage, see http://www.jeffreyshawcompendium.com/.
7. For details and footage, see *Landscape One*, http://www.fondation-langlois.org/html/e/page.php?NumPage=127 and https://vimeo.com/5995405; *Panoscope*, http://panoscope360.com/ and https://youtu.be/jRGjImyhuR8.
8. For details and stills, see http://www.ellardjohnstone.com/assets/WallofDeath.html.
9. For details and stills, see http://on1.zkm.de/zkm/stories/storyReader$6140.
10. From Deleuze's two cinema books, 1986 and 1989, charting a fundamental shift from classical pre-WWII cinema (movement-image) to post-WWII cinema (time-image).
11. See note 5, this chapter.
12. For footage and more information, see *Place: A User's Manual*, at http://www.jeffreyshawcompendium.com/portfolio/place-a-users-manual/, and *Place-Ruhr* at http://zkm.de/en/media/video/place-ruhr-2000-place-urbanity-2002 and http://www.jeffreyshawcompendium.com/portfolio/place-ruhr/.
13. For footage and more information, see http://alive.scm.cityu.edu.hk/projects/related/place-hampi/.
14. For footage and more information, see http://alive.scm.cityu.edu.hk/projects/alive/pure-land-inside-the-mogao-grottoes-at-dunhuang-2012/.
15. For footage and more information, see http://alive.scm.cityu.edu.hk/projects/alive/pure-land-ii-2012/.
16. For more details, see http://stefangreuter.info/?page_id=248.
17. See https://www.weforum.org/agenda/2015/09/can-3d-save-our-cultural-heritage/.
18. See, for example, Bruno 2002, where she explores the psycho-geography of cultural life, mapping connections between film, architecture, and the body. She highlights the embodiments and emotional terrains first evidenced in historical mapping, extrapolating from the *Carte du pays de Tendre*—a geography of love from Madeleine de Scudey's novel *Clelie* (1654). See also, of course, Debord 1955.
19. In the Bio-mapping project (which commenced in 2004), participants were connected to innovative galvanic skin response (GSR) devices which provide indicators of emotional arousal in conjunction with their geographical location (Nold 2004). See also http://www.blasttheory.co.uk.
20. See Clifford Lucas, "Deep Mapping," http://cliffordmclucas.info/deep-mapping.html.
21. Research and development of the *Atlas* for immersive display is supported by the Australian Research Council grant, "Transforming Visualisation in Museums: Deep Mapping for Narrative Coherence," 2016–19.
22. For more information on the project collaborators, see http://ecai.org/projects/MaritimeBuddhism.html.
23. See https://www.jauntvr.com/lobby/StoryLobby.
24. See http://veeah.com/.

## CHAPTER 11

1. "La photographie accueille d'abord avec enthousiasme a soulevé, comme toute invention, une multitude de critiques. Des esprits, bien intentionnés sans doute, ont voulu voir dans cette admirable découverte un péril pour l'art; ils ont craint que la main humaine devint inhabile, sachant qu'une machine était là qui travaillerait pour elle. Cette crainte n'a rien de fondé. . . . La photographie est à l'art ce que le famulos Wagner est à Faust. Elle compile, elle commente, elle fournit les textes, elle garnit de documents les murs du cabinet, mais elle ne conclut pas et elle attend la décision du maitre. Que signifie tout cela? Elle rient sait rien—Cela est, voilà tout" (Gautier 2015).
2. See, for example, the opinions by Stanley Fish, professor of Humanities and Law at Florida International University (Miami), in the *New York Times*. In particular, see "The Digital Humanities and the Transcending Mortality," January 9, 2012, http://opinionator.blogs.nytimes.com/2012/01/09/the-digital-humanities-and-the-transcending-of-mortality/.
3. The Portuguese Restoration War began with the Portuguese revolution of 1640 and ended with the Treaty of Lisbon in 1668, when Spain acknowledged the independence of Portugal.
4. Voltaire published *Poème sur le désastre de Lisbonne* in 1756 and *Candide ou L'optimiste* in 1759, expressing his critical views on Leibniz's assertion that ours is "the best of all possible worlds."
5. Information on the projects is available at http://www.museudelisboa.pt/exposicoes-actividades/detalhe/news/museu-de-lisboa-palacio-pimenta.html?tx_news_pi1%5Bcontroller%5D=Event&tx_news_pi1%5Baction%5D=eventDetail&cHash=82af80a3a48909027d1099d5c3fcba9e.
6. See https://lisboastorycentre.pt./.
7. See http://lisboaemazulejo.fcsh.unl.pt/.
8. Engraving by Jacques Philippe le Bas (1757), from drawings by M. Paris and T. Pedegache, *Ruins of the Opera Royal House* (at National Library of Portugal); cross section and plan of ground floor of an Opera House (at National Academy of Fine Arts, Lisbon).
9. See https://operadotejo.org/sobre/.
10. See https://lisbon-pre-1755-earthquake.org/en/about//.

## CHAPTER 12

1. At present, Vienna's population is growing by approximately 30,000 each year.
2. These include the car industry's powerful lobbying, the necessarily traffic-generating planning concept of separating urban functions (living, work, recreation), as in classical modernism (and older urban planning concepts, such as the Garden City, Ciudad Lineal, the Decentrists, the Ville Radieuse, etc.), and last but not least, suburbanization, which has its roots in economic, but also in military tactical considerations and that led to the extension of road infrastructure outside urban areas (cf. Lewis 2004; Plunz 1990, 278; Psenner 2015c).
3. See Psenner 2004. In identifying criteria for public-space perception, I focus on architecture as a determining factor. The aim of the process is to study the interrelation between street space and buildings.
4. Psenner 2004.
5. 3D-UPM is a newly developed method of surveying and representing in three-dimensional form all relevant data on the street-level environment (see "Methodology").

6. Detailed articles by the author on this issue are Psenner 2015b, 2015c; 2014a, 2014c; 2013; 2012b; 2011a, 2011b; 2005; 2004.
7. The pilot study surveys the exemplary *Stadtparterre* of a typical Gründerzeit ensemble in a street in one of Vienna's inner districts.
8. Cf. Peters 1990, 30; Peters 1999, 779f.
9. RBTC = Residential Building Type Catalogue (classification of dwellings)
10. CGPSs in fact often cover different floor levels—in most cases, the ground floor, basement, and a standard upper floor. Some also include a typological register, specifying the particular building type. The register is based on recurring elements of the ground floor and its geometrical structure. Cf. Caniggia 1963, Muratori 1960, Malfroy 1986.
11. Plans are kept on file in the database of the Vienna Municipal Building Inspection Department. Unlike in other countries (e.g., Italy and Switzerland), detailed plan data for individual buildings, which contain precisely the information needed here, are considered private under data-protection laws in Austria.
12. Building height is the factor that determines the extent and type of shading of the street space and lower floors. In relation to street width, Gründerzeit buildings are rather tall, which leads to disadvantageous daylighting conditions particularly on the ground floor. On the other hand, the typical Viennese multiple-story apartment building has exceptional ceiling heights (3.6 to 5 m), which allows for specific and highly flexible types of use. For more on this, see detailed studies: Psenner 2014c, 2012a, 2011c.
13. Revit is a standalone application that includes architectural design, MEP and structural engineering, and construction features. The program was developed as a specific building information modeling (BIM) tool which facilitates a coordinated and consistent model-based approach. It supports easily controllable transparency and fade-in/-out effects based on flexible keys for all objects. Cf. Campagnol 2014.
14. The Ponticelli/Naples ZGA includes axonometric plans, which, however, only show building skins (Fioravanti, quoted in Caniggia 1986, 336).
15. Cf. Psenner 2014a, 2014b, 2013, 2012a, 2012b.
16. This brings to mind other urban cultures, where the *StadtParterre* still functions this way, as, for example, in Rome. It is interesting to observe that the historical practice of numbering every window or door in the street, apart from having the usual house numbers, is still in use there. This means firmly inscribing a flexible small scale on the constantly changing structure of the city; which may start a discussion on the value of flexible small scales.
17. Relevant data were collected from participant observation, since official data about this frequency factor are of course not available.
18. Integration is societal participation, both by members of minorities and by the majority. It is not a static condition but, rather, an ongoing interactive process that happens in everyday life through reciprocally oriented and interpreted actions (Psenner 2011, 203).
19. Cf. (Scheuvens and Schütz 2013); see also the artistic-scientific work of the "Gehsteig-Guerrilleros" (sidewalk guerrillas), www.gehsteigguerrilleros.net.
20. Sidewalks are part of the street space, which is why, under §82 RTRA, approval is mandatory for any nontraffic use of streets: "(1) The use of streets, including the air space above them, insofar as it must be considered with respect to traffic safety, for purposes other than street traffic, e.g. for commercial activities or advertising, requires prior approval under this federal law, notwithstanding other

applicable legal provisions. The same applies for activities that are likely to cause crowds to gather or impair the attention of vehicle drivers." Pursuant to the law, approval can only be granted if there is no "impairment whatsoever of the safety and easy flow of traffic" (§83 cl. 1).

21. The point is realizing and utilizing the advantages of this distinctive structural feature. The spaces available hold an enormous potential for a growing city.
22. In view of the shortage of resources and under the general heading of "scarcity," urbanists today sometimes call for the downscaling of our present-day apartment sizes.
23. Today, interior courtyards mostly accommodate garbage cans or dumpsters. More intensive, diversified uses of this part of the *Stadtparterre* are rare (Psenner 2014c, 2015b).
24. In order to develop a sustainable solution to the various problems that afflict the Vienna street-level environment, it will be necessary to employ a systemic view of urban structures. One objective of this study is an in-depth analysis and representation of these interrelations.

## CHAPTER 14

1. See Seaman 2014. See also documentation video http://athanasius.trinity.duke.edu/projects/I_E/i_e.html.
2. Seaman 2014.
3. Verbauwhede et al. 2002. The principal Investigators were Dr. Ingrid Verbauwhede, Department of Electrical Engineering, University of California, Los Angeles; Dr. Bill Seaman, Head, Graduate Digital Media Department, Rhode Island School of Design; Dr. Mark Hansen, Department of Statistics, University of California, Los Angeles; and graduate students Shenglin Yang, in electrical engineering at UCLA, and Fabian Winkler, in design and media arts at UCLA.

The manufacturer's authorised representative in the EU for product safety is Oxford University Press España S.A. of El Parque Empresarial San Fernando de Henares, Avenida de Castilla, 2 – 28830 Madrid (www.oup.es/en or product.safety@oup.com). OUP España S.A. also acts as importer into Spain of products made by the manufacturer.

Printed in the USA/Agawam, MA
May 15, 2026

905072.186